Commanding Lincoln's Navy

Commanding Lincoln's Navy

UNION NAVAL LEADERSHIP DURING THE CIVIL WAR

Stephen R. Taaffe

Naval Institute Press
ANNAPOLIS, MARYLAND

Naval Institute Press
291 Wood Road
Annapolis, MD 21402

Library of Congress Cataloging-in-Publication Data

Taaffe, Stephen R.

Commanding Lincoln's navy : union naval leadership during the Civil War / Stephen R. Taaffe.

p. cm.

Includes bibliographical references and index.

ISBN 978-1-59114-855-5 (alk. paper)

1. United States—History—Civil War, 1861–1865—Naval operations. 2. United States. Navy—History—19th century. 3. Leadership—United States—History—19th century. 4. Command of troops. I. Title.

E591.T33 2009

973.7'5—dc22

2008054724

Printed in the United States of America on acid-free paper

16 15 14 13 12 11 10 09   9 8 7 6 5 4 3 2

First printing

For Cynthia

Contents

Acknowledgments

MANY PEOPLE HAVE OBSERVED that writing is often a lonely and tedious process. In my case, however, I was fortunate enough to receive valuable assistance and support from numerous sources. Stephen F. Austin State University's Office of Research and Sponsored Programs gave me a Faculty Research Grant so I could take a summer off from teaching to write. The Naval Historical Center's Edwin B. Hooper Research Grant provided additional money for research. Two of my friends and colleagues, Allen Richman and Philip Catton, read through the manuscript and gave me sage advice, and the folks at the Naval Institute Press were patient and helpful throughout the publishing process. My old friend Ken Wilson generously agreed to create the maps for this book. Finally, my wife Cynthia and our three children served as happy reminders that there is more to life than the Union Navy.

Introduction

THE FIRST STREAKS of light illuminating the predawn skies off of Fort Fisher, North Carolina, on 13 January 1865 revealed the largest collection of U.S. Navy warships ever gathered in one place up to that time. The Union armada contained more than sixty vessels in all, riding in formation in the waters off Cape Fear. It was an eclectic conglomeration consisting of squat ironclads roiling in the morning swell, screw frigates and sloops towering majestically beyond them, and numerous smaller vessels straining their boilers to maintain their positions. What the fleet lacked in grace and homogeneity, it more than made up in firepower. The warships sported 627 artillery pieces, and over the course of the next two and a half days their heavy Dahlgren and Parrott cannon systematically demolished the Confederate fortifications in and around Fort Fisher, blanketing the area in a haze of smoke and dust. By the time Union soldiers stormed the fort and raised the Stars and Stripes above its battered ramparts on 15 January, the Navy had fired a precisely tabulated 19,682 projectiles at its target, a total of 825 tons of metal. It was no wonder that Rear Adm. David Dixon Porter, the commander of the North Atlantic Squadron and the author of the Navy's successful barrage, later wrote, "I don't suppose there ever was a work subjected to such a terrific bombardment, or where the appearance of a fort was more altered. There is not a spot of earth about the fort that has not been torn up by our shells."[1]

Porter's presence as commander of the naval expedition was one of many remarkable things about the Fort Fisher operation. Porter had begun the war as a lowly lieutenant, but had risen rapidly through the ranks until he became one of the Navy's seven wartime rear admirals on the active list. He did so with a combination of daring, resourcefulness, skill, and not a little duplicity and guile. Along the way he had overcome questions of his loyalty, the hostility of many of his brother officers, naval tradition, and the mistrust and doubts of the president and secretary of the navy. Porter's meteoric ascent, inconceivable in peacetime, epitomized the dramatic changes that the Civil War wrought on the Navy's command and personnel structure.

The Union Navy played a vital role in winning the Civil War. Although it contained only 1,300 officers, 7,600 sailors, and 42 steam-propelled warships when the conflict began, its growth over the next four years was impressive. By the time Porter's fleet bombarded Fort Fisher, the Navy possessed more than 650 vessels of all kinds, manned by approximately 6,700 officers and 51,000 sailors. Moreover, its contributions to Union victory justified its phenomenal expansion. Most prominently, the Union Navy blockade of Confederate ports gradually starved the rebels of many of the supplies, equipment, and weapons they needed to continue the fight. In addition, naval expeditions seized strategic points along the Confederate coast and established bases there, depriving blockade-runners of harbors from which to operate and permitting Union warships to remain on station much longer than would have otherwise been the case. These lodgments along the Confederate periphery also exposed the Confederate heartland to attack by army forces. On the high seas, the Navy protected Union merchant ships by hunting down Confederate commerce raiders. Finally, Union warships operated on the numerous big rivers that ran through the Confederacy, spearheading and succoring army offensives deep into rebel territory. The Navy was certainly not the deciding factor in suppressing the rebellion, but it is impossible to discount its accomplishments.

The Navy's changes during the Civil War were more than just arithmetic. As Porter's experience indicates, the Navy also enacted reforms that permitted qualified officers to rise to some of its most important posts. In terms of its personnel system and command structure, the antebellum Navy was an inflexible, inefficient, close-minded, and tradition-bound organization in which creativity and competence often counted for little. Its quarrelsome and clique-ridden officer corps was led by elderly men past their prime who ascended through the ranks by strict seniority. Instead of cooperating with each other for the good of the service, the bureau chiefs who oversaw the Navy's daily operations frequently competed with each other for scarce resources. Navy secretaries came and went, and during their often short tenures they were usually unable to understand and address the Navy's numerous personnel problems. During the Civil War, however, Secretary of the Navy Gideon Welles worked with Congress to develop a new ranking system and flexibility in promotions and assignments. These changes did not fix all the Navy's flaws, but they temporarily alleviated the worst of them. By doing so, they contributed significantly to the Union Navy's Civil War successes by enabling Welles to place capable men in key positions.

Nothing exemplifies these organizational changes like the selection of the various squadron commanders. To wage the conflict, the Navy Department ultimately divided the theater of war into six stations—North Atlantic, South Atlantic, East Gulf, West Gulf, Mississippi River, and West India—and assigned a squadron of vessels to each one. Because these squadrons did almost all of the Navy's fighting, their commanders had enormous responsibilities. Naval officers sought squadron commands for several reasons. For one, they were the most prestigious posts in the Navy, attracting more attention from the press and presenting more opportunities for glory than other positions. In addition, squadron command gave officers a chance to ply their craft in an important way and influence the war effort by planning and conducting operations. Finally, squadron commanders received a portion of the prize money allotted from the seizure and sale of blockade-runners, enabling them to accumulate nice nest eggs for their postwar years.

Selecting squadron commanders was Secretary of the Navy Gideon Welles' job. To be sure, President Abraham Lincoln, as commander in chief of the country's armed forces, had the final say on all naval appointments, but he rarely tried to influence Welles' decisions. Although Welles appreciated the confidence in him that this implied, he also complained that more consultation with the president and the cabinet on personnel issues would have benefited the Union war effort. Lincoln, however, had his reasons for his hands-off attitude that was in such marked contrast to his relationship with the Army. First, he trusted Welles' judgment in personnel matters, once noting that the navy secretary made good use of the human material at his disposal. Moreover, the Navy simply was not as important as the Army in terms of size and mission. Whatever the Navy's contributions to Union victory, it was the Army that destroyed the Confederacy and physically occupied the seceded Southern states. This was as obvious to Lincoln as to everyone else, so it made sense for him to devote more attention to War Department matters. Finally, the Navy was not as politically useful to Lincoln as the Army. To cement various constituencies to his administration and its war, Lincoln nominated scores of prominent citizens to be army generals. Not even the most egotistical politician, though, pretended to be skilled enough to captain a warship or lead a squadron, so there were no naval equivalents to the often incompetent political generals that plagued the Army. In short, Lincoln usually had little to gain, politically or otherwise, from meddling in Welles' choices for squadron commanders.[2]

Left to his own devices, with minimal interference from above, Welles was free to develop and apply his own criteria for selecting his squadron commanders. Despite this autonomy, however, he was still subject to the bureaucratic and cultural pressures that permeate any organization. One of the Navy's most powerful influences was seniority. Promotion in the prewar Navy was almost always based on longevity, meaning that the next available commission went to the most senior eligible officer. The highest-ranking officers, of course, usually received the most important and prestigious assignments too. There was a logic behind this practice. Theoretically, the most experienced officer was probably the most capable, and consequently the most deserving. Seniority was also predictable, so every officer knew where he stood, and where he was likely to stand in the future. Finally, promotion by seniority could reduce congressional meddling and officer string pulling. Welles actually had little respect for seniority, recognizing early in the war that longevity did not necessarily translate into competence and ability. Still, so strong was its hold on the Navy that he felt obligated to take it into account when choosing squadron commanders. Blatantly and publicly disregarding it would cause an uproar among the tradition-bound officer corps, and Welles had no desire to disrupt the service more than absolutely necessary when the country's survival was at stake. Although Welles always kept seniority in mind, he also worked diligently with Congress to enact legislation to undermine its impact. As Welles explained after the war, "Seniority had its influence, but was not always satisfactory."[3]

Personal and political ties also influenced Welles' choice of squadron commanders. The officer corps was a small, close-knit community whose members had often known each other for decades. They had almost all started as adolescent midshipmen and served together on distant stations under the tedious and occasionally dangerous circumstances that brought out the best and worst in people. Over the years, they had formed themselves into quarreling and interconnected cliques of friends who helped each other secure choice assignments. Many officers had also developed and cultivated relationships with various politicians who looked after their best interests. The war did not change this. Throughout the conflict, officers and their patrons lobbied Welles directly and indirectly for squadron command. The upright Welles was hardly the kind of man to publicly condone such activity, and in fact he wrote to one naval officer, "The Department disapproves the efforts of any officer to influence its actions by outside political influences."[4] Welles was also a politician, however, who understood the need to placate and appease the constituency that made it possible for him to do his job. He never appointed an officer to squadron command solely on the basis of his connections, but his willingness to read their petitions and hear their pleas demonstrated that he took those connections into account when he made his personnel decisions.

Welles also considered availability when assigning squadron commanders. Early in the war, many officers theoretically eligible for squadron command because of their rank and seniority were actually too old or sick for duty afloat. Although on paper they possessed good records, by the Civil War they were beyond their prime, so Welles frequently passed them over. Later on, as more and more officers gained battle experience and promotion, there were often plenty of good candidates for squadron command. The problem, however, was that many of them were already gainfully employed elsewhere. Welles recognized that sending capable men from one important post to another still left a position unfilled, so he therefore often looked for underutilized officers to fill his vacant squadrons.

Finally, Welles weighed an officer's birthplace and commitment to the Union cause when he chose his squadron commanders. In the war's first hectic months, Welles doubted the fidelity of Southern-born officers, and intentionally sidelined them until he could be sure of their loyalty. Even after everyone had chosen sides, Welles still believed that the Navy contained a number of men who, though nominally loyal, were not sufficiently enthusiastic about waging war against their Southern brethren. He noticed that such officers sometimes tried to secure transfers to remote regions far from combat, or, if unable to do so, followed the letter of their orders, but would not go much beyond that. Their hearts, Welles felt, simply were not in the war. Welles was determined to keep such officers out of squadron command and other important positions, regardless of their seniority, prewar record, or abilities. As he later explained about his choices for squadron command, "The important question of earnest, devoted loyalty to the Constitution of the Union was of course a primary consideration."[5]

Seniority, connections, availability, and loyalty had their places, but for Welles, ability—either demonstrated in battle or suspected from his evaluation of an officer's character and record—mattered more than anything else in selecting his squadron commanders. As a prewar newspaper publisher, editor, and writer, Welles was a keen judge of men, and he recognized that it took a certain kind of person to lead a naval force effectively. He preferred resourceful, dynamic, driven, and independent-minded officers who generated results as his squadron commanders. He usually wanted an officer to prove himself in battle before giving him more responsibility, and he learned to trust his instincts unless he received specific evidence to the contrary. Indeed, Welles was even willing to give commands to men he personally disliked, as long as he believed them capable of delivering victories. He also tried to marry the right officers to the right jobs. Welles understood that each squadron required a different kind of leader depending on its geography, available resources, and mission, and he sought to find the best qualified officer for that particular position. Such considerations may seem commonsensical, but they flew in the face of decades of naval tradition and procedure.[6]

Nineteen men served as commanders for the six relevant squadrons during the Civil War. This high number might call into question Welles' ability to pick capable officers, but a closer look shows that most lost their posts for reasons other than incompetence. Of the nineteen, five were holding their positions when the war ended. Although most squadron commanders saw their share of combat, sickness took a greater toll on these men than injuries and wounds. Commanders often remained on board their flagships for months at a stretch, and were subject to poor food, lack of exercise, tedium, never-ending and mind-numbing paperwork, the stress of responsibility, and unhealthy climates. It is small wonder that some squadron commanders surrendered their commands because they were debilitated by disease. One commander lost his job when Welles dissolved his squadron, a few were transferred without prejudice, and several were temporarily leading their squadrons for extended periods until supplanted by their permanent replacements. In all, only a half-dozen officers were relieved of their commands or quit under pressure because they failed to live up to Welles' expectations, mostly early in the war. These disappointments, though, should not detract from the majority of squadron commanders who performed creditably, or the few who rendered outstanding service to the Union war effort. While Welles' selections were not perfect, on the whole he chose well from the material available to him.

The Civil War was the defining event in nineteenth-century U.S. history, one with enormous consequences that reverberated for generations. Moreover, its scale matched its significance. For four years, hundreds of thousands of men fought from the Virginia Tidewater to the New Mexico desert in battles that killed and wounded at a rate never before seen in American history. Despite its advantages in population and industrial capacity, there was nothing inevitable about Union victory. It is possible to discuss the conflict in terms of impersonal social, political, military, and economic forces, but in reality the war was waged by individuals whose decisions could impact unforeseen events weeks, months, and even years in the future. Fortunately for the Union, Welles overcame serious bureaucratic obstacles in the Navy and selected capable squadron commanders who contributed greatly to defeating the Confederacy.

Commanding Lincoln's Navy

CHAPTER ONE

Scraping the Barnacles

Gideon Welles and the Navy Department

IN LATE WINTER 1869, visitors, sightseers, and hangers-on crowded into Washington to participate in Ulysses Grant's presidential inauguration. Even amid the nonstop parties and festivities, though, the federal government's administrative machinery continued to operate as departing cabinet officials struggled to clear their desks of old business before their replacements took over. Secretary of the Navy Gideon Welles was among those hard at work. Welles had little patience, constitutionally or otherwise, for the unrestrained revelry around him, but he accepted it as one of the unpleasant facets of republican government. On 3 March, a Wednesday, Welles and his bureau chiefs walked across the street to the White House to pay their respects one last time to outgoing president Andrew Johnson. Welles formally introduced each bureau chief to Johnson, starting with the most senior, Rear Adm. Joseph Smith, head of the Bureau of Yards and Docks. As Johnson cordially shook Smith's hand, Welles remembered a similar White House scene eight years earlier. Back then, the war that had scarred them all one way or another was still hypothetical, and the new president, Abraham Lincoln, remained untested. Smith warned Lincoln of the approaching upheaval, and added sternly, "We will do our duty, and expect you to do yours." Lincoln had lived up to Smith's expectations, and paid for it with his life, which was why Johnson now held the position to which Lincoln was twice elected. Now, however, the country was united and at peace, and Smith saw no need to resort to foreboding and admonition. After the ceremony, Welles and his bureau chiefs returned to the Navy Department building. Everyone there—bureau chiefs, messengers, clerks, etc.—one by one entered Welles' office to say their good-byes to him. Many of them had worked for Welles for eight years through trying and difficult times, so the emotion some displayed was understandable. Summing up his day in his diary, Welles concluded: "It was past four when, probably for the last time and forever, I left the room and the building where I had labored earnestly and zealously, taken upon myself and carried forward great responsibilities, endured no small degree of abuse, much of it unmerited and undeserved; where also I have had many pleasant and happy hours in the enjoyment of the fruits of my works and of those associated with me." He returned to his Hartford, Connecticut, home, where he lived out his last remaining nine years until his death in February 1878.[1]

Abraham Lincoln's decision to appoint Welles his secretary of the navy was based primarily on political considerations. The young Republican Party was a disparate organization filled with people who often had little in common beyond a commitment to restrict slavery's expansion into the Western territories. Abolitionists, old Free-Soilers, disenchanted Democrats, former Whigs, and rootless Know-Nothings all vied with one another for power and influence in the new party. This competition became even fiercer after Lincoln won the 1860 election and gained the right to appoint his supporters to federal jobs. As a skilled politician, Lincoln well understood that he could use this patronage to glue the various Republican factions to his administration. The problem, however, was that there were never enough offices to go around. This was especially true of the limited number of cabinet positions. One of Lincoln's challenges, therefore, was to construct a cabinet that best represented and satisfied the Republicans' innumerable geographic and ideological divisions. To Lincoln, Welles was one piece in an incredibly complicated political puzzle he had to put together before he assumed office.

In the weeks and months leading up to his inauguration, as seven Southern states left the Union one after another in response to his election, Lincoln received all sorts of solicited and unsolicited advice on his cabinet appointments. He had, however, quickly come to some tentative conclusions about whom he wanted in his administration. In a conversation with Hannibal Hamlin, his vice president–elect, Lincoln noted that he hoped to name a New Englander as his secretary of the navy, mentioning Nathaniel Banks, Charles Francis Adams, and Welles as possibilities. Banks was a powerful Massachusetts politician who had served as Speaker of the House of Representatives and later as state governor. Unfortunately for Banks, he was a well-known political trimmer who had accumulated his share of influential enemies in the course of his rise through the political ranks, including both of Massachusetts' Republican senators. Besides, he had recently accepted a job as president of the Illinois Central Railroad and moved to Chicago, so he no longer represented New England's interests. As for Adams, the son and grandson of presidents, people spoke well of his character, but warned that he lacked sufficient practical experience to run the Navy Department. Although some suggested Amos Tuck, a prominent New Hampshire Republican, his candidacy went nowhere. This left the Connecticut-born Welles, whose numerous supporters, including Hamlin and some prominent regional editors, were loud and persistent in their enthusiasm for him.[2] Moreover, many of Banks' detractors jumped on the Welles bandwagon too to keep the department out of Banks' hands.

On the other hand, Welles had his opponents, including secretary of state designate William Seward and his New York patron, Thurlow Weed. Weed was an old political foe of Welles, and, during a meeting with Lincoln, he warned the incoming president that he would do just as well to make a ship's wooden prow secretary of the navy as the staid Welles. Lincoln, though, was unconvinced. He liked and respected Welles, and appreciated his steadfast support. Besides, Lincoln wanted a former Democrat such as Welles to represent New England in the cabinet. Finally, just days before his inauguration, Lincoln asked Hamlin to send to Hartford, Connecticut, for Welles. While Welles was well aware of the lobbying on his behalf, he was still surprised by the sudden summons. He caught a train for Washington and arrived disheveled and exhausted at Willard's Hotel on 2 March 1861. At a meeting next morning, Lincoln asked Welles if he preferred to become secretary of the navy or postmaster general. Welles opted for the former, Lincoln agreed, and the Senate confirmed the nomination two days later.[3]

Welles' appointment may have had more to do with politics than anything else, but Lincoln was a shrewd judge of men, and he no doubt recognized that Welles could bring more to his cabinet than mere political support and geographic balance. Born in 1802, Welles came from a deeply rooted Connecticut family. Even though his parents were prosperous enough to provide him with an education, he was initially unable to find his niche in the world. He dabbled in writing fiction, traveled to Pennsylvania, and studied law before he stumbled upon his calling as part owner and editor of the *Hartford Times*. Not surprisingly, newspaper editorship segued into politics, and Welles became a stalwart Jacksonian Democrat committed to states' rights and the strict construction of the Constitution. His regional clout and political connections secured him a variety of government jobs over the years, the most important of which was chief of the Navy's Bureau of Provisions in President James Polk's administration during the Mexican War. This position introduced him both to the Navy's way of doing things and to some of the officers with whom he would later work during the Civil War. Despite his political influence, electoral success eluded him. Although he served eight years in the Connecticut state legislature in the 1820s and 1830s, he ran unsuccessfully over the years for the House, the Senate, and the governorship. He broke with the Democratic Party over the slavery issue in the mid-1850s and became a Republican. Along the way he met Lincoln a number of times and was impressed with his simple and clear logic. In 1860 he led the Connecticut delegation to the Republican national convention in Chicago and helped Lincoln secure the presidential nomination.

Physically, Welles cut an almost comic figure. He wore a badly fitted wig, parted in the middle, which did not match the color of his remaining hair, or that of his bushy white beard. His demeanor, however, belied his appearance. To most people, he was a distant, austere, and curmudgeonly man with a limp handshake and a placid countenance. One person compared him to a ghost that inhabited the Navy Department whose presence was always felt but only rarely seen. Indeed, Welles shunned the Washington social scene, preferring to spend his evenings scribbling baleful and wary observations in his diary. His aloofness helped to insulate him from outside pressures, but it also aroused the antipathy of journalists and politicians who resented his inaccessibility, so he suffered from poor press throughout the war. In fact, one critic noted, "[He] possessed a strong intellect, but manifested little warmth of feeling or personal attachment for any one. He was a man of high character, but full of prejudices and a good hater. He wrote well, but was disposed to dip his pen in gall."[4] Although he and Lincoln respected each other, they were not socially intimate. Welles firmly believed in hard work and duty, and spent almost every day, including a good many Sundays after church, at his desk in his second floor office in the Navy Department building. Neither periodic illness nor the death of his young son, Herbert, in 1862, deterred him from his responsibilities. In fact, he rarely left the capital during the war, except for occasional inspections of naval yards and trips home to Connecticut to see his family. He was a man of immense integrity and moral courage, with sound judgment and a ferocious commitment to the Union. Indeed, on several occasions during the war he sacrificed relationships in order to do what he believed was best for the Navy, explaining, "Individual feelings, partialities, and friendships must not be in the way of public welfare."[5] Pessimistic by nature, he was neither elated by success nor overly depressed by failure. As a former editor, he understood the value of information, and accumulated as much of it as possible. For example, he kept close tabs on the crushing volume of Navy Department paperwork that crossed his desk, and carefully scrutinized the newspapers. He recognized that the Navy was no better than its officers, so he went to considerable lengths to watch and evaluate them through correspondence and personal interviews. Unlike some, Welles realized from the start that the sectional conflict would be protracted and bitter, so he did his best to prepare the department in his charge for the long war ahead.[6]

Welles may have become secretary of the navy in name, but, as he quickly learned, this did not necessarily mean that he controlled the Navy. In the early days of Lincoln's administration, Secretary of State William Seward convinced himself that he should serve as some sort of premier for the inexperienced and backward president, with authority over the various cabinet secretaries. His early attempts to give orders to other cabinet secretaries in Lincoln's name aroused their resentment and suspicion. To make matters worse, Seward and Lincoln did not quite see eye to eye on policy toward the Southern states. Seward advocated a more conciliatory approach in order to maintain the loyalty of the vital Border States, whereas Lincoln and the rest of the cabinet were willing to take a harder stance. As part of his efforts to bring the president around to his way of thinking, Seward tried to subvert Welles' position in the Navy Department.

On the evening of 1 April 1861, Welles was eating dinner at Willard's Hotel when Lincoln's personal secretary, John Nicolay, delivered to him a package from the White House. Upon opening the parcel, Welles discovered among other things an executive order from Lincoln assigning Capt. Samuel Barron to the newly created Bureau of Detail, with wide latitude to act as he saw fit. A flabbergasted Welles understood immediately that this directive reduced him to a figurehead, so he rushed over to the White House to protest. He found Lincoln in his office working alone. When the president looked up and saw Welles' distressed look, he asked, "What have I done wrong?" After Welles read the document to him, Lincoln admitted that he had no memory of issuing it, but he had without looking signed a bunch of papers earlier in the day that Seward submitted to him. Welles quickly explained that Lincoln lacked the power to establish a new naval bureau without congressional consent, that it violated tradition to place a naval officer instead of a civilian in such a position, and that he doubted Barron's loyalty. The real issue, though, was Welles' authority over his department, which Lincoln instantly grasped. The president said that he would never knowingly undermine Welles, and he agreed to rescind the order immediately.[7]

If Welles believed that his late-night appeal to Lincoln would put an end to Seward's machinations, he was sorely mistaken. In his inaugural address, Lincoln had pledged to retain all federal property in the seceded Southern states, including Fort Sumter, which protected the harbor of Charleston, South Carolina. The small and isolated Union garrison there was running out of supplies, and its commander, Maj. Richard Anderson, made it clear that he would soon have to surrender the post. Lincoln and his cabinet were initially divided on their response, but, as March turned into April, an increasing consensus emerged to try to hold onto the fort. The question, however, was how. Speaking for the Army, Winfield Scott, the general in chief, and chief engineer Joseph Totten argued that relief was impossible. The naval officers whom Welles consulted had serious doubts as well, though a few of them were at least willing to try something. Confronted by this phalanx of skeptics, Lincoln turned to Gustavus Fox. Fox was a former naval officer who presented a plan to run supplies to Fort Sumter at night in shallow draft steamers, protected by warships. Lincoln appreciated his enthusiasm and gave his approval for the mission. Despite Fox's irregular position, Lincoln placed him in charge of the transports and ordered Welles to provide the necessary warships. Welles put the Navy's bureaucratic wheels into motion, which included orders to the acting commandant of the Brooklyn Navy Yard, Cdr. Andrew Foote, to prepare the side-wheel steamer USS *Powhatan* for action.[8]

Within days, however, Welles began to suspect that something was amiss. Foote telegraphed cryptically that he assumed that Welles knew that the "Government" was sending new orders for *Powhatan*. Welles did not, even after a 4 April meeting with Lincoln, so just to be safe he ordered *Powhatan*'s departure delayed while he sought more information. At around 11:00 PM on 6 April, Seward and his son Frederick called upon Welles at Willard's to complain that Welles was interfering with a presidential directive dispatching *Powhatan* under Lt. David Porter's command to Union-held Fort Pickens outside of Pensacola, Florida. Seward believed that Lincoln should make his stand there instead of at emotionally charged Fort Sumter, whose evacuation he was advocating, and had persuaded the president to reinforce the place. A shocked Welles responded that he had instructed Capt. Samuel Mercer, not Porter, to take *Powhatan* to Fort Sumter as part of Fox's expedition, and that he knew nothing of any effort to strengthen Fort Pickens. Seward suggested that perhaps Lincoln had told Capt. Silas Stringham, Welles' naval adviser, about the operation, and that Stringham had neglected to inform him. Welles immediately summoned Stringham, who on his arrival denied any knowledge of Seward's scheme. Thoroughly exasperated by now, Welles insisted that everyone trudge over to the White House to see Lincoln and get to the bottom of the matter. Despite the late hour, the president was still awake. Lincoln was perplexed, and his only explanation was that he had probably confused *Powhatan* with another warship, the side-wheel steamer USS *Pocahontas*. He said nothing about the propriety of using the Navy's resources without Welles' knowledge, but he was obviously embarrassed by the situation. He immediately placed *Powhatan* back under Welles' control, and refused to discuss the expedition to Fort Pickens. By that time, however, Porter had already taken the warship to Florida, having persuaded Mercer that his orders from the president trumped Welles' instructions.[9]

As humiliating as this episode was for everyone involved, indicating as it did the new administration's incompetence, greenness, and ineptness, it probably did little to change Fort Sumter's fate. The leading elements of Fox's expedition arrived off Charleston on 12 April, just before the impatient Confederates finally opened fire on Anderson's little garrison, and could do little but watch helplessly as rebel artillery battered the defenders into submission the next day. Fort Sumter's fall persuaded the Northern public to wage war to crush the rebellion, but Lincoln's decision to ask the state governors to call out their militias to do so convinced four of the Border States—Arkansas, Tennessee, North Carolina, and Virginia—to secede from the Union and join the Confederacy. For Welles, the intrigue surrounding Fort Sumter had an equally cathartic effect. Seward was so mortified by his midnight visit to the White House that he promised Welles to never again interfere in another department's business. Being meddlesome by nature, Seward was unable to quite live up to his pledge, but his subsequent transgressions were not as blatant or serious as those in April 1861. As for Lincoln, Welles noted that he thereafter always spoke regretfully of his collusion with Seward, and attributed it to his inexperience and inattention. He never again treated Welles so cavalierly. Whatever distress Welles may have felt about the onset of hostilities, he could at least take comfort in the fact that he went to war with full authority over the department in his charge.[10]

Treason in the Air

Several days after the Confederates fired on Fort Sumter, Capt. Charles Wilkes visited the Navy Department building to see a friend. As he entered one of the first-floor offices, he spotted eight or ten officers, Southerners all, discussing whether they should resign their commissions and join their seceded states. Although the cantankerous Wilkes was not a very popular man in the service, the officers hailed him and called for his views. Wilkes had hoped to withdraw without attracting any attention, but now, put on the spot, he later asserted that he exclaimed, "I am not of your opinions and have only to say—while there is a plank to float on and a spar to hoist the old flag on, I will always be found under its folds. I am as true as steel and shall be as long as my country exists."[11]

The issue may have been crystal clear to Wilkes, but the same could not be said for a good many Southern-born naval officers. Indeed, of the 571 captains, commanders, and lieutenants in the service in March 1861, 138 either resigned their commissions or were dismissed from the service. In most cases, these men found their way into Confederate service and played important roles in the rebel war effort on both land and sea. As far as most of them were concerned, their primary loyalty was to their states, and they felt that surrendering their commissions absolved them of any commitment to the federal government. As one remarked, "Civil War has begun and I cannot draw my sword against my native state, where my early years were passed and the bones of my father and mother lay—and I do not think it honorable to hold a commission with any reservation as to the service I may be directed to perform."[12] Others quit because they disliked the Republicans and their agenda, succumbed to peer pressure, did not want to fight against their friends and family, or were alienated by the Navy Department's increasingly strident attitude toward them. They did not consider themselves traitors, but instead saw themselves as honorable men coping as best they could with difficult circumstances. Most left the Navy only after considerable angst and soul-searching, some eventually regretted their decision, and one, Maryland-born Cdr. Edward Tilton, was so distressed by his competing loyalties that he committed suicide.[13]

There were, however, other Southern-born officers who remained faithful to the Union. They believed that they had sworn an oath to the federal government, not to their individual states. As officers, their job was to obey orders regardless of the political implications. Long years at sea had diluted their sectional ties and simultaneously strengthened their attachment to the nation as a whole, and they had no desire to see the United States weakened in the international arena in which they operated. Summing things up, one loyal South Carolinian, Cdr. Charles Steedman, wrote, "I am now, as I have always been, what I intend to live up to, a union man. I know no North or South in this unfortunate and deplorable affair. All that I know is my duty to country and flag which I have served for the last thirty years and seen respected upon every sea."[14] Although these men often sympathized with the emotional tug-of-war some of their fellow officers underwent, they themselves had few doubts where their loyalties lay.[15]

When Welles assumed his office, he was appalled by the attitudes he encountered in Washington in general and in the Navy Department in particular. As Wilkes' story indicated, many Southern-born officers were amazingly frank and open in their discussions of their loyalties. They continued to fulfill their duties and draw their pay while publicly contemplating when and if they should resign their commissions and join the Confederacy. Welles felt that like so many Southerners in the capital in the days surrounding Lincoln's inauguration, these officers saw secession as a partisan political issue between Republicans and Democrats, not as treason. Welles, however, had no such illusions. In his estimation, those officers who advocated or sympathized with secession were traitors who conspired against their government. He believed that these men were a threat to the Navy because they undermined morale, sowed mistrust and defeatism, and eroded efficiency. In his first weeks as navy secretary, Welles received numerous reports questioning the loyalty of some of the Navy's highest-ranking and most distinguished officers. These included

Samuel Barron, who accepted a commission in the Confederate navy days before Seward intrigued to impose him on Welles; bureau chiefs George Magruder and Joseph Smith; Matthew Maury, superintendent of the Naval Observatory; and Franklin Buchanan, commandant of the Washington Navy Yard. Not all of them ultimately joined the Confederacy, but enough did to lend validity to rumors that Welles was surrounded by secessionists.[16]

Welles was neither alone nor incorrect in his assertion that wavering Southern-born officers were hindering the Navy's ability to fulfill its responsibilities. Indeed, he had firsthand evidence of the disruption they caused. On 20 April Welles had a half-hour conversation with Maury and Magruder in his office about the riot the previous day in Baltimore by pro-Confederate locals against some Massachusetts troops en route to Washington. Although Welles was well aware of continuing rumors that both men intended to resign their commissions, neither gave evidence of planning to do so during their discussion. When Welles returned from his supper at Willard's, however, he learned that the Naval Observatory was abandoned, and Maury and his family were gone. He sent for Magruder, who arrived with the observatory keys that Maury had left behind. Magruder assumed that Maury had headed south, and admitted that he was not surprised. Magruder himself waited two more days to submit his resignation as chief of the Ordnance Bureau, but he at least had the decency to personally inform Welles of his decision.[17]

On 22 April, the same day Magruder said good-bye to Welles, Capt. Frank Buchanan quit too. Buchanan commanded the Washington Navy Yard and was one of the Navy's most respected officers. In the chaotic days after the Baltimore riot, Washington was isolated from the Northern states and virtually defenseless. The naval yard, with its valuable property, was particularly vulnerable to Confederate seizure. Buchanan did not believe in secession, but he was a Marylander who expected his state to leave the Union at any time. He had no desire to fight his friends and family, so he decided to resign his commission. Like Magruder, Buchanan felt obligated to explain his reasoning to Welles personally, noting that this was the hardest decision he had ever made. Welles heard him out, warned him that he would regret his choice, and asked him sardonically who had paid his salary all these years. A couple of days later, Buchanan officially surrendered his command at a formal ceremony at the naval yard flagpole in front of a silent and tearful crowd of workmen. After he said a few words, the workmen offered up three cheers. Buchanan, however, had acted prematurely; despite widespread Southern sentiment, Maryland did not secede from the Union. After reevaluating the situation, he wrote to Welles and asked to be reinstated, but Welles refused. He felt that Buchanan had abandoned his post in a time of crisis, and it did not matter if he did so with great dignity and without immediately jeopardizing the naval yard's security. As Welles explained, "Confidence once destroyed would never again be restored." With a heavy heart, Buchanan thereupon cast his lot with the Confederacy and against Maryland and the Union.[18]

Considering his experiences with Magruder, Maury, and Buchanan, it was hardly surprising that Welles sometimes felt like the only loyal man in the Navy Department building. Throughout that spring, Southern-born officers resigned their commissions by the dozens, and Welles could never be sure if the remaining ones would be faithful to the Union. It was possible that an officer harboring Southern sympathies would use his authority to bring his warship into a Confederate port, or assist the rebels in seizing a naval yard. More realistically, a temporizing Southern-born officer might hurt the Union war effort by showing insufficient enthusiasm for the cause. It was therefore imperative that the Navy Department identify and remove those officers who constituted real or even potential security risks. Despite his time as chief of the Bureau of Provisions during the Mexican War, Welles was still an outsider unfamiliar with the officer corps. He simply did not know which officers to trust with the Navy's most important positions.

Fortunately, Welles' long career as an editor had taught him the value of cultivating credible sources. If he did not know on whom he could rely, then he would seek help from those who did. Happily, several such men were available. One was Andrew Foote, assistant commandant of the Brooklyn Navy Yard. Foote and Welles had attended school together as children, and had remained friends ever since. Before the Confederates fired on Fort Sumter, Foote traveled to Washington to advise Welles on personnel matters. Foote was only a commander, however, so he lacked sufficient rank to provide the kind of assistance Welles wanted. For this, Welles turned to Capt. Silas Stringham. Not only was Stringham one of the Navy's highest-ranking and most respected officers, but, of equal importance, Welles had known him for five years. With Lincoln's consent, Welles asked Stringham to serve as his adviser on all navy matters, including personnel. In the few weeks in March they served together, Stringham gave Welles valuable advice. It was quickly apparent, though, that Stringham lacked the talent and inclination for the paperwork his new post required, so Welles obliged him when he asked for command afloat.[19]

Foote's and Stringham's advice was important, but Welles' mainstay in personnel matters during the first months of the war was Capt. Hiram Paulding. Welles appointed Paulding his detailing officer, or the person responsible for distributing assignments. Born in 1797 in New York City to one of the men who captured John André during the Revolutionary War, Paulding entered the Navy as a midshipman in 1811 and fought bravely at the Battle of Plattsburgh Bay during the War of 1812. In the ensuing years he saw action against the Barbary pirates, graduated from the Norwich Military Academy in Vermont, and served throughout the Pacific Ocean region not only in routine patrol duty but also in hunting mutineers, confronting hostile natives, and carrying dispatches across the Andes Mountains to Simón Bolívar. He was promoted to captain in 1844, and put in charge of the Home Squadron a decade later. Southerners roundly condemned him for seizing William Walker and his filibusters in Nicaragua in 1857, which cost him his command. He was on court-martial duty in Washington when Welles took office, and in the weeks before the Confederates fired on Fort Sumter he impressed the navy secretary with his intelligence, solid judgment, obvious patriotism, and sound advice. Welles asked him several times to become detailing officer, but Paulding agreed only when he learned that Lincoln wanted him to take the job too. There was much about Paulding that people liked and respected. He was a tall man, straight as an arrow, with gentle blue eyes. Although noted as a strict disciplinarian while shipboard, just beneath his rough surface he was kind and sympathetic, a combination that invariably appealed to his crews. Off duty Paulding liked nothing more than puttering around his Long Island farm in an old Panama hat and brown linen jacket while enjoying the company of his wife and six children. Welles hoped that these traits would enable him to handle perhaps the Navy's most sensitive job.[20]

Paulding worked long hours, oftentimes from six in the morning until midnight, with only a couple quick breaks for meals. The paperwork quickly became overwhelming, so he brought in Cdr. Charles Davis, an old friend, to help establish order out of the chaos. Usually the Navy had more officers than it needed, but not now. As the Navy rapidly expanded, Paulding had to assign officers to warships, naval yards, recruiting stations, the bureaus, and so on. He also had to tactfully decline offers from enfeebled senior captains seeking commands commensurate with their rank. It was a tiresome and delicate job that required him to quickly transfer to new duties men with big egos who were very jealous of their honor, reputation, rank, seniority, and prerogatives. Taking offense was practically a naval tradition, and the haste with which Paulding had to act made complaints even more likely than usual. Fortunately, Paulding had the right combination of frankness and cordiality to soothe the majority of the ruffled feathers he created, and most observers agreed that he acted as scrupulously as he could. In doing so, he often placed more emphasis on seniority than ability, however, which eventually caused problems when it came time to choose the first squadron commanders.[21]

Whatever their differences in background and temperament, Welles and Paulding both shared a suspicion of Southern-born officers. Indeed, Paulding had expressed his concern the previous January, months before Lincoln assumed office. Even though his oldest son was named after a close friend who had joined the Confederate navy, Paulding still concurred with one officer who urged him to "put the screws to all such people."[22] Many Southern-born officers had already resigned their commissions, but Welles and Paulding had grave doubts about the commitment of the rest. In an effort to separate the lukewarm from the steadfast, Welles did several things after the war began. First, the Navy Department began labeling officers who resigned their commissions as dismissed from the service, thus attaching a permanent black mark to their records for everyone to see. It also required all officers and men to swear an oath of allegiance to the Union. Hopefully, these actions would smoke out the disloyal before they could do damage to the service. Finally, and most significantly, Welles and Paulding systematically removed all Southern-born officers from important positions, especially those commanding warships returning from foreign stations. They acted even against officers whose only transgression was their Southern birthright. It was ruthless and unfair, as even Welles later admitted to one victimized officer.

Among the most painful and trying duties that have devolved upon me in administering the affairs of this Department, has been that of deciding what course of policy to pursue toward officers from regions that are in insurrection who were intrusted [*sic*] with positions of responsibility in distant stations. Not knowing their opinions and feelings on questions involving the integrity of the Union and the existence of the Government, unable to ascertain them, except indirectly, and by inference and conclusions drawn from their associations and other circumstances, neither satisfactory nor reliable, I was compelled to act in some instances with apparent harshness, and with a severity I did not feel toward gentlemen sensitive as regarded their honor and professional obligations. . . . Without any means in my possession to discriminate between the faithful and faithless, I was compelled to act, and in doing so I have done yourself and others an injustice.[23]

Welles, however, figured that the mistreated officers, if truly loyal to the Union, would understand that their country's needs came before their own. It is hard to measure the effectiveness of this strategy, based as it was on prevention, but by autumn Welles no longer worried about perfidious officers, and in fact explained to a friend, "Some of the Southern officers are very loyal and true, and when there are such I honor and respect them."[24]

Loyal Southern-born officers responded to Welles' purge in a variety of ways, depending upon their circumstances and personalities. Some resented a policy based on the presumption of guilt, especially when their Northern-born replacements were men of inferior rank and ability. Most of them, however, did their duty and obeyed orders. Capt. Cornelius Stribling, for example, commanded the East India Squadron when the war began. Stribling was a South Carolinian whose son had resigned his commission the previous December to join the Confederate navy, factors guaranteed to arouse Welles' suspicions. Stribling knew that his head was on the chopping block, but he had no intention of letting that interfere with his responsibilities. When he realized that a good many of his officers in his small Hong Kong–based squadron entertained secessionist sympathies, he issued orders forbidding any discussion on the subject, and called for them to remain loyal to the Union. To Welles he wrote: "I can give you no pledges, I can offer you no security. And if you should think proper to supersede me in my command, I feel that the circumstances of the country would justify you in doing so. . . . All I can do, is to promise, if you continue to express confidence in my honor, patriotism, and fidelity, that the flag of the Union shall not be tarnished in my keeping, and will faithfully deliver the ships under my command, to the same authority that gave it, when the cruise is up, or whenever required to do so."[25] Welles did in fact order Stribling home and gave his flagship, the screw sloop USS *Hartford,* to someone else. However, his forthright letter no doubt appealed to Welles because Stribling ended the war as a squadron commander.[26]

Command and Personnel

In terms of its command structure, the prewar Navy Department was among the most inefficient divisions of the federal government's bureaucracy. As Welles' experience indicated, presidents usually selected their navy secretaries for political reasons, not because the appointees possessed any real naval expertise. As a result, navy secretaries often had little understanding of the Navy's culture and problems, let alone sufficient insight to present realistic solutions. Worse yet, the turnover rate for navy secretaries was very high, especially during the 1840s, so there was often not much continuity at the top. This might not have been so debilitating if navy secretaries had possessed a well-trained professional staff to offer planning and advice, but this was not the case. In fact, there was not even a naval equivalent to the Army's general in chief. The general in chief may have been mostly a figurehead with little authority over the Army, but at least he was available to provide informed counsel to the secretary of war. As a result, many navy secretaries were frequently isolated from and ignorant of the very department in their charge. Real power lay with the chiefs of the five bureaus—Navy Yards and Docks; Construction, Equipment, and Repairs; Provisions and Clothing; Ordnance and Hydrography; and Medicine and Surgery—who ran the Navy's day-to-day affairs. All too often, however, these men, whatever their abilities, were more interested in defending their own bureaucratic turf than in promoting the service as a whole. To inoculate themselves from outside pressure, they usually cultivated alliances with congressmen and other naval officers, making it extremely difficult for navy secretaries to remove them or to implement necessary reforms that might impinge on their prerogatives.

Unfortunately, the Navy's officer corps was equally dysfunctional. The Naval Academy did not open until 1845, so almost all of the high-ranking Civil War officers entered the service as teenage midshipmen who learned their trade shipboard through on-the-job training. They had almost no formal education in strategy, tactics, international relations, history, logistics, technology, management, or politics. Promotion was by strict seniority, not merit, so longevity counted for more than ability and accomplishment in ascending the naval register. Moreover, until the 1850s there was no provision for retirement, so officers stayed in the service as long as possible in order to draw their pay, thus clogging the uppermost ranks with superannuated men who contributed little to the Navy's efficiency. The only way to get rid of the aged, incompetent, physically infirm, alcoholic, and insane—short of resignation or death, that is—was through courts-martial, but even this was rarely effective because officers frequently used their congressional allies to pressure the navy secretary or president to set the verdicts aside. Instead, navy secretaries learned that there were certain useless officers whom they should never assign to sea duty, but these men continued to receive their salaries and rise through the chain of command. Needless to say, this bizarre system demoralized capable officers stuck for years in lower ranks and seemingly unable to make much of a difference. One young officer, future Spanish-American War hero George Dewey, later remembered, "Such a system was killing to ambition and enterprise. It made mere routine men to face a crisis in which energy and initiative were needed. No subordinate was expected to undertake any responsibility on his own account."[27] Not surprisingly, many officers resigned their commissions out of frustration to take more lucrative civilian jobs. Others turned to drink, or channeled their resentment into the petty quarrels and cliques that plagued the Navy before the Civil War. The most remarkable thing about the prewar officer corps was not the innumerable worthless men who infected the service, but the many proficient officers who stayed, did their duty well, and fought for needed reforms. Thus, when the Civil War began, Welles' challenge was to identify these talented officers and find ways to give them positions of authority.

Actually, the Navy had already attempted a significant and partially successful stab at personnel reform several years before Welles took office. In 1854 Cdr. Samuel "Frank" Du Pont persuaded one of the more dynamic prewar navy secretaries, James Dobbin, to take action against the deadwood in the officer corps. At Dobbin's urging, the following year Congress authorized the creation of a fifteen-man Efficiency Board to review all officers and weed out those unable to contribute to the Navy ashore or afloat. Those officers found wanting would be put on a reserve list at partial pay—retired, in effect—or dismissed from the service altogether. They would be replaced from the lower ranks in order of seniority. The board—whose members included Du Pont, Cornelius Stribling, Andrew Foote, Frank Buchanan, and Samuel Barron—met throughout

June and July of 1855 and ruthlessly performed its duty. When it finished, it recommended the removal of 201 officers. Some of the Navy's most prominent men were among those discharged, including Capt. Charles Stewart, the most senior captain in the service. Not surprisingly, considering the high stakes involved, the board's findings provoked a fierce and immediate backlash. Targeted officers and their congressional allies complained that the board had kept no official record and provided no due process. Feeling the heat, Congress backpedaled and in 1857 gave dismissed officers the opportunity to appeal their cases to a court of inquiry that ultimately reinstated sixty-two of them. Even so, the Efficiency Board was not a complete failure. It rid the Navy of some of its worst officers, led to the promotions of some capable men, and familiarized Congress with the Navy's complicated personnel problems.[28]

One of Welles' problems was that the Navy's personnel system contradicted everything in which he believed. As a Jacksonian Democrat from the Connecticut middle class, Welles had little respect for aristocracies of any kind, especially one as blatantly undemocratic and arbitrary as the Navy's. While Welles conceded that seniority had some value in maintaining order and harmony in the tradition-bound officer corps, he was not convinced that it provided the best means for selecting officers for responsible positions. Nor did he like all the bickering cliques that permeated the service. Some of the worst of them disintegrated when their members resigned their commissions and went south, but others remained. Welles thought that these cliques protected mediocre officers from the scrupulous evaluation that the Navy so desperately needed. Welles' instincts, honed by years of experience in the cutthroat world of newspaper publishing and politics, told him to judge officers not by their rank or friends, but by a rigorous and thorough examination of their character, abilities, record, and fidelity. Only then could he identify the right officers for the right tasks.[29]

Whatever changes Welles hoped to make in the Navy's command and personnel structures, he would have to do without Hiram Paulding. Despite Paulding's sympathy with the seniority system, Welles continued to respect him for his hard work, solid judgment, and loyalty. Most observers agreed with Welles that Paulding had used considerable discretion and tact in assigning officers to their posts, but there were some malcontents. Chief among them was the powerful Blair family of Maryland. Patriarch Francis Preston Blair was an adviser to presidents going back to Andrew Jackson, and two of his sons, Frank and Montgomery, were equally influential, the former being a prominent Missouri politician and the latter serving in Lincoln's cabinet as postmaster general. The Blairs did not believe that Paulding had given one of their kinfolk, Cdr. Samuel Phillips Lee, the command they thought he deserved, so they conspired to have Paulding replaced with Frank Du Pont, now a captain. Although normally as politically astute as they were ferocious, the Blairs made the mistake of asking Seward to raise the issue with Welles in late May. Welles was hardly in the mood to tolerate more meddling from the secretary of state, so he frostily rejected the proposal, and Lincoln backed him up. In the end, however, it was overwork and exhaustion that drove Paulding from his post. Paulding had found his job distasteful from the start, and the weeks of endless grind took their toll. Sometime during the spring, he asked Welles to reassign him as soon as he could be spared, and Welles finally did so in September. Welles gave Paulding the plum assignment of commandant of the Brooklyn Navy Yard, which, after a vacation with his family at his Long Island farm, Paulding took up in late October. He spent the remainder of the war there capably running the naval yard, sitting on various committees, and supporting Welles against his enemies. In gratitude for valuable services rendered, Welles later persuaded Lincoln to nominate Paulding as one of the original rear admirals on the retired list.[30]

By the time Paulding left Washington, Welles had already found—or, more accurately, had thrust upon him—the person who served as his right-hand man throughout the war: Gustavus Fox. Although the Blairs may have disliked Paulding, they had nothing against Welles. In fact, Francis Blair and Welles were longtime political allies and friends. Montgomery Blair in particular believed that Welles needed a professional at his side to offer advice and share the workload, so he recommended Fox as the Navy Department's chief clerk. Lincoln, who appreciated Fox's energy and forthrightness during the Fort Sumter crisis, liked the idea too. Welles, though, was not so sure. He respected Fox also, but he had already slated William Faxon, an old newspaper colleague from Connecticut with lots of administrative experience, for the job. One morning, however, Fox showed up at the Navy Department building with a note from Lincoln asking Welles to appoint him chief clerk. Although Welles normally did not like such interference in his department, he initially acquiesced and awarded the post to Fox on 9 May, much to Faxon's dismay. Fortunately for everyone, in the end Welles found a way to retain the services of both men. In July Congress created the position of assistant secretary of the navy, which Lincoln at Welles' behest conferred on Fox on 1 August. Welles then reappointed Faxon his chief clerk. Although Welles was satisfied with the arrangement, Faxon never completely reconciled himself to it because Fox's gregariousness annoyed him. Nevertheless, together these three men—Welles, Fox, and Faxon—formed the triumvirate that capably operated the Navy's administrative machinery throughout the war.[31]

The Massachusetts-born Fox was a Navy man, having entered the service as a midshipman in 1838. He fought in the Mexican War and later commanded mail steamships. Like many young capable officers, however, he realized that the Navy's stultifying seniority system provided him with few opportunities for timely advancement, so at his new wife's prompting he resigned his commission in 1856 to work in the wool industry as an agent for the Bay State Mills Company in Massachusetts. He prospered in civilian life, learning much about the business and technological worlds that would later serve him well in the Navy Department. The Civil War brought him back into government employment. Kissing his wife a hurried good-bye, he rushed to Washington from New York City to offer his help to the Lincoln administration in the tense days following the president's inauguration. Although Fox's expedition to rescue the beleaguered Union garrison at Fort Sumter failed, his can-do attitude impressed everyone, including Lincoln and Welles. He was related to the Blairs by marriage—he and Montgomery Blair both wed daughters of Levi Woodbury, a former navy secretary during the Jackson administration—so it was hardly surprising that Montgomery's thoughts turned to his brother-in-law when he concluded that Welles needed some help running his department, and it was at his urging that Congress created the position of assistant secretary.[32]

In fact, Fox impressed people for many reasons. He was a witty, affable, and direct man full of humorous stories and fond of good food and wine, traits that appealed to the president, with whom he frequently socialized. Fortunately for the Union, there was more to him than mere personality. He possessed an encyclopedic knowledge of the Navy and its way of doing things, but he also understood the service's innumerable organizational deficiencies. These factors, combined with a keen mind, boundless energy, and a determination to get things done, contributed enormously to the Navy's Civil War accomplishments. He had his flaws, however. In the first part of the war, he enjoyed planning more than execution, so he sometimes got bogged down in the preparatory details of important operations. Although he was usually friendly with the press, his occasional officiousness toward congressmen and naval officers generated unnecessary anger and resentment against the Navy Department. His bonhomie occasionally obscured serious disagreements instead of clarifying them, leading to misunderstandings and hard feelings later. He disliked the Army, which complicated interservice planning, and he sometimes played favorites with officers whose interests he looked after. Finally, the constant pressure and tension of his office took a toll on his health, so he became increasingly rotund and pasty-faced as the war progressed. Taken as a whole, though, it is hard to see how anyone could have better filled his position.[33]

Despite their obvious differences in personality, the extroverted Fox and cantankerous Welles got along extremely well. Their offices were close together on the second floor of the Navy Department building, and they consulted each other frequently on problems big and small. As Welles confided in his diary after the war, "We have worked together with entire harmony, never in a single instance having a misunderstanding. I have usually found his opinions sensible and sound. When I have had occasion to overrule his opinions, he has acquiesced with a readiness and deference that won my regard."[34] Indeed, the two men complemented each other's strengths, permitting them to divide up the Navy Department's workload. Fox's nuts-and-bolts understanding of the Navy enabled him to concentrate on the minutiae of outfitting and equipping warships, supervising the development of new technology, lobbying Congress, dealing with the press, and engaging in strategic planning. Although Lincoln once noted that Fox really ran the Navy Department, this was a gross simplification. Instead, Fox served as the navy secretary's de facto chief of naval operations, allowing Welles to focus on the big picture by overseeing the implementation of departmental policies.[35]

Their relationship, while good, was not trouble free. The president was one source of difficulty. Since Lincoln and Fox were close socially, the president often consulted with Fox, not Welles, on naval matters. Lincoln defended this practice by claiming that he did not want to waste Welles' valuable time, so he only sought out Fox when he needed information, not decisions. Welles, however, did not like this one bit. In addition, Fox sometimes issued orders in his own name after he had secured Welles' approval, giving the impression that he was making the final decisions. This practice, along with his occasional overbearing attitude, led naval officers, bureau chiefs, and congressmen to complain to Welles about Fox's highhandedness. Welles was aware of these problems, but, after weighing the pros and cons, concluded that Fox's sins were minor compared to the vast benefits he brought the department.[36]

Fox also played an important part in selecting squadron commanders. Like Welles, Fox believed that the Navy needed to overcome its obsession with seniority, and instead rely on merit in assigning officers. Finding the best available men for the right positions, however, was not easy. In an inbred, clique-bound organization like the Navy, where courts-martial were convened for the most trivial of reasons, officers were often reluctant to tell Welles and Fox the unvarnished truth in official dispatches that might become public. This made it difficult for Welles and Fox to evaluate officers accurately. To circumvent this obstacle, Fox, with Welles' consent, cultivated an unofficial correspondence with dozens of officers that did not go into the Navy Department's files. In his letters Fox encouraged these officers to explain their problems, including those of personnel, in a forthright and honest manner so he could help overcome them. Doing so enabled Fox to tap into the Navy's extensive gossip network. Naturally, some officers used their conduit to Fox for self-promotion at the expense of others, but Fox and Welles were willing to run that risk. Fox's unofficial correspondence provided Welles with an additional tool with which to screen officers for important positions, and it undoubtedly contributed to his ability to choose good squadron commanders.[37]

The war's outbreak brought about a rapid increase in the Navy's strength as Welles' agents scoured the eastern seaboard to purchase vessels capable of mounting a few cannon and cruising along the Confederate coast. Someone had to command these improvised warships, so the Navy experienced the unusual problem of having too many berths for its available officers. For once almost everyone on the naval register had an assignment. Gratifying though such full employment might seem, implying as it did the complete utilization of available resources, there was a downside to it. Many of the captains were physically or mentally unfit, but they still wanted positions worthy of their station. Not surprisingly, they took a dim view of the idea of serving under men with less seniority and rank, or of having their applications rejected altogether. This greatly complicated personnel assignments because the Navy contained no grade above captain to which to appoint capable officers. It did not take Welles long to conclude that this seniority-based system was unworkable, and that the best solution was to get Congress to change the law. Although Congress had usually shown little inclination to grapple with the Navy's complicated personnel structure in peacetime, the impetus of war put its members in a different frame of mind. Unfortunately, Welles and Fox did not get along with the chairman of the Senate's Committee on Naval Affairs, New Hampshire's John Hale, who at times seemed more interested in waging war with the Navy Department than with the rebels. As far as Welles was concerned, Hale was a hypocritical, buffoonish, and dishonest man. In fact, Hale provoked Welles into scrawling one of the few humorous entries in his diary: "Oh, John P. Hale, how transparent is thy virtue!" Instead, Welles and Fox worked closely with Charles Sedgwick, Hale's counterpart in the House of Representatives, and Senator James Grimes of Iowa, who served on Hale's committee and was close to the assistant secretary. Together, these men crafted several laws that revolutionized the Navy's command system during the war.[38]

The first, enacted on 3 August 1861, established a board of naval officers to review the active list and recommend for retirement or dismissal those incapable of performing their duties. This time, however, Congress gave officers the right to a full and fair hearing before the board. Although the board convened on 18 October, subsequent events overtook its work. With so many officers off fighting the rebels, it was simply impractical to provide the kind of due process the law demanded. Instead, Welles resorted to a simpler and more draconian solution. In his annual report in early December, Welles called for the mandatory retirement of all officers who had served more than forty-five years in the Navy, but with the proviso that the president could recall these retired officers back to active duty if necessary. Welles also asked for official sanction of the hitherto courtesy title of flag officer for squadron commanders so they would outrank all the captains who would otherwise have to be transferred elsewhere because of their seniority. Congress responded favorably, and on 21 December Lincoln signed legislation more or less in line with Welles' recommendations, except that it retired officers not only after they had served forty-five years, but also if they were at least sixty-two years old. It also permitted retired officers to hold ship or squadron command only with the Senate's consent.[39]

This legislation was significant for a couple of reasons. For one thing, it pensioned off all but fourteen of the sixty-five remaining captains on the active list. Welles could ask Lincoln to recall them to active duty if he thought they could contribute to the war effort, but they would serve at his pleasure where he assigned them. Their only legal recourse to return to duty, therefore, was to convince Welles of their usefulness, which guaranteed their cooperation and subservience. In addition, the establishment of the flag officer rank permitted Welles to appoint any captain, or even commander, to squadron command, regardless of the rank and seniority of his new subordinates. The law, then, freed Welles from the seniority system's restrictive bonds and gave him substantial flexibility in selecting squadron commanders based on his own personal criteria. Seniority, in short, now had only as much importance as Welles chose to attach to it.

Considering the firestorm that earlier legislation to reform the Navy's personnel structure had ignited, the officer corps' response to these laws was unexpectedly muted. Not surprisingly, those officers retired against their will were most upset, even though Welles had Lincoln recall many of them to duty. Some believed that the new policy was part of a conspiracy by younger men such as Frank Du Pont to supplant them. Others associated their high rank with ability, so, using this logic, they argued that removing them to make way for more junior officers meant placing less competent men in positions of power. Finally, some complained that the Navy was discarding them like so much trash after years of devoted service to live out their remaining years in poverty on their inadequate pensions. As Capt. Louis Goldsborough explained to Fox, "The rule . . . is certainly a slashing one, and carries with it not only no discrimination, but sheer cruelty." During wartime, though, it seemed almost unpatriotic to protest too loudly. Goldsborough, for example, continued, "I shall not say one word to oppose it, however, until after I am kicked adrift and consigned to poverty, nor shall I allow the probable fate that awaits me to interfere a solitary iota with my duty to my country, and the sacred cause in which it is now engaged."[40] Besides, there was considerable support for the law among junior officers, if for no other reason than it gave them the greater opportunity to rise to important commands.[41]

This legislation solved many but by no means all of the Navy's command troubles. The war required considerable interservice cooperation, but the Army's and Navy's different ranking scales made effective coordination difficult. The Army had generals, but there was no naval equivalent. Indeed, some naval captains complained that even army regimental colonels tried to boss them around. Although the establishment of the flag officer alleviated some of these problems, Fox in particular recognized that the Navy could never deal with the Army from a position of parity until it had a rank comparable to an army major general. He and Welles therefore lobbied Congress for an appropriate remedy, and Congress responded in July 1862 by passing An Act to Establish and Equalize the Grade of Line Officers in the United States Navy. The law abolished the flag officer position and replaced it with two new ranks, commodore and rear admiral, that were analogous to the Army's brigadier general and major general. Congress called for the creation of up to eighteen rear admirals, half from the active list and half from the retired list. It authorized the president to appoint the retired rear admirals, with the Senate's consent, from those pensioned captains who had "given the most faithful service to their country." As for active list rear admirals, the president would also nominate them for the Senate's approval, but he had to choose them only from the commanders, captains, and commodores who had received Congress' official thanks for distinguished services.[42]

Although one officer hailed the law as the most important piece of naval legislation in a half century, this did not mean that it met with universal acclaim. Some complained that requiring a congressional vote of thanks for active-list rear admirals would turn the new rank into a political football manipulated for partisan reasons. Moreover, they doubted Congress' ability to recognize and determine true merit. Others noted that creating rear admirals without full admirals was as illogical as having a vice president without a president.

Less than a month after the law was enacted, Welles submitted his recommendations for the nine retired-list rear admirals. He did not rely on strict seniority, but instead omitted retired captains whose enthusiasm and competence he doubted. Not unexpectedly, some of the overlooked officers protested the humiliation Welles inflicted on them. Hardly anyone, however, commented on the power the law bestowed on the executive branch. It said nothing about seniority in selecting rear admirals, so the president—or, in practice, Welles—could use almost any criteria he wanted in nominating officers for the position. While Welles was quick to fill the retired admiral slots, he only appointed seven men as active-list rear admirals during the war.[43] He deliberately left several vacancies as incentives for officers to perform well. As he commented in his diary, "The higher appointments must be kept open to induce and stimulate our heroes."[44] Instead, he made some squadron commanders acting rear admirals until they had proven themselves worthy of a permanent commission. From now on, rising to the highest positions in the Navy was no longer contingent on longevity and seniority, but rather on pleasing the austere navy secretary.[45]

The Strategy

Alexander Dallas Bache had much in common with his great-grandfather, Benjamin Franklin. Like Franklin, he was a scientist and educator who devoted his life toward advancing both fields. Although he graduated from West Point at age nineteen without accumulating a single demerit, he resigned his commission in 1828 to become a professor of natural philosophy and chemistry at the University of Pennsylvania. In addition, he reorganized Philadelphia's public schools and served on the Lighthouse Board. Finally, in 1843 he was appointed superintendent of the federal government's Coast Survey, whose mission, as its name implies, was to chart the American shoreline. The Coast Survey worked so closely with the Navy that the Navy often lent out its officers for service on Coast Survey vessels. The Civil War's outbreak, however, threatened Bache's beloved Coast Survey with neglect and irrelevance. Fortunately, Bache had also inherited some of his great-grandfather's bureaucratic savvy. So when Fox repeatedly asked for data on Confederate coastal waters, Bache recognized an opportunity to secure for the Coast Survey a role in the war.[46]

Like any good bureaucratic infighter, Bache knew the importance of relationships in getting things done. He was already part of Fox's extensive social network, and he had agreed with Cdr. Charles Davis, who was still helping Paulding assign officers, to eat dinner together every Tuesday evening. Bache also approached Welles, but while the navy secretary liked Bache personally, he made it clear that he was simply too busy to invest the necessary time in a friendship. To all these men, Bache in mid-May suggested the establishment of a board that would examine the Confederate coastline and recommend the best way to conduct a naval blockade of it. Such a study would obviously require Coast Survey participation because it possessed much of the necessary data. Both Welles and Fox liked the idea. The president had already on 19 April ordered the seceded Southern states blockaded, but such a daunting task required serious planning, which Bache's proposed board could provide. As for the board's composition, Bache was an obvious choice. Welles and Fox also wanted help from the Army, so they secured the services of its chief engineer, Maj. John Barnard. For the Navy, Fox recommended Capt. Frank Du Pont, one of the Navy's rising stars. Du Pont was currently commandant of the Philadelphia Navy Yard, but he agreed to chair the board. He was a convivial man who enjoyed the Washington social scene, and he wanted to be near the center of the Union war effort when Congress convened that summer. Moreover, he believed that the blockade would play a vital role in winning the war, and he hoped to contribute to its effectiveness. Finally, the job of board secretary went to Charles Davis. Davis felt as though he already had enough work on his plate just then, so he was unhappy with yet another assignment, though he welcomed the opportunity to renew his friendship with Du Pont.[47]

Du Pont's committee, variously called the Strategy Board or Blockade Board, met almost daily at the Smithsonian Institution throughout the summer of 1861 to pore over the vast amount of data the Coast Survey possessed. Du Pont at first expected that the board would require only a few weeks to finish its assignment because Welles' official orders called for it to simply select two ports along the Atlantic Coast that could serve as coaling stations for the Union blockading fleet. Bache, however, wanted the board to write a more elaborate memorandum, and he carried the day. Davis was initially optimistic that Washington policy makers would take the board's reports seriously, but his enthusiasm waned as the summer dragged on, interest lagged, and events overtook the board's mission. On 25 July, though, Welles and Fox summoned the board members to the Navy Department building to review their work. They had only completed their survey of the Atlantic Coast, but Welles and Fox were still impressed enough to forward their conclusions to Lincoln, the cabinet, and Winfield Scott, general in chief of the Army. The board did not issue its last report until late September, and by then Du Pont and Davis were too busy with a new project to write a planned summary.[48]

The Strategy Board ultimately produced six major reports—three on the Atlantic Coast and three on the Gulf Coast—full of mind-numbing details about currents, tides, shoals, topography, the local population, nearby natural resources, and so on. Although the Navy had already undertaken efforts to implement its blockade of the Confederacy, the Strategy Board's memoranda underscored the enormity of the Navy's mission and offered some suggestions to fulfill it. For one thing, it clarified the geographic complexities that the blockade entailed. The Confederate seaboard stretched 3,550 miles from the Potomac River to the Rio Grande, and that was not counting the innumerable island chains that hugged the shoreline. There were 189 harbors and inlets in which blockade-runners could take refuge, though fortunately only a few possessed rail-

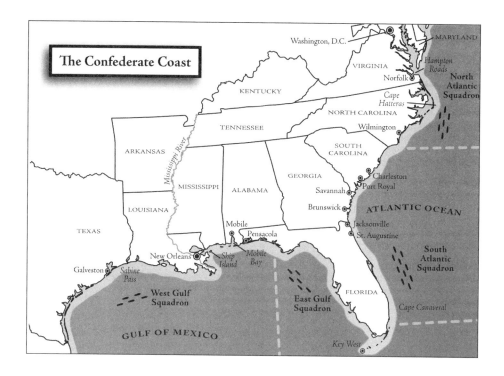

roads connecting them with the interior. Port cities such as Wilmington, North Carolina, Charleston, South Carolina, Savannah, Georgia, St. Augustine, Florida, Mobile, Alabama, New Orleans, Louisiana, and Galveston, Texas, studded the Confederate periphery. Local geography surrounding these and other towns varied greatly. The North Carolina littoral, for instance, contained low and sandy islands that separated the stormy Atlantic Ocean from large and comparatively placid inner sounds. Georgia's coastline reminded the board members of Holland's intricate island network. West Florida's shores were barren, sterile, and unpopulated. And Louisiana's coast was a labyrinth of swamps, bayous, and lakes bisected by the expansive Mississippi River delta. Such geographic particulars were more than enough to demoralize even the most stalwart naval officer charged with blockading the Confederacy.

Fortunately, the Strategy Board did more than just point out difficulties; it also offered a number of solutions to these seemingly intractable problems that ultimately found their way into the Union Navy's strategic plans. As

Welles had already realized, the Navy's steam-propelled warships needed supply bases along the rebel coast to take on coal. Without them, warships would have to devote most of their cruising time to steaming back and forth to bases up north. The Strategy Board called for the occupation of isolated Confederate harbors that could be easily seized, held, and converted into logistical bases. Doing so would not only extend the amount of time Union warships could remain on their blockading stations, but it would also mean one less harbor for the Navy to watch. The Strategy Board also strongly implied that the best way to seal off large Confederate port cities such as New Orleans was to simply conquer them at some point. All these amphibious operations, however, required army cooperation. Welles understood this too, noting, "As for the Navy it has the hardest and most thankless practice of the war. There is no commerce on the other side to enrich it—no way from which to win laurels—we cannot go and bombard Charleston or any place but by cooperation with the Army. Yet we are blamed for not doing all these things."[49] Although the Strategy Board also acknowledged this, it provided no guidance as to how such complicated interservice coordination would work. Nor did it recognize the strategic possibilities that occupying stretches of the rebel coastline presented the Army. In fact, the Navy's inability to persuade the Army of the value of working together to seal the Confederacy off from the outside world was the Union's biggest strategic mistake of the war. Finally, the board failed to appreciate the Mississippi River's value to the Confederacy. Even so, the Strategy Board's suggestions provided Welles and Fox with the rough blueprint they needed to construct the Navy's blockade.

Ready for War

The Navy was a deeply flawed organization in many ways when the Confederates fired on Fort Sumter. Fortunately, Welles was familiar enough with the institution from his days as a bureau chief to recognize many of its problems from the start of his tenure as navy secretary, and confident enough in his instincts to propose the necessary remedies. Welles was not the first navy secretary to realize that the Navy's command and personnel systems in particular were fundamentally faulty, but he had a number of advantages that enabled him to implement the proper reforms. These included a supportive president, a cooperative Congress, a dynamic and knowledgeable assistant secretary, and sympathetic officers. Most important of all, however, was the war itself, which focused attention on the Navy's difficulties like nothing else could have. To support a Navy capable of waging war effectively, the president and Congress were willing to sacrifice long-enshrined but counterproductive principles like seniority. Thus, in the space of a little more than a year, Welles succeeded in purging the Navy of disloyal and unenthusiastic officers in responsible positions, neutralizing the seniority principle, restructuring the ranking system, and formulating the outlines of a strategy for his commanders to follow. All that Welles now needed to do was to select the right squadron commanders and make sure they had the proper tools to complete their missions. But this, as Welles discovered, was more easily said than done.

CHAPTER TWO

Atlantic Storms

Silas Stringham and the Atlantic Squadron

THOSE UNIONISTS who believed that the Lincoln administration could crush the rebellion in a matter of weeks received a rude shock on 21 July 1861. That day, the Confederates defeated a Union army marching on the rebel capital of Richmond, Virginia, at the Battle of Bull Run, and drove it back in confusion to Washington. Confederates were of course elated by their victory, which seemed like conclusive evidence that one Southerner could whip any number of Yankees. In reality, though, Bull Run obscured other events that spring, summer, and fall that taken together gave the Union the edge in the first months of the war. In Maryland, Lincoln stretched the law to the breaking point by imprisoning pro-Confederate legislators to prevent the state from seceding from the Union. Rebel victories in Missouri at Wilson's Creek and Lexington did not change the fact that Unionists had already seized control of the state's political machinery. Finally, the ill-conceived Confederate occupation of Columbus, Kentucky, in early September drove that hitherto neutral state into the Union camp. The rebels may have won the war's first major battle, but the Lincoln administration had succeeded in securing the strategically vital Border States that were collectively worth a dozen Bull Runs.

Facing as yet no real opposition afloat, the Navy managed to avoid any Bull Runs of its own. It had its fair share of problems though. Cabinet

intrigue and disloyal officers were bad enough, but they paled in comparison with the debacle at the Norfolk Navy Yard in Virginia a week after
Fort Sumter fell. In addition to possessing one of only two granite dry
docks in the country, the Norfolk Navy Yard also contained more than
one thousand artillery pieces of various sorts, a hospital, magazines,
foundries, and machine and boiler shops. It had, in short, everything a
blockading fleet required to maintain itself. There were several vessels laid
up there as well. Most were relics from the bygone era of sail, but one of
them, the screw frigate USS *Merrimack,* was among the most powerful
warships afloat. Unfortunately, Welles failed to give Capt. Charles
McCauley, the yard's superannuated and befuddled commandant, clear
and timely orders to safeguard the federal government's property at all
costs because he did not want to do anything to provoke Virginia into
seceding from the Union. As it was, Virginia left the Union on 17 April,
and rebel forces began gathering outside the naval yard soon after.
McCauley did not believe that he had sufficient manpower to protect the
warships, so on 20 April he ordered the anchored vessels, including
Merrimack, burned and scuttled. McCauley's panic proved contagious.
Welles had already dispatched Hiram Paulding there to consult and advise
McCauley. When he arrived, Paulding discovered the naval yard in a state
of chaos. He therefore directed the naval personnel to destroy anything
else of value to keep it out of Confederate hands. Paulding and his men
worked throughout the night before evacuating the yard just before dawn,
but they lacked sufficient time to do a thorough and systematic job. When
rebels occupied the place the next morning, they recovered nearly twelve
hundred cannon—including three hundred modern Dahlgren guns—that
eventually found their way into rebel warships and fortifications throughout the Confederacy. They also gained the dry dock, vast stores of ammunition, 130 gun carriages, valuable machinery—and the opportunity to
raise, refloat, and refit *Merrimack* in a way those who sank her could not
have imagined.[1]

The Norfolk Navy Yard fiasco mortified Welles, but it did not stop
him from doggedly prosecuting the war. Lincoln had proclaimed a blockade of the seceded Southern states on 19 April, presenting Welles with his
first major wartime personnel decision. To command the blockade of the
Confederacy's Atlantic Coast, Welles turned to Capt. Silas Stringham. On
the surface, Welles' reasoning made perfect sense. He had known
Stringham for nearly five years, and Stringham had performed faithfully

and dutifully as Welles' naval adviser that spring. Stringham was also one of the most respected officers in the Navy, with numerous supporters from his home state of New York who vouched for him. He possessed plenty of experience, having served in the Navy for fifty years, twenty of them as a captain. He was tenth on the list of captains, and at sixty-four years old was younger than his seniors, almost all of whom were too enfeebled one way or another for sea duty. Stringham was in Boston, readying the steam frigate USS *Minnesota* for action, when Welles chose him as squadron commander. On 1 May Welles ordered Stringham to begin collecting, organizing, and outfitting the vessels his newly named Atlantic Blockading Squadron would need to conduct a blockade from Chesapeake Bay to Key West, Florida.[2]

Stringham had entered the service as a twelve-year-old midshipman back in 1809, and was on board the frigate USS *President* when she pummeled the British sloop HMS *Little Belt* in 1811. He participated in the War of 1812, the expedition to Algiers, the campaign against West Indies pirates, and efforts to suppress the African slave trade. He was promoted to captain in 1841, ran the Brooklyn Navy Yard for three years, and during the Mexican War commanded the ship of the line USS *Ohio* during the siege of Veracruz. Afterward he led the Brazilian and Mediterranean squadrons, and was on court-martial duty in Washington when Welles summoned him to the Navy Department building to become his adviser. Ruddy, clean-shaven, with thinning graying hair and a pleasant but tired smile, Stringham was an upright and honorable gentleman, but he could also be wordy and ponderous. Although a perfectly acceptable officer during peacetime, he was ill-suited for a modern conflict like the Civil War. Stringham was fully capable of captaining a warship, but leading a squadron of dozens of disparate vessels deployed over hundreds of miles of ocean in wartime was beyond him. He had quit as Welles' naval adviser because he did not possess the talent and temperament for the bureaucratic wrangling the position required, but squadron command involved much of the same kind of tedious paperwork. Worst of all, however, Stringham lacked the killer instinct that motivated aggressive and hard-driving men to come to grips with the enemy at every opportunity. It was Stringham's misfortune to be an ordinary officer when success called for extraordinary abilities.[3]

To implement the Atlantic Coast blockade, the Navy Department initially put at Stringham's disposal a small flotilla operating on the

Potomac River, Capt. Garrett Pendergrast's tiny Home Squadron, which had recently returned from Mexican waters, and fourteen additional vessels hurriedly scraped together. Stringham arrived off of Hampton Roads on 13 May in his flagship, *Minnesota,* and began assigning warships to patrol Confederate shores. Welles had given him considerable latitude to act according to his best judgment, but soon Stringham complained to the Navy Department about the burdens placed upon him. He grumbled that he lacked sufficient strength to carry out his mission, and that those warships he did have were in poor condition. He also protested his deficiency in and high turnover rate of his younger officers. While there was considerable truth to Stringham's grievances, their validity did little to endear him to Welles and Fox. After all, the Navy Department was purchasing, equipping, and manning vessels as rapidly as possible, but there was nothing easy about it. Shortages were endemic in almost everything except obligations. Welles and Fox, however, expected their officers to make do with what they had, not constantly grouse about what they needed. They were increasingly concerned with the blockade's apparent ineffectiveness, for which they blamed Stringham. Welles also suggested that Stringham seize some point along the rebel coast as a supply depot, but Stringham made no immediate plans to do so. Stringham resented such criticism, noting in particular that it was hard to conduct an efficient blockade when prize courts freed even ships flying Confederate flags. He summed up his frustration in a stiff letter to Fox: "You, Sir, as an officer of some experience must appreciate the very unpleasant position in which I am placed."[4]

The long, narrow, sandy, and barren islands that composed North Carolina's Outer Banks served as sentinels protecting the state's mainland. Several gaps, or inlets, in the island chain provided blockade-runners with opportunities to dart out to sea and through the blockade. Worse yet, the stormy waters off Cape Hatteras in particular made it difficult for Union warships to remain on station there. The Strategy Board had called Welles' attention to Hatteras Inlet in one of its reports, and recommended sealing it by sinking stone-filled vessels, called blockships, there. The rebels, however, had constructed two hastily erected earthworks, Fort Hatteras and Fort Clark, to defend the inlet, and garrisoned them with 580 men. Welles wanted to assault the forts, but by now he had serious doubts that Stringham was the man for the assignment. He approached Frank Du Pont, who was still in Washington chairing the Strategy Board, and offered him the job. Du Pont declined because he would not have as much control as he wanted, or so he later told a friend. In all likelihood, he was also waiting for more important duty elsewhere. Welles thereupon summoned Stringham to the capital in late July to explain the mission, but Stringham's lack of enthusiasm worried the navy secretary. At this point, though, Maj. Gen. Benjamin Butler, the commander of Union troops at Fortress Monroe, lent a hand. Butler was one of Lincoln's political appointees, and as such he was always looking for ways to get his name in the newspapers and erase the stain of some previous defeat—in this case, the Battle of Big Bethel on 10 June—so he persuaded Winfield Scott to send him and 860 men to help the Navy. Butler's contribution gave the operation an impetus it had hitherto lacked, prodding Stringham into action. Bad weather delayed the expedition's departure for nearly two weeks, but finally, on 25 August, Stringham's fleet of seven warships and two troop-laden steamers towing a couple of blockships left Hampton Roads for Hatteras Inlet.[5]

The advent of steam power changed warfare in all sorts of ways that Union naval officers only dimly understood at the conflict's onset. In particular, it freed warships from their reliance on wind and tides, giving them a predictable and controllable mobility they had previously lacked. This made it harder for land-based cannon to target them, as the Confederates in Fort Clark and Fort Hatteras discovered. The Union fleet arrived off Cape Hatteras on 27 August, and Stringham bombarded Fort Clark the next day. Union cannon had greater range and firepower than the rebel guns, so the outcome was practically predetermined. The rebels abandoned the fort that afternoon, even though the heavy surf swamped and wrecked the Army's landing craft after only a few hundred men had reached the shore. Next day at dawn, Union warships turned their attention to Fort Hatteras, which surrendered sometime around 11:00 AM after a hellish barrage. Union forces captured more than seven hundred prisoners, including reinforcements who arrived during the battle, while suffering no casualties of their own. At this bargain price, the Union Navy won its first victory of the war.

Although Butler played little role in the battle, he rushed back to Washington to claim credit for delivering the good news and to secure retroactive permission to leave his soldiers behind to garrison the captured rebel forts. Arriving in the capital from Annapolis after midnight on 31 August, Butler spotted from his rented carriage Montgomery Blair and Fox talking in

Blair's study. Eager to share his information, he stopped to tell them about Stringham's success. Fox was so elated that he insisted on taking Butler across the street to the White House immediately to inform Lincoln. Butler doubted the propriety of disturbing the president so late at night, but Fox argued that Lincoln would sleep better with the intelligence than without it. The watchman ushered them into the cabinet room, and a few minutes later a nightshirt-clad Lincoln joined them. Without formality, Fox blurted the most welcome news. He and Lincoln threw their arms around each other—awkwardly, though, because Fox was as short and portly as Lincoln was tall and lanky—and danced an impromptu jig. To Butler, Lincoln expressed sentiments that might have applied just as well to Fox and the Navy: "You have done all right. You have done all right."[6]

Welles exaggerated when he told Stringham that the operation's "brilliant" results could "scarcely be overestimated," but the capture of Fort Clark and Fort Hatteras did yield the Union some strategic benefits. Butler's decision to leave his troops behind to garrison both forts gave the Union a permanent foothold in the region. In the aftermath of the Union triumph, the Confederates evacuated other positions in the Outer Banks, enabling the Union to cordon off much of the North Carolina coastline. Stringham's victory also exposed Pamlico and Albemarle sounds, as well as all the rebel cities and towns on their shores, to Union attack, although the Navy was not yet ready to take advantage of the opening. In addition, it raised Union morale in the depressing weeks after the Battle of Bull Run, and showed naval officers that contrary to the conventional wisdom, their steam-powered warships could successfully assail rebel coastal fortifications.[7]

Unfortunately for Stringham, his success at Hatteras Inlet did little to restore his dwindling credibility in the Navy Department. Welles left Washington on 6 September to inspect naval yards and visit his family in Connecticut, leaving the Navy Department in Fox's hands. While Welles' frustration with Stringham was rapidly approaching the boiling point, he continued to like him as a person. Fox, on the other hand, had scant regard for Stringham professionally or personally, in part because of Stringham's lack of enthusiasm for Fox's plan to relieve Fort Sumter the previous spring. As far as Fox was concerned, Stringham epitomized the slow-witted and ossified officers who infected the Navy. Stringham had steamed to New York City in *Minnesota* after his victory to deliver rebel prisoners, and he stopped in Washington on his way back to Hampton Roads. Welles had already left town, but Stringham's conversation with

Fox did nothing to change the assistant secretary's low opinion of him. Nor did subsequent reports that Stringham had left warships idle at Hampton Roads while several rebel ports went unblockaded. Thoroughly disgusted by now, Fox expressed his displeasure in a series of terse telegrams to Stringham on 14 September whose collective tone was more chastising than their content.[8]

Obeying orders was of course important to Stringham, but so was protecting his personal honor. He had labored long and hard to establish the blockade from scratch, but believed that if the department was as displeased with his performance as Fox's telegrams implied, then he had no choice but to step down. After mulling things over for a couple days, he submitted his resignation on 16 September. In doing so, he noted tartly that he had seen it as his duty to inform the department of the obstacles he encountered in his efforts to fulfill his mission. Although Welles had by then returned from his trip up north, he made no effort to persuade Stringham to rescind his resignation. Welles had been dissatisfied with Stringham for some time, so he was not unhappy with the chain of events that transpired in his absence. Even so, he was gracious in his 18 September telegram to Stringham accepting his decision. He regretted any hurt feelings on Stringham's part, attributing the department's—meaning Fox's—reproaches to overzealousness in the pursuit of victory. Welles concluded, "I deem it a duty, as it is a pleasure, to express my high appreciation of your patriotism and zeal in the cause of the country which you have so long honored and served."[9] Stringham hauled down his flag on 23 September, and then took a three-month furlough that Welles granted him. When it was over, he was retired because of age. Welles always respected Stringham's service as his naval adviser, though, so he not only eventually appointed him commandant of the Boston Navy Yard, but he also made sure that Lincoln nominated him as one of the retired list rear admirals. He kept his post until 1866, after which he served on various boards, and eventually became port admiral of New York City. He died in 1876.[10]

Frank Du Pont: The Finest Officer in the Navy

A quarter of a century after the war had ended and Frank Du Pont had gone to his eternal reward, one naval officer who served under him wrote, "It will not be considered adulatory to those who knew him to say that no

officer in our navy within the past half century was gifted with a more distinguished appearance or exulted character."[11] Such praise was hardly
unusual among those who worked with Du Pont. He was a handsome,
well-proportioned man, over six feet tall, with a graying beard and mustache surrounding a hairless chin. His courtly and outgoing manner made
him a favorite in Washington society, and helped him navigate the federal
government's bureaucracy in pursuit of his objectives. People admired his
obvious intelligence, clear-sightedness, and coolness under pressure, traits
that enabled him to identify and resolve the most intractable of problems.
Although he was undoubtedly ambitious, he camouflaged it with a sincere
desire to improve the Navy and a deep belief that he was acting in accord
with God's divine will. Considering these attributes, it was hardly surprising that he was a Navy Department star, with Paulding and Fox among
his many patrons and friends. At the same time, he was generous and kind
toward his subordinates, who repaid him with fervent loyalty and support.
Viewed under a different light, however, Du Pont's shine took on a more
sickly hue. There was a thin line between charm and manipulation, and
some saw Du Pont's ingratiation as that of a fawning courtier currying
favor. Similarly, the fidelity he inspired among his subordinates could just
as easily be construed as an effort to construct a clique around him for his
own egocentric purposes. Hostile observers interpreted his sense of honor
as overweening pride, his ambition as selfishness, his self-assurance as arro-
gance, and his dispassion as indifference. As with much of life, evaluation
depended on perspective.[12]

Whatever people thought of Du Pont, it was hard to argue with the
success he enjoyed in the prewar Navy. Born in 1803 into a French émi-
gré family that eventually settled in Delaware, Du Pont entered the Navy
as a midshipman in 1815. For thirty years there was nothing remarkable
about his career trajectory. He served on a variety of warships, usually in
the Mediterranean, and engaged in the customary petty quarrels with his
brother officers. During the Mexican War, he commanded the sloop of
war USS *Cyane* off the Pacific Coast and led a contingent of sailors and
Marines in the occupation of Mazatlán, Mexico. After the conflict, however, his career took off. Stationed on shore throughout the early and mid-
1850s, often in Washington, he had the opportunity to apply his
intellectual, social, and bureaucratic skills on numerous administrative
tasks. He sat on committees that revised the Naval Academy's curriculum,
established the Lighthouse Board—of which he became a member—and

rewrote U.S. Navy rules and regulations. It was his participation on the 1855 Efficiency Board, though, that cemented his reputation as a reformer. Many praised his efforts to eliminate some of the Navy's ineffective officers, but the board's victims and their congressional allies severely criticized him. Even so, he was promoted to captain in 1856, and the following year he took the new steam frigate *Minnesota* on a two-year cruise to East Asia. After he returned, he was appointed commandant of the Philadelphia Navy Yard in December 1860. Although he hailed from something approximating a Border State and had little sympathy with abolitionism, Du Pont had no doubts as to his loyalties. During the secession crisis, he wrote to Capt. William Shubrick, one of his patrons and friends, "I may be all wrong, but I never believed I was serving *two* masters. I have been nurtured, clothed, and fed by the general government of my whole country, [and] have had the honor of carrying her flag abroad and representing her sovereignty on the ocean."[13]

In one of its reports, the Strategy Board recommended the division of the Atlantic Squadron into two at the border of the Carolinas. Its rationale was geographic. The Virginia and North Carolina coasts were so dissimilar to those of South Carolina, Georgia, and Florida that blockading them required two different approaches. Welles agreed, adding that the Atlantic Squadron was growing too large for one man to lead effectively anyway. Fortunately, Stringham's departure made a separation much simpler because it spared the Department the embarrassment of reducing his responsibilities. To lead the new South Atlantic Squadron, Welles chose Du Pont. For Welles, it was an easy decision; Du Pont seemed to epitomize the energetic, cooperative, and can-do attitude the navy secretary was seeking in his officers. Du Pont had impressed Welles by assisting the Army in sending reinforcements through Maryland to Washington after the Baltimore riot, and later as chairman of the Strategy Board. Welles saw a good bit of Du Pont in the summer of 1861, and found him intelligent, congenial, and deferential. Besides, Fox and Paulding both spoke highly of him. Welles concluded that Du Pont had the discretion, courage, and organizational ability a squadron commander needed, so he issued the necessary orders on 18 September. The only fly in the ointment was that Du Pont was so far down on the naval register, ranking just fifty-first out of sixty-five available captains. Not surprisingly, his appointment to such an important job angered many of those above him. In fact, some senior captains saw Du Pont's selection as a Navy

Department plot to force their retirement. Du Pont dismissed these concerns, writing to a friend, "The theory gotten up is funny—that the real object of going down so low on the list for flag officers is that all above them should be retired."[14] The officers' fears, however, were not groundless; Welles' mind was already cultivating such ideas, which would bear fruit with congressional legislation in December retiring all officers over sixty-two years old. In picking Du Pont, Welles sent his high-ranking officers the clear message that they could no longer count solely on their seniority to secure prominent positions.[15]

As it was, Du Pont's new title dovetailed nicely with a job Welles had given him back on 3 August. The Strategy Board had emphasized the Navy's need for a supply base somewhere along the Confederate coast to succor Union warships on blockade duty, so Welles had put Du Pont in charge of organizing and leading an expedition to do so. Based on the Strategy Board's reports, Welles identified four possible targets—Bull Bay, Port Royal, and St. Helena Bay in South Carolina; and Fernandina on the Florida coast—and ordered Du Pont to occupy two of them. After a series of meetings in August with general in chief Winfield Scott and his advisers, the Army agreed to contribute 13,000 soldiers under Brig. Gen. Thomas Sherman's command. Scott also recommended that the Navy postpone the operation until October, when Union forces would be less susceptible to the tropical diseases that plagued the rebel coast in summertime. Although Du Pont and Fox saw the suggestion's merit and agreed, Fox in particular was unhappy with the delay. Du Pont traveled to New York City to start preparations, but in late September he slipped away to his Delaware home for a few days of rest. There he received an urgent summons from Fox to report to Washington for a hastily called meeting with Lincoln, who was urging the Navy to follow up on its success at Hatteras Inlet as soon as possible. Du Pont hurried to the capital, and on 1 October he and Fox traveled to William Seward's house for the conference. Lincoln, Seward, and several others were there—Du Pont was disturbed when Seward absentmindedly blew cigar smoke in the president's face—but Secretary of War Simon Cameron, Sherman, and Army of the Potomac commander George McClellan were not and had to be sent for. Although Lincoln was normally very good with faces, he failed to recognize Du Pont until Fox introduced him. All this was disconcerting enough, but there was also obvious confusion over Du Pont's mission and another in the works for an assault on the North Carolina sounds, so

Lincoln chastised everyone for their poor coordination. The two undertakings were competing for resources, and Fox worried that Du Pont's would get shortchanged. After some discussion, however, everyone agreed that Du Pont's force should have priority, and in fact they determined it should get under way in just four days.[16]

Du Pont believed that such a timetable was unrealistic. He had been working since early September to ready his expedition, but there was nothing easy about it. He was starting practically from scratch, and outfitting, manning, and supplying the necessary vessels was an enormous undertaking. Indeed, he confessed to his wife that he felt overwhelmed by his responsibilities, and feared he lacked the temperament for the task. Moreover, although the Navy Department expected a lot from him, it was not providing the kind of assistance he wanted. In mid-September Du Pont traveled to Washington to meet with Fox. He was encouraged by Fox's usual optimistic and ebullient manner, and even more so by his help in personnel matters. Du Pont was unhappy with the poor quality of some of the officers the Navy Department had assigned him, so Fox agreed to let him select his chief subordinates. In response, Du Pont plucked Charles Davis out of his Navy Department office and gave him the choice of captaining a warship or serving as his chief of staff. Davis claimed that twenty years earlier he had predicted that he would someday become Du Pont's chief of staff—or "fleet captain," in contemporary parlance—so he opted for the latter position. Du Pont also recruited Cdr. Percival Drayton as his ordnance officer. Drayton had proven invaluable in arming the motley conglomeration of vessels the Navy was purchasing, but Du Pont was initially reluctant to include him in the expedition. Drayton was a Union-loyal South Carolinian, and Du Pont worried that he might have qualms about attacking his home state. Drayton assured Du Pont, however, that as an officer it was his duty to obey orders, and he felt it would be dishonorable for him *not* to participate in the operation. Finally, Du Pont brought on board the Rodgers cousins—Cdr. John and Lt. Raymond. Raymond Rodgers had been commandant of midshipmen at the Naval Academy, and Du Pont put him in charge of his flagship, the steam frigate USS *Wabash*. John Rodgers, fresh from a stint of duty on the Mississippi River, took over the converted steamer USS *Flag*, but he spent most of his time as Du Pont's very effective troubleshooter and aide. Du Pont quickly molded these men into a loyal and close-knit team, so that, as one officer later noted, "It has been said that not even Nelson had the more devoted service of his officers."[17] Together, they got the job done, and Du Pont's fleet was able to leave New York City harbor on 16 October for Hampton Roads. There it rendezvoused with the transports carrying Sherman's soldiers from Annapolis. Unfortunately, a weeklong storm delayed its final departure until 29 October, and reduced a seasick Du Pont to a liquid diet. Even so, by now Du Pont was confident enough to write Welles an upbeat letter thanking him for his staunch support and predicting ultimate success.[18]

The storm and his seasickness gave Du Pont plenty of time to determine his target. In his meetings with the Army the previous July, he had tentatively decided to assault Bull Bay and Fernandina. Davis and John Rodgers urged him to do so, arguing that the fleet could easily occupy both places. Du Pont, however, was no longer so sure. By now the expedition's size had grown so large that he doubted that the Lincoln administration and Union public would be satisfied with such a small return on their collective investment. Seen in this light, Du Pont acceded to Fox's suggestion that Port Royal was the proper objective, even though he heard it was well protected. Strategically located between Charleston and Savannah, amid a maze of palmetto-filled islands, Port Royal, unlike Bull Bay and Fernandina, contained a deepwater harbor big enough to handle as many vessels as the Union cared to build. It was the perfect supply base for the blockading fleet, making its conquest a fitting goal for Du Pont's force.[19]

Du Pont and *Wabash* arrived at Port Royal on 3 November, two days after a horrific storm off Cape Hatteras scattered his fifty-ship fleet. When it finally re-formed, Du Pont learned that although only one vessel had sunk, the gale damaged valuable supplies and forced several small steamers necessary to land the soldiers to return north. This meant that if Du Pont wanted Port Royal, he would have to take it without the Army's help. Confronted with the unpleasant choice of assailing Port Royal alone or returning north empty-handed, Du Pont opted to continue the operation. After crossing the sandbar and reconnoitering the bay, Du Pont, Davis, and the Rodgers cousins formulated their plan. It was a collaborative effort, which permitted all of them to claim credit for it afterward. To defend the bay, the Confederates had deployed a tiny three-boat flotilla and constructed two forts—Beauregard and Walker—containing twelve hundred men, on either side of the sound. At Davis' urging, Du Pont decided to steam through the middle of the sound with his warships divided into two parallel columns. After bombarding both forts, the northernmost column would peel off up the sound to take on the rebel flotilla and protect the transport ships. The southernmost column, with Du Pont's powerful *Wabash* in the van, would turn around and rake the forts again, before circling back to repeat the process. Du Pont tailored his tactics to take advantage of the fleet's firepower and mobility, but it would also expose it to fire from two directions. *Wabash*'s officers understood this as well as anyone when Raymond Rodgers explained their role to them, and, after silently mulling over their chances of surviving the coming engagement, one asked another in a stage whisper, "Isn't the cake and wine coming around?"[20]

Poor weather and the temporary grounding of *Wabash* and the side-wheel frigate USS *Susquehanna* delayed the Union attack for two days, but on the morning of 7 November Du Pont's fleet finally pushed into Port Royal Sound. Although the two lines of warships passed between the rebel forts as intended, after that the Union plan quickly unraveled. One of the warships in Du Pont's southern column, Cdr. Sylvanus Godon's sloop USS *Mohican*, without orders pulled out of the line to take an advantageous position from which she could effectively pound Fort Walker. Except for *Wabash* and Capt. James Lardner's *Susquehanna*, the remaining vessels in the southern line followed *Mohican*, throwing the Union fleet into confusion. The northern column of vessels, led by Cdr. Charles Steedman's side-wheel gunboat USS *Bienville*, rushed back down the sound to support *Wabash* and *Susquehanna* after throwing enough shells at the tiny rebel flotilla to keep it in check. On the surface Du Pont remained perfectly calm and collected as rebel shells tore through and around *Wabash*, but he was distressed by the collapse of his carefully laid design. He issued without much luck numerous signals to try to straighten out the mess. In frustration he called out to Steedman, "How is it that I can't get my signal obeyed and my orders carried out?"[21]

In the end, though, such mistakes did not matter much. The Confederates fought gamely, but increasingly heavy and accurate Union firepower carried the day. *Wabash* and its accompanying warships steamed between the rebel forts five times in two and a half elliptical patterns, pummeling the Confederates with each pass while Godon and the other warships continued to rain destruction on Fort Walker. By early afternoon the rebels at Fort Walker had had enough and evacuated the place. John Rodgers led a party ashore to take possession of the abandoned fort, and the sailors in the fleet gave three cheers when they saw the Stars and

Stripes raised over it. The rebels also slipped away from Fort Beauregard, though Union forces did not occupy it until the next morning. Thus, at the bargain price of only eight killed and twenty-three wounded, and with minimal damage to his vessels, Du Pont had seized Port Royal.

That evening most of the ship captains, as well as a good number of army officers, came aboard *Wabash* to celebrate the victory. Davis later complained to his wife that they overstayed their welcome, but their jubilation and exuberance were certainly understandable. Percival Drayton was among those most satisfied with the day's events. Drayton's ship, *Pocahontas,* had barely survived the heavy storm that dispersed the fleet, and she reached Port Royal halfway through the battle. Despite a lack of orders, Drayton immediately engaged the enemy batteries. When Drayton returned from the party on board *Wabash,* a young officer, future naval theoretician Alfred Thayer Mahan, asked him if he had learned who commanded the Confederate forces. Drayton smiled slightly and responded that it was his brother, Brig. Gen. Thomas Drayton.[22]

Coming as it did at a time when Union victories were few and far between, Du Pont's triumph was especially welcome to the Lincoln administration. The normally stoic Welles was among the most effusive in his praise. He ordered a salute fired from each naval yard in honor of the victory, and he wrote to Du Pont: "The results of the skill and bravery of yourself and others have equaled and surpassed our highest expectations."[23] The victory also stifled whatever feeling there was among the officer corps that Du Pont owed his position to his connections, not his ability. Du Pont of course appreciated all the commendations, but he also wanted to make sure that some of his subordinates got the credit he felt they deserved. He saw no need to publicize the fact that the battle did not go anywhere near according to plan, no doubt at least in part because doing so would reflect poorly on him. In fact, he did not even mention Godon's unauthorized actions in his official reports. In his correspondence with Welles and Fox, Du Pont extolled John Rodgers and Drayton for their usefulness. He also appreciated Lardner's steadfastness during the battle. When Du Pont concluded that Lardner was not getting proper recognition for his contribution to the victory, he issued a general order to the officers and crew of *Susquehanna* acknowledging it. Charles Davis, however, received Du Pont's greatest applause. In a letter to Welles, he noted Davis' calmness and courage, and added, "In the organization of our large fleet before sailing, and in the preparation and systematic arrangement of the details of our contemplated work, in short, in all the duties pertaining to the flag-officer, I received his most valuable assistance. He possesses the rare quality of being a man of science and a practical officer."[24] Du Pont, it seemed, had found a winning team for his future campaigns.[25]

No doubt conquering Port Royal was an important achievement for the Navy. Doing so gave the South Atlantic Squadron a nearby base that Du Pont eventually filled with machine shops, storage facilities, and much else that Union warships needed to remain on station. On the other hand, Du Pont and Sherman missed an opportunity to exploit their success by attacking weakly held Charleston and Savannah. In their defense, the War Department had little interest at that time in a major effort in that part of the war, so it did not reinforce Sherman sufficiently to take advantage of the opening Du Pont developed. Despite this failure to seize the main chance, Du Pont did not remain idle. Instead, he and Sherman used their available resources to seal off as much of the rebel coast as possible from blockade-runners. Fortunately, the Confederates decided not to seriously contest these operations, but instead to concentrate their resources on fortifying Charleston. In March Drayton and Raymond Rodgers led a naval task force that occupied Fernandina, Jacksonville, and St. Augustine in Florida. At about the same time, Godon took Brunswick, Georgia, without a fight. Most important of all, on 10 April 1862, Fort Pulaski, on Cockspur Island, surrendered after a thirty-hour bombardment by army and naval artillerymen, thus severing Savannah's links to Europe and the rest of the world and rendering it useless as a refuge for blockade-runners. Such accomplishments not only tightened the blockade, but also disrupted the Confederate home front. Union forces entering Beaufort, South Carolina, for example, discovered that the white population had fled, leaving their black slaves behind to loot the stately old homes and contemplate their newfound freedom. Shortly thereafter, Drayton visited his mother's plantation on Hutchinson Island, and observed prophetically, "My own brother will probably like so many others lose everything, but although sorry I think he richly deserves it."[26]

Thanks to these efforts, by the summer of 1862 Charleston was the only major Confederate port left for the South Atlantic Squadron to blockade. It was enough of an achievement to persuade Congress to vote Du Pont its thanks in February 1862. While Du Pont appreciated the accolades that came his way, they did not make his job any easier. As he and other naval officers were discovering, guarding enemy harbors was more difficult than anyone had expected. Patrolling hundreds of square miles of trackless ocean was complicated enough, but Charleston harbor's shifting channels, treacherous shoals, and defending rebel batteries kept Union warships a respectful distance from the shore. These circumstances made it possible for the sleek, shallow-draft, steam-powered vessels the Confederates increasingly employed to sneak through the Union cordon on dark nights, bringing in the weapons, matériel, and supplies the rebels needed to prosecute the war. Moreover, Du Pont lacked enough warships, especially fast gunboats, to completely clamp down on such activity. In early 1862 newspaper reports that sixty-five blockade-runners had slipped into Charleston since the war began prompted Senator John Hale, chairman of the Naval Affairs committee and Welles' archenemy, to initiate a congressional investigation of the blockade's effectiveness. Welles acknowledged that the Charleston blockade was not as efficient as he wanted, but he defended Du Pont by stating that he was doing a good job under adverse circumstances. Fortunately for Du Pont, Welles' support was neither rhetorical nor solely for public consumption. Although he urged Du Pont to find ways to tighten the blockade, he accompanied his exhortations with effusive praise that he rarely gave his other squadron commanders. On 28 July 1862, for instance, Welles wrote to him, "Most earnestly do I congratulate you on the continued excellent command and service of the South Atlantic Squadron. This I do the more cheerfully, for while I have felt compunction that you were not better supplied, no complaint or censure has come from you for this omission. While we have really done all we could, I feel that we have not done half enough."[27] Fox too remained steadfast in his support, and in fact referred to Du Pont as the Navy's best officer.[28]

Although his flagship, *Wabash,* was usually anchored at Port Royal, Du Pont rarely stepped on dry land. Like any good naval officer, he felt that his proper place was shipboard, and anyway going ashore was more trouble than it was worth for a busy squadron commander swamped with the details of overseeing dozens of warships spread out across hundreds of miles of ocean. He spent as much time as he could on deck, inhaling the salt air to maintain his health. Indeed, he was in fine physical shape for a man approaching sixty, but the burdens of leadership wore him down mentally. To be sure, he was proud of his recent accomplishments, and in his more optimistic moods he even predicted that the blockade's constricting coils might crush the Confederacy by autumn. Overseeing a naval squadron, however, was one headache after another. Du Pont lacked sufficient warships to fulfill his missions, and the machinery of those he had was constantly breaking down. He believed that he needed at least twenty vessels off Charleston to effectively seal off the harbor, but he rarely had more than eight on station at any given time. Du Pont visited the blockade several times to consult with the officers there and make some organizational changes, but he could only do so much with the resources at hand. Welles' memory was either faulty or selective because Du Pont did in fact grumble to the Navy Department on several occasions that he required more warships. He was, however, a tactful and ingratiating man, so he was always careful to phrase his complaints in ways that did not offend the touchy navy secretary. Besides, he knew that he was one of Welles' favorite officers, and his backdoor channel to Fox kept him informed and up-to-date on the Navy Department's inner workings. Finally, Welles was always tolerant of his successful squadron commanders, and Du Pont had delivered one of the Navy's biggest victories to date. Whatever Du Pont's concerns, he could at least take comfort in the fact that he retained Welles' and Fox's support.[29]

Unhappily for Du Pont, not everyone in his squadron shared the Navy Department's confidence in him. Although Du Pont had a knack for inspiring loyalty among his subordinates, the growing size of his squadron made it difficult for him to give everyone his personalized attention. As a result, some of his officers, such as Cdr. John Marchand and Cdr. Charles Steedman, groused among themselves that he was playing favorites. As they saw things, Du Pont gave special treatment to Davis and the Rodgers cousins at the expense of everyone else. Du Pont permitted these men to lead the high-profile expeditions against weakly defended Confederate ports while Marchand, Steedman, and others were relegated to tedious and thankless blockade duty off Charleston that promised little by way of combat or glory. Instead, the constant tension and frustration among the blockaders led to petty bickering. Marchand and James Lardner, for instance, got into a row because Marchand failed to understand one of Lardner's signals while blockading Charleston. Steedman's grievances were mostly professional, but Marchand's were of a more personal nature. In March 1862 Marchand received permission to participate in the occupation of Fernandina. After the town surrendered, he asked to send word to his wife that he was okay, but Davis refused because Du Pont wanted to announce Fernandina's fall in his official dispatches first. When Marchand persisted, Du Pont angrily thrust a piece of paper at him and told him to write his letter and he would see that his wife received it. Du Pont seems to have forgotten the incident, but not Marchand, who increasingly questioned Du Pont's competence. Such discontent did little to promote the Union Navy's war effort.[30]

Du Pont was actually aware of the dissension within his squadron because Cdr. John Missroon told him about it, but he did not think that this sentiment was justified or fair. While he had his favorites, he was hardly blind to their faults. In fact, he had come to some conclusions about Charles Davis that would have surprised Marchand and Steedman, had they known about them. Du Pont's opinion of his old friend and chief of staff changed a good bit after they planned and executed the Port Royal expedition together. The more Du Pont thought about it, the more he concluded that Davis was an odd fellow falling short of the mark. He freely acknowledged Davis' composure, devotion, and strong astute opinions, but he also believed that Davis was too lazy and passive to cope with the crushing paperwork and routine that were part and parcel of a fleet captain's responsibilities. To Du Pont, this indolence was symptomatic of an untidy and unprofessional mind that should disqualify Davis from ever attaining squadron command. Du Pont shared his evaluation only with his wife, however, not Welles and Fox. On 10 February 1862, the Navy Department ordered Davis back to Washington for new duty. Du Pont protested the transfer, but in such a way that precluded Welles from changing his mind. He wrote to the navy secretary, "You are sufficiently acquainted with the merits of this officer not to require any additional testimony from me. His withdrawal from my command is a great loss to it and to me, but the Department is quite right to avail itself in a larger sphere of his professional abilities and loyal devotion to his country's cause."[31] No doubt Du Pont did not want to hurt his old friend's career, but by endorsing Davis against his better judgment, he placed personal loyalty ahead of his country's welfare. It was a mistake that Welles for one never would have made.[32]

Du Pont was less effusive, but more honest and sincere, in his praise for James Lardner. Lardner captained *Susquehanna,* and during the Battle of Port Royal had obeyed orders and stuck close to Du Pont and *Wabash* when everyone else in the column followed Godon. He later commanded the naval force stationed outside of Charleston harbor. Du Pont could have blamed him for all the problems associated with blockading that port, but he did not. Although Lardner was not part of his inner circle, Du Pont still admired him for his zeal, efficiency, experience, and devotion to duty, and he said as much to Welles. In April the Navy Department detached Lardner and *Susquehanna* from the South Atlantic Squadron and sent them to Norfolk. Du Pont regretted losing not only a powerful warship, but also such a dependable subordinate. [33]

The officer who impressed Du Pont the most, however, was Raymond Rodgers, who became the South Atlantic Squadron's fleet captain in August, and a commander three months later. The forty-two-year-old Rodgers had seen extensive prewar service throughout the world in the Navy and the Coast Survey, and served in the Mexican War on the Gulf Coast. To Du Pont, Rodgers was everything that Davis was not. Rodgers not only anticipated Du Pont's needs and made helpful and timely suggestions, but he also possessed abundant energy, orderliness, and ability. Moreover, he supplied Du Pont with considerable inside information on Navy Department thinking during a trip to Washington in July. Rodgers did much to alleviate the constant pressure under which Du Pont labored, so it was small wonder that Du Pont referred to him as "the most perfect officer that I have ever met in [my] life." [34]

Du Pont was not much for pomp and ceremony, but even he felt humbled and impressed by the pageant that unfolded before him at Port Royal around noon on 9 August 1862. In response to congressional legislation the previous July, Welles had asked the president to appoint Du Pont one of the first four active-list rear admirals. Although the official commission had not yet arrived, Raymond Rodgers insisted on going through with the commemoration because so many officers had to return to their stations in the next few days. Du Pont did not want to disappoint his officers, so he gave his assent. He let Rodgers handle all the details, confident that he would do a tasteful and appropriate job. Now, as he stood on the barge that carried him to *Wabash*, he saw the crews of the nearby vessels aloft in their masts. A Marine guard greeted him when he boarded *Wabash*, and he shook hands with all the officers before going aft to witness the raising of the square rear admiral's pennant on the mainmast. Then all the warships in the harbor fired their salutes, followed by the cannon at the fort at Hilton Head. For all its simplicity, it was a moving experience for Du Pont, Rodgers, and the officers present. After all, none of them had ever before seen an American admiral. The ceremony symbolized the Navy's growing power and prestige, as well as the fact that Frank Du Pont was a man at the very pinnacle of his profession.[35]

"Old Guts"

Five hundred miles north of Port Royal, off Hampton Roads, another naval officer had also recently received word of his promotion to rear admiral. For Louis Malesherbes Goldsborough, however, the news was bittersweet because he had recently asked to be relieved of his command. It was a peculiar situation, but no more peculiar than the fact that Goldsborough had risen to such a high position in the first place. Goldsborough was sixteen months younger than Du Pont, having been born in 1805, but he possessed more seniority, ranking forty-fifth out of the sixty-five captains at the war's start. He was born in Washington, the son of a longtime Navy Department clerk, and as a young boy he was a favorite of then Secretary of the Navy Paul Hamilton. These factors no doubt opened as many doors for him as did his marriage to the daughter of former attorney general and Anti-Masonic Party presidential candidate William Wirt. When Hamilton wrote out Goldsborough's midshipman's warrant, he backdated it to the day Congress declared war on Great Britain, giving the youth several years of unearned seniority. In fact, Goldsborough did not enter the service until 1816. His performance thereafter was impressive. He studied in Paris in the mid-1820s, served as head of the Depot of Charts and Instruments, and saw action on land and sea in the Seminole War in Florida. Like Du Pont, he captained *Ohio* during the Mexican War. Afterward he was a member of a commission that explored recently acquired Oregon and California, and then he became superintendent of the Naval Academy. When the war began, he was finishing a two-year stint as commander of the Brazil Squadron.

Everything about Goldsborough conveyed size and mass. He was a large man, topping three hundred pounds, with broad shoulders, a thick neck, heavy black beard, and square head. His personality was outlandish, extravagant, and pretentious. He was a big, loud, and long-winded talker who punctuated his conversations with profanity. A stickler for the rules, he had no patience for those who broke them. He was not terribly thoughtful, but he did have a certain amount of common sense and a prodigious work ethic. Although sometimes fawning toward his superiors, he often exhibited a violent and coarse temper to his subordinates. His brusqueness, however, masked the fact that his bark was a good deal worse than his bite. One officer remembered that while superintendent at the Naval Academy, Goldsborough often screamed and shouted at the midshipmen's misbehavior, but in private he chuckled at their transgressions. The best way to get along with Goldsborough was not to take his tantrums very seriously, but this was hardly the kind of relationship a good squadron commander wanted with his officers.[36]

Welles' reasons for choosing Goldsborough for squadron command are unclear. Although Goldsborough had his supporters, Welles did not know him well, and Fox had never exhibited much partiality toward him. It is likely, though, that Goldsborough's selection was by process of elimination. Congress did not pass legislation creating the flag officer rank until December, so Welles had to pick Stringham's replacement from the captains' list only. Of these, some were too physically or mentally unfit to lead a squadron. Others were Southern-born officers whom Welles still mistrusted. Finally, many had already been assigned important jobs elsewhere. Welles wanted Stringham removed as soon as possible, so he did not wish to wait long for his replacement to arrive from some distant station. Goldsborough met Welles' basic requirements; he was healthy, of undoubted loyalty, and readily available. Moreover, he had experience leading a squadron, albeit a peacetime one, and Welles thought him capable enough. On 18 September 1861 Welles ordered Goldsborough to take charge of all the warships along the Atlantic Coast until Du Pont was ready to establish his South Atlantic Squadron, which did not happen until 12 October. There was, however, one last obstacle. Four of the officers commanding warships in the Atlantic Squadron—William Nicholson, Joseph Hull, John Chauncey, and Samuel Mercer—were captains senior to Goldsborough. Welles did not think much of any of them, so he was no doubt happy for an excuse to transfer all of them to shore duty, where they could do less damage to the Union cause. On 23 September Goldsborough arrived at Hampton Roads and designated *Minnesota* as his flagship.[37]

With his customary enthusiasm and energy, Goldsborough immediately went to work on the numerous problems squadron commanders encountered day in and day out. For instance, Welles had directed him to sink blockships in the major inlets along North Carolina's Outer Banks to prevent their use by rebel blockade-runners. Although Goldsborough fulfilled his orders, it was more difficult than anyone expected, and it did not

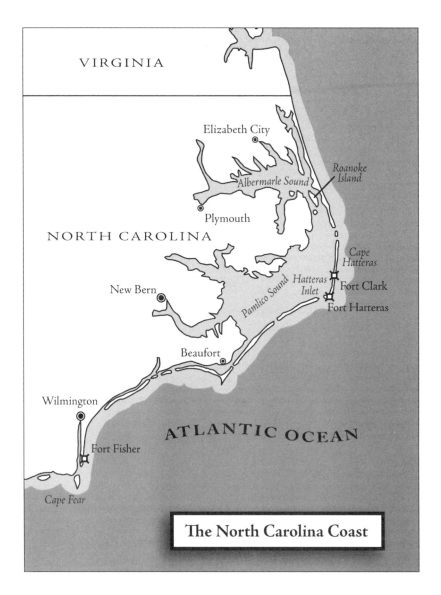

The North Carolina Coast

accomplish much because the blockships failed to obstruct the channels. Goldsborough for one felt that the Navy should aim a little higher. As he saw things, the Navy had to take the lead to win the war, and that required sounder strategic thinking than anyone had hitherto exhibited. After scanning his maps and charts, Goldsborough proposed to Welles that his squadron seize control of Pamlico and Albemarle sounds off the

North Carolina coast. Doing so would not only tighten the blockade, but also provide Union troops with access to the North Carolina and Virginia interiors. The first step in such a campaign was to occupy Roanoke Island, situated in the channel between the two sounds. In Union hands, it would serve as a staging base for further operations in the region. Goldsborough was confident that with moderate naval reinforcements he could sweep aside any Confederate seaborne opposition, but conquering the island itself required army assistance. Fortunately, such help was forthcoming. In October the War Department approved Brig. Gen. Ambrose Burnside's proposal to recruit a force of New Englanders to fight along the Confederate coast. After some interservice negotiations, the Army and the Navy agreed to cooperate in an invasion of Roanoke. By early January 1862, Burnside had gathered about 13,000 bluecoats at Annapolis, ready to board the motley collection of eighty vessels the War Department had hastily purchased to transport them, their supplies, and their equipment down to North Carolina.[38]

The combined Army-Navy expedition weighed anchor on 11 January 1862 at Hampton Roads and steamed south, arriving off Cape Hatteras two days later. In preparation for the shallow-water work he knew would come, Goldsborough had not only transferred his flag to the side-wheeler USS *Philadelphia,* but had also established a naval division of twenty gunboats under Cdr. Stephen Rowan. Once off the North Carolina coast, however, a terrible storm scattered the fleet for two weeks. As it slowly reformed, Goldsborough discovered to his shock that Hatteras Inlet was only seven and a half feet deep. Most of the vessels required a good deal more clearance than that. Goldsborough and Burnside immediately got their men to work lightening, pulling, and kedging their ships to get them across. Some vessels even used their propellers to carve deeper channels out of the sandy bottom. It was a long, hard, and frustrating job that involved a good many collisions, mishaps, and lost tempers. Although at one point Goldsborough doubted that he would ever come to grips with the enemy, in early February all the vessels were finally in Pamlico Sound and ready for action.[39]

On 5 February the fleet headed north toward Roanoke Island, with Rowan's gunboats in the lead. Despite Roanoke's obvious strategic significance, the rebels had fewer than three thousand men there. They had also deployed a naval force, called the Mosquito Fleet, of seven small warships mounting a gun apiece, in front of a double line of piles driven into the muddy bottom that extended from the island to the mainland. Heavy rain delayed the Unionists for a couple days, but on 7 February Burnside began landing his troops at Ashby's Harbor, halfway up the island's west coast. Rowan's gunboats initially supported the invasion force, but they quickly engaged the Mosquito Fleet when it came out to fight. Rowan not only had more cannon than the Confederate warships, but their range was much greater, so he had little trouble driving them off. By the end of the day, the Mosquito Fleet was down to five badly damaged warships, and it withdrew to Elizabeth City, North Carolina, to refit. In the meantime, Burnside's troops were able to secure Roanoke and capture twenty-six hundred rebels in two days of sharp fighting. Naval losses amounted to six killed and nineteen wounded.

Goldsborough and Burnside did not rest on their laurels, but instead took advantage of the opportunities their victory at Roanoke created. On 9 February Goldsborough sent Rowan and fourteen of his gunboats to Elizabeth City to finish off the Mosquito Fleet. The Irish-born Rowan came to Ohio at ten years of age to join his parents after sustaining serious burns in a fire, entered the Navy as a midshipman in 1826, and later served in the Coast Survey, during the Seminole War, and along the California coast during the conflict with Mexico. Now he was determined to finish the job he started the day before. His warships were short of ammunition, so he ordered his officers to ignore enemy fire from the battery at Cobb's Point and instead steam directly toward the seven vessels the rebels had scraped together for their last stand. Rowan's warships had thirty-seven cannon in all, so the outcome was never in serious doubt. By the time the smoke cleared, only two of the Confederate ships had managed to escape through the Dismal Swamp Canal to Norfolk. The Mosquito Fleet's destruction now exposed the entire North Carolina coast to Union attack. On 14 March a combined Army-Navy force took New Bern on the Neuse River, and a week after that nearby Washington fell. Most important of all, Fort Macon surrendered on 26 April after a daylong bombardment, securing the Union occupation of Beaufort.

On the morning of 14 February, Goldsborough's private secretary rushed into the Navy Department building to inform Welles of Roanoke's fall. More good news from the region followed, and within two months the Union had sealed off almost the entire North Carolina coast to Confederate shipping, enabling the Navy to twist the blockade that much tighter. Unfortunately, Burnside lacked the resources to strike inland and sever the vital Weldon Railroad leading to Richmond, but this was not Goldsborough's fault. Indeed, both Welles and Fox warmly congratulated their North Atlantic Squadron commander for accomplishments that rivaled Du Pont's to the south. To his credit, Goldsborough magnanimously gave his chief subordinates the recognition they deserved, especially Rowan, whose assistance in planning and executing the campaign Goldsborough termed "invaluable." Congress agreed by voting Rowan its thanks in February. Goldsborough had to wait until July for his congressional endorsement, but by then Welles and Fox had a much less positive view of his abilities.[40]

In fact, it was a season of Union successes. Out in the remote Transmississippi, Brig. Gen. Samuel Curtis' victory at the Battle of Pea Ridge on 7–8 March 1862 secured Missouri for the Union and opened the door for further operations in Arkansas. In February Brig. Gen. Ulysses Grant's forces seized Fort Henry and Fort Donelson on the Tennessee and Cumberland rivers, thus bringing most of Kentucky into the Union fold and exposing Tennessee to Union invasion. In the wake of Grant's triumph, the rebels abandoned Nashville, lost Island Number Ten on the Mississippi River, and surrendered New Orleans. Although the Confederates launched a ferocious surprise assault on Grant's army at Pittsburg Landing in the Battle of Shiloh on 6 April, Union troops beat off the attack in two days of bloody fighting and began a slow but ultimately successful advance toward the vital Confederate railroad junction of Corinth, Mississippi.

Union armies, it seemed, were moving forward to victory everywhere except in what many felt was the most important theater of the war. Throughout early 1862, Maj. Gen. George McClellan's powerful Army of the Potomac remained quiescent around Washington, despite the president's best efforts to prod it into action. McClellan's plans to begin his campaign against the Confederate capital, though, were complicated by events at Hampton Roads on 8–9 March. In an effort to use technology to compensate for shortcomings in just about every other area, Confederate Secretary of the Navy Stephen Mallory had decided to construct an ironclad warship. The rebels raised *Merrimack* from the bottom of the Norfolk Navy Yard, sheathed her in iron, and added a 10-cannon battery and an iron prow. Mallory gave the ship to Flag Officer Frank Buchanan, who was now, thanks to Welles' unwillingness to let bygones be bygones, fighting against the state he had been willing to sacrifice his career to protect.

The refurbished *Merrimack*, now renamed *Virginia*, was launched on 13 February. Despite her deep draft and creaky engines, Buchanan determined to attack the collection of Union warships anchored at nearby Hampton Roads. These included some of the most majestic vessels in the Union fleet, such as *Minnesota* and her sister USS *Roanoke*, as well as the old sailing warships USS *St. Lawrence*, USS *Cumberland*, and USS *Congress*. On the morning of 8 March, *Merrimack* steamed out of the Elizabeth River toward *Cumberland* and *Congress*. Because of her painfully slow speed, the Union sailors had plenty of time to clear their decks and prepare for action. Unfortunately, their broadsides ricocheted off the iron-plated rebel warship. *Merrimack*, oblivious to the barking Union cannon, rammed *Cumberland* and then fired repeated volleys that ripped though the old vessel, tore her apart, and ultimately killed 121 of her 376 crewmen. After *Cumberland* settled on the seabed, her mastheads still poking above the water, *Merrimack* slowly turned her attention to *Congress*. *Congress* tried to get into open water to maneuver better, but she instead grounded on a shoal. Once again the invulnerable rebel ironclad devastated her wooden opponent, which blew up that night. By now it was getting late and Buchanan had been wounded, so he decided to call it a day and fall back to Sewell's Point.

Next morning *Merrimack*, now under Lt. Catesby Jones, steamed out into Hampton Roads to finish the job so promisingly begun the previous day. Before she could do so, though, the Union's own new ironclad, USS *Monitor*, intervened. *Monitor*, under Lt. John Worden, had arrived the night before from New York City in time to survey *Merrimack*'s handiwork. Worden was determined to prevent a repeat of that performance. For four hours the two ironclads pounded each other without much success, except to underscore to everyone that a new era of naval warfare had dawned. Although *Merrimack* eventually withdrew to Norfolk, the battle was a tactical draw. Strategically, however, it was a Confederate defeat because Worden managed to preserve the vital Union blockade.

Goldsborough was at Roanoke planning the attack on New Bern when a fast steamer brought word of *Merrimack*'s rampage. He immediately returned north, arriving at Hampton Roads on the morning of 12 March. Fox was already there, and the two compared notes. Neither was surprised by *Merrimack*'s existence; Union intelligence had kept close tabs on the warship since the Confederates began transforming her. Goldsborough had originally boasted that he would easily destroy her if she ventured out of the Elizabeth River, but that was before he observed the shattered hulks of *Cumberland* and *Congress.* McClellan was just now beginning an amphibious assault on the peninsula between the York and James rivers in Virginia, and Goldsborough shuddered to contemplate the havoc *Merrimack* could wreak upon the hundreds of unarmed vessels busily transporting the Army of the Potomac's men, supplies, and equipment if she evaded or blasted through the Union warships at Hampton Roads. The consequences for the Union blockade would be just as serious. Mulling things over, Goldsborough decided that the best thing to do for now was to remain on the defensive and contain *Merrimack.* This did not, however, imply passivity on his part. If *Merrimack* emerged from her Norfolk lair to challenge the Union fleet, Goldsborough planned to throw everything he had at Hampton Roads at her. In particular, he gave orders for several fast steamers to sacrifice themselves by ramming the rebel ironclad. Although the press criticized him for his apparent lassitude as the days stretched into weeks and the Union warships remained inert at Hampton Roads, Goldsborough was increasingly confident that he could destroy *Merrimack* if the Confederates would just indulge him by coming out to fight. Except for an occasional feint, however, the rebel ironclad stayed put. Patience was not an integral part of Goldsborough's makeup, so waiting came hard for him, but he was buoyed by his belief that he retained the support and confidence of both the Navy Department and his fellow officers.[41]

In fact, Goldsborough did not know that his standing in the Navy Department was a good deal less secure than he realized. In praising his North Atlantic Squadron commander for his plans to deal with *Merrimack,* Welles was not being altogether honest. News of *Merrimack*'s attack had caused considerable consternation in Lincoln's cabinet, and Welles later referred to it as his most stressful time of the war. Although *Monitor*'s timely arrival prevented a setback from blossoming into a catastrophe, it did not end the threat *Merrimack* posed. Aggressive as ever, Welles wanted the Confederate ironclad destroyed, not contained. As March stretched into April, Welles was increasingly convinced that Goldsborough's apparent inaction was due to cowardice. Indeed, Welles even concluded that Goldsborough had gone down the coast to assail Roanoke because he was afraid to confront *Merrimack.* This accusation was unjust for many reasons, all the more so because the previous November Goldsborough had warned Fox that *Merrimack* might be formidable, and that she might sortie out while he was preoccupied in the North Carolina sounds. Furthermore, while Goldsborough's decision to remain on the defensive meant surrendering the initiative to the enemy, it was probably the most sensible course of action under the circumstances because storming *Merrimack* in her lair might cost the Union its only oceangoing ironclad.[42]

Goldsborough did not recognize that some of his colleagues questioned his judgment too. Down at Port Royal, Du Pont could not understand why Goldsborough had left those antiquated and vulnerable sailing vessels at Hampton Roads in the first place. Closer to home, there were pockets of discontent even within the North Atlantic Squadron. American naval officers valued calmness and dispassion, but dealing with *Merrimack* sapped Goldsborough's limited supply of these traits. On one occasion, when it looked as if *Merrimack* might emerge to engage the Union fleet, he sent so many signals to James Lardner, still commanding *Susquehanna,* that Lardner told the acting master to ignore them. That same day, a young officer approached Goldsborough with a requisition order, and Goldsborough bellowed at him, "Damn you, sir, go back to your ship, sir, don't you see that I am expecting the *Merrimac[k]*?"[43] Such intemperate behavior did not sit well with some officers. In the end, however, it was the Army of the Potomac that resolved the *Merrimack* problem. As McClellan slowly approached Richmond, the Confederates decided to abandon their outlying positions and concentrate on defending their capital. On 11 May they evacuated Norfolk. *Merrimack* drew too much water to go up the James River to safety, so the rebels destroyed her that day. Union naval supremacy remained intact, but Goldsborough's credibility had suffered irreparable damage.[44]

Keeping an eye on *Merrimack* was merely one of Goldsborough's many responsibilities. He also had to maintain the blockade and assist McClellan's Army of the Potomac in its advance on Richmond. McClellan wanted as much naval support as possible for his march up the peninsula between the York and James rivers. Regrettably, Goldsborough had little to spare after he accounted for all the other demands on his resources. Although he sent what vessels he could, it never completely satisfied McClellan. Norfolk's fall and *Merrimack*'s destruction freed up more Union warships for McClellan, but that did not help much. For one thing, it did not dramatically improve the military situation. Goldsborough ordered John Rodgers to take five vessels, including *Monitor* and the new ironclad USS *Galena,* up the James River toward Richmond, but the rebels drove them back at Drewry's Bluff on 15 May. For another, these new naval assets could not salvage Goldsborough's deteriorating relationship with McClellan. McClellan made plenty of demands on the Navy, but he rarely consulted with Goldsborough. Goldsborough grumbled that McClellan and his generals tried to boss around his officers as if the Navy was subordinate to the Army, not an equal and independent branch of the American military. Goldsborough was a stickler for protocol, so he resented the fact that McClellan did not even address him properly. With his usual volubility, he complained about the perceived slight in a letter to McClellan that began, "I am neither particularly fastidious about unimportant concerns nor prone to verbo-casuistry, but there is a matter to which I wish to invite your attention and ask for its correction, because it is just that I should do so."[45] Goldsborough's protests, however, had little impact on McClellan; he continued to call insistently for the Navy's help, especially after the rebels defeated the Army of the Potomac at the Seven Days' Battles in late June and drove it to Harrison's Landing on the James River.[46]

Welles could sympathize with Goldsborough's troubles with army officers such as McClellan, having since the war began experienced more than his share of frustration in dealing with that particular branch of the military. Even so, this did nothing to alter his declining opinion of Goldsborough. He now saw Goldsborough as a blowhard who was all talk and no action, and he regretted appointing him to squadron command. To Welles, Goldsborough was a product of the Navy's old seniority system that he had had to adhere to the previous September, before new congressional legislation made it possible for him to select the most talented men. For example, Goldsborough seemed eager to attack Fort Caswell, which guarded Wilmington, the last Confederate-held port in North Carolina. He bragged that he could do so easily and without army assistance, but nothing came of his strong talk. Fox was more forbearing. In fact, one of the reasons the assistant secretary was so keen to assault Fort Caswell was to give Goldsborough an opportunity to prove his worth. Fox, however, did not speak for many of Goldsborough's officers. These men, including Lardner and Rowan and John Rodgers, were increasingly fed up with Goldsborough's outbursts and inaction. They referred to him derisively behind his back as "Old Guts," and mocked his size and his temper.[47]

McClellan's defeat during the Seven Days' Battles and retreat to Harrison's Landing sent shock waves through the Lincoln administration. Even worse were reports that the Army of the Potomac might have to evacuate the peninsula or even surrender. To facilitate the Army's departure, should that be necessary, Lincoln wanted to be sure that the Navy was capable of keeping the James River open. The president had met Goldsborough several times, most recently the previous May during his trip down the Virginia coast to witness the occupation of Norfolk. Goldsborough's cautiousness on that occasion had not especially impressed Lincoln, so he believed that someone else should oversee the naval effort on the James River. Fortunately, Lincoln knew just the man for job, and he was right there in the capital. Cdr. John Dahlgren was an ordnance expert—he invented the massive bottle-shaped Dahlgren cannon that graced the Navy's warships—and commandant of the Washington Navy Yard. He was also the president's close personal friend. As far as Lincoln was concerned, Dahlgren possessed the can-do attitude a naval officer required to secure the James River, so on the morning of 5 July he and Secretary of War Edwin Stanton sent for him at once. Dahlgren readily agreed with Lincoln's reasoning. He was an ambitious man who yearned for a flag command, and he recognized an opportunity when he saw one. Unhappily for Dahlgren, Welles demurred because he believed that Dahlgren was too valuable where he was. Although Dahlgren protested strenuously, Welles and Fox would not budge, and Lincoln refused to overrule them.[48]

Despite this rebuff, Lincoln still liked the idea of putting some high-ranking competent naval officer in command on the James River. He and Seward huddled and suggested to Welles that he assign the post to Capt. Charles Wilkes. As commander of the steam sloop USS *San Jacinto*, Wilkes had the previous November boarded the British passenger steamer *Trent* and without orders seized two Confederate diplomats, John Mason and John Slidell, as contraband. The British threatened war for this alleged violation of their neutral rights, and in the end Lincoln backed down and released the rebels. Although Wilkes became a national hero for standing up to John Bull, Welles was at a loss about what to do with him. He believed that Wilkes was an overly ambitious, willful, disobedient, trouble-some, conceited, rash, and indiscreet man intensely disliked by his fellow officers. He was too dangerous for squadron command, too tactless for a desk job, and currently too important to captain a single warship. Now, out of the blue, Lincoln and Seward had found a satisfactory place for the Navy's white elephant. On 6 July Welles wrote out orders placing Wilkes in charge of the newly designated James River Flotilla. Unfortunately, though not unsurprisingly, there was bad blood between Wilkes and Goldsborough, so Wilkes did not want to answer to him. Welles was not normally very tolerant of those who let personal relationships interfere with their duty, but he had already committed himself to the scheme, so he agreed to let Wilkes report directly to him. Besides, doing so would help Welles keep close tabs on McClellan and the Army. He failed, though, to clearly explain this to Goldsborough.[49]

Goldsborough had his share of woes and grievances that spring and summer. He resented newspaper criticisms of his handling of *Merrimack,* as well as army efforts to take credit for his actions in occupying Norfolk. He wanted to assault Fort Caswell, but he could not until McClellan seized Richmond and freed up the naval resources necessary for such an undertaking. Finally, he missed his family, and a painful rheumatic wrist sometimes made writing them, or anyone else, impossible. All this was bearable, though, because he still believed he had the Navy Department's confidence. On 6 July he learned from Welles about Wilkes' new James River Flotilla. Welles wrote that the flotilla would be an independent division within the North Atlantic Squadron, but Wilkes had a different story when he called on Goldsborough three days later. Wilkes explained that he would be reporting directly to the Navy Department, not to Goldsborough. Somewhat puzzled, Goldsborough took advantage of Lincoln's visit to Harrison's Landing on 8–10 July to clarify his standing. He secured a few minutes with the president, and, after reviewing the military situation, asked if Wilkes would answer to him. Lincoln responded that that was his understanding, so clearly Wilkes' orders had originated with Welles. After digesting this implied rebuke for a couple days, Goldsborough wrote to Welles to acknowledge this latest directive. He stated that he assumed that Welles was doing what he thought was in the country's best interests, but added that permitting a junior officer such as Wilkes to bypass his authority hurt his pride, made him look bad in the eyes of the public and officer corps, and exposed him to attacks from the press. Nevertheless, he was willing to sacrifice everything short of honor to win the war. When Welles did not reply to this message, Goldsborough probably interpreted the silence as evidence that the Navy Department lacked confidence in his leadership. On 15 July he wrote again to Welles. This time he noted that the negative press Wilkes' appointment had generated toward him had, as he feared, damaged his credibility in the eyes of the public and his brother officers, staining his honor. Under such circumstances, he believed it would be best for Welles to remove him from his command.[50]

Welles waited nearly a week to respond to Goldsborough's request. He made no effort to dissuade or reassure his North Atlantic Squadron commander, but instead icily acknowledged his decision and stated that he would dispatch a replacement as soon as possible. Doing so took longer than Welles expected, so Goldsborough did not haul down his flag until 6 September. Ironically enough, this was a week after Welles dissolved the James River Flotilla due to the Army of the Potomac's evacuation to northern Virginia, and transferred Wilkes to command the Potomac River Flotilla. In all likelihood, Goldsborough's decision to quit had little to do with newspaper criticism or homesickness—in fact, he derided the press in his correspondence—but rather with his belated realization that he did not enjoy Welles' confidence. He was right about that; Welles had been dissatisfied with Goldsborough for a long time, and was happy to take this opportunity to be rid of him. Welles believed that Goldsborough was an inefficient, pompous, and cowardly officer who had accomplished next to nothing since he took over the North Atlantic Squadron. He even contemplated withdrawing his nomination to rear admiral, but eventually reconsidered. In thinking this way, Welles was not being fair to Goldsborough. Du Pont, for example, readily acknowledged Goldsborough's shortcomings, but added after learning of his removal, "A kinder-hearted man does not live, nor one who is more ready to oblige than he is—so conscientious in the discharge of his duty, so incorruptible in his integrity as a man and officer. There is no one who is going to be made an admiral who could aid the government more in all those matters connected with the national defense of the country than he—it is a thousand pities every way."[51] This was true enough, but Welles was not interested in character except to the extent that it contributed to Union victory, and Goldsborough had not delivered enough of that particular commodity to suit the exacting secretary of the navy. Although Goldsborough's days as North Atlantic Squadron commander were over, his role in the war was not. Welles eventually put him in charge of Navy rules and regulations, a job for which he was well suited. At the end of the conflict, Welles assigned him to command the European Squadron for two years. After that he served on a variety of boards and in a number of positions, eventually ending his career as commandant of the Washington Navy Yard. He died in Washington in 1877.[52]

Conclusions

Stringham, Goldsborough, and Du Pont all contributed to the Union Navy's successes along the Atlantic seaboard in the first sixteen months of the war. By the summer of 1862, large chunks of the Virginia, North Carolina, South Carolina, Georgia, and Florida coastlines were under Union control, so that only Wilmington and Charleston remained in Confederate hands and open to blockade-runners. Goldsborough had also beaten back the Confederacy's one major attempt to wrest naval superiority from the Union by neutralizing *Merrimack*. Despite these important accomplishments, two of these squadron commanders resigned their positions when they realized that they no longer had the Navy Department's support. Welles lost confidence in these men because they failed to exhibit the consistent aggressiveness he believed victories required. As far as he was concerned, Stringham and Goldsborough spent too much time complaining about one thing or another, and not enough prosecuting the war. Du Pont had his grievances too, but he accompanied his with a steady stream of victories. Moreover, Du Pont was tactful in a way that Stringham and Goldsborough were not. Goldsborough's blustery demeanor especially rubbed Welles the wrong way. Welles attributed his failure to find a good North Atlantic Squadron commander to the counterproductive seniority system under which he had to operate when he made his selections, which limited his choices to the sixty-five mostly overage and feeble captains. There was some truth to this; most of the captains were unfit for squadron command. There were, however, some capable men among them who went on to give good service in the war, but Welles initially overlooked them. In fact, Welles was learning his way just like everyone else, and was bound to make mistakes in the process. Fortunately, congressional legislation granting Welles the authority to appoint rear admirals and pick squadron commanders from lower-ranking officers gave him greater opportunity to choose competent officers—and fewer excuses if he chose wrong.

CHAPTER THREE

The Mighty Mississippi

Andrew Foote Breaches the Barricades

THE MISSISSIPPI RIVER originates in the northern Minnesota wilderness at Lake Itasca. From there, it takes on the reddish tint of the soil as it carves its way through the pine, cedar, and maple trees indigenous to the region. Pushing southward, it grows in size and power by absorbing thousands of waterways, including large rivers such as the Platte, Illinois, Missouri, Ohio, Arkansas, White, and Red. During the Civil War, its course after merging with the Ohio River became increasingly meandering as it wound and twisted its serpentine path toward the Gulf of Mexico. Springtime floods often changed its route, leaving isolated bayous and oxbow lakes behind as reminders of its former presence. By the time it reached the sea, 2,320 miles from its headwaters, it was nearly a mile across and up to two hundred feet deep. Since it carried more than 159 million tons of sediment annually—or, from a shorter perspective, 436,000 tons daily—its enormous delta poked its way far into the Gulf. American settlers moving westward across the Appalachian Mountains reached the river in the late eighteenth and early nineteenth centuries and quickly grafted it onto their way of life. Towns such as St. Louis, Memphis, Vicksburg, Natchez, Baton Rouge, and New Orleans developed beside it, and sugarcane and cotton fields sprouted along its unpredictable banks, crowding out the swamps and woods. The river became the young country's main economic artery, carrying Midwestern products out to the Gulf of Mexico and to the rest of the world on ships of all shapes, sizes, and nationalities. By the time Lincoln was elected, the Mississippi River was as much a part of the American experience as the slave system that soon brought unimagined violence and bloodshed to its muddy waters.

The Navy had operated on inland waters during the War of 1812 and the Seminole War of 1835–42 in Florida, but by the time the Confederates fired on Fort Sumter there was little institutional memory of these events. Most naval officers spent their careers afloat serving on the high seas, so it was hardly surprising that their thoughts turned in that direction when the conflict began. Indeed, the Strategy Board considered the Mississippi River only as part of New Orleans' local geography, not as a highway into the Confederacy's heartland. The Navy may not have initially understood the Mississippi's strategic value, but fortunately others did. Winfield Scott, for example, suggested using gunboats on the river to help isolate the Southern states as part of his Anaconda Plan. Edward Bates, Lincoln's attorney general, was even more astute. Bates was on friendly terms with a fellow Missourian named James Eads, who had made a fortune salvaging wrecked vessels and their cargoes from the Mississippi's muddy bottom. If anyone knew the river, he did, so Bates recommended him to the Navy Department. Eads traveled to Washington and explained to Welles the importance of establishing some sort of naval presence on the river. Welles, however, was reluctant to divert scarce resources away from the Atlantic and Gulf coasts. Besides, he at first considered the Mississippi and its tributaries the Army's bailiwick, so he forwarded Eads' report to the War Department for its consideration. Still, Welles wanted to be cooperative. In mid-May 1861, just after the war began, Secretary of War Simon Cameron asked him for an officer to assist the Army in arming the vessels it was purchasing on the Ohio River, and Welles agreed to send Cdr. John Rodgers to help out.[1]

John Rodgers was part of American naval aristocracy. His father and namesake gained fame as an officer in the Quasi War with France, the Tripolitan War, the War of 1812, and finally as head of the Board of Naval Commissioners. It was therefore hardly surprising that his son entered the Navy as a fourteen-year-old midshipman in 1828. Although over the years Rodgers saw the usual naval service in the Mediterranean and off the African and South American coasts, he also had more eclectic duties. He participated in the Seminole War, for example, and later the Navy sent him to Pittsburgh to help with the construction of one of the new steamships. He spent most of the 1850s with the North Pacific Exploring and Surveying Expedition, of which he became commander in

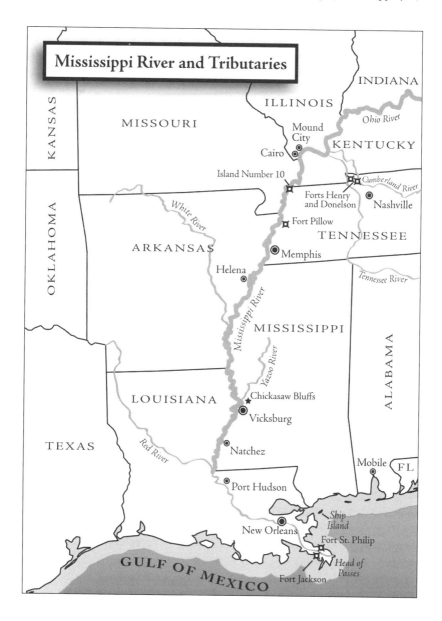

1854. There he faced hostile natives, harsh working conditions, and scarce resources without much help from or contact with the Navy Department. When the Civil War began, he was in Washington preparing his expedition's final report for publication.

Rodgers was not high on the naval register, ranking just sixty-seven out of its eighty-two loyal commanders. His new assignment out west under the Army's jurisdiction was a thankless and dreary task that no ambitious officer would solicit or covet. Welles and Paulding probably gave Rodgers the job because of his previous experience with steamships in Pittsburgh, but they were fortunate in that he had other positive qualities that served him and his country well, although Welles did not initially appreciate them. Most prominently, he was accustomed to operating on his own with minimal resources, far from Navy Department supervision. He knew how to get things done, and believed it was easier to obtain forgiveness from superiors than permission. He also had an enlightened attitude toward leadership, once explaining to his wife: "I am fair, I think, I have no humbug. I prefer to let the commanders know what I am doing and what they are doing instead of making automatons of them. I think by this more intelligent service is rendered—and more willing service—and I love to give credit where credit is due. At the same time I do not think that I neglect any points necessary to be observed. I do not care for popularity, and I make no effort to attain it. As I am amiable, fair, and I hope efficient enough to command respect, popularity has come I think as a matter of course."[2] Naval officers praised him for his energy, work ethic, and judiciousness. Finally, he possessed a wry sense of humor that belied the exasperated look in his eyes and determined set of his jaw. In another letter to his wife later in the war, he explained to her in literary deadpan, "The responsibilities of my position are no doubt very great, but I do not myself know in what."[3]

Rodgers traveled to Cincinnati to join army efforts already under way to buy and construct a gunboat fleet for riverine operations. Rodgers' instructions were to arm and crew these vessels, but he exceeded his authority by doing much more than that. Upon observing the organizational chaos around him, he moved instinctively to establish order and get things done. He contracted for gun carriages, ammunition, anchors, chains, clothing and bedding, cooking utensils, and even entire vessels. Sometimes he used his own money to pay for them. He also appointed officers for these warships, fixed their pay, and assigned their duties. None of this was strictly legal, but he rationalized his actions by explaining, "I am forced to do irregular things."[4] In the process, however, he alienated Maj. Gen. John Frémont, commander of the Department of the West, by rejecting a ship that Frémont wanted to purchase. Besides, Frémont believed that the Navy should send somebody to actually command the flotilla Rodgers was organizing, not serve merely as a liaison and adviser to the Army. Frémont wrote to both Fox and Montgomery Blair and asked them to use their influence to persuade Welles to replace Rodgers with someone with more legitimate authority. By now Welles recognized that the Navy's role on the Mississippi would be greater than he had originally anticipated, and would require someone with more rank to direct it, so he was receptive to such a change. On 30 August 1861, therefore, he ordered Capt. Andrew Foote to go west and assume command of all naval forces on the Mississippi and its tributaries.[5]

Welles realized that Foote would have his work cut out for him. He would have to cooperate with army officers who had little understanding of or respect for the Navy and its ways, sort out the contracts Rodgers had entered into with or without proper authority, and manage a fleet of tatterdemalion warships and untrained crews without access to a proper naval yard. On the surface, Foote was an unlikely choice for such a difficult assignment. Most obviously, he had only recently been promoted to captain, so there were plenty of officers in his grade with more seniority. Welles, however, had reason to believe that Foote would be the right man for the job. Although Foote was several years younger than the navy secretary, the two had attended Cheshire Academy together as children and had remained friends ever since. When Welles took over the Navy Department, he looked to Foote for information on which officers he could and could not trust. As assistant commandant of the Brooklyn Navy Yard, Foote had also impressed Welles by quickly readying *Powhatan* for its planned role in relieving Fort Sumter. In June Foote had lobbied Welles for promotion and command of a warship, but added that he was willing to serve wherever the navy secretary thought best. Now Welles proposed to take him at his word by giving him this difficult job. Welles did so with great reluctance because he did not want to jeopardize his friend's career, but, as usual with the navy secretary, duty trumped all else. Welles had faith in Foote's zeal and judgment, and hoped that he would infuse life into the Navy's new enterprise out west. Just to make sure that Foote understood the stakes involved, Welles summoned him to Washington to explain his decision. While Welles downplayed some of the difficulties he expected Foote to encounter, he made it clear to his friend that he would confront some unique problems. Foote promised to do his best, and on 5 September he reported to Frémont in St. Louis as commander of the newly dubbed Western Flotilla.[6]

Like Welles, Foote was Connecticut born and raised, the son of a prominent politician who served as a congressman, senator, and governor. With his father's influence, he secured an appointment to West Point and even attended the academy briefly, but he ultimately opted to accept a midshipman's warrant in 1822. In the ensuing years he saw duty in the Caribbean, Pacific, and Mediterranean. While on board USS *Natchez* in the Caribbean in 1827, a fellow officer converted him to Christianity. Foote did not embrace the distant and dispassionate deity his Congregationalist ancestors had worshipped, but rather a personal and loving God who called for His flock to put their faith into action by helping their fellow man. Foote initially wondered if the Navy was now the proper profession for him, but his father assured him that if the seas needed policing, it was best for Christians to do it. In fact, Foote concluded that since discipline and morality went hand in hand, the Navy was the ideal place for him to spread the gospel. During the war, one observer remembered that Foote ordered his entire crew to turn out for Sunday services so he could read scripture to them, and he did not seem to mind when many of them fell asleep and disrupted his discourse with their snoring. With a zealot's enthusiasm, Foote threw himself into a variety of causes in that reformist era. After witnessing a drunken riot on board USS *Cumberland*, he organized a shipboard temperance society and persuaded all but one of the crew to give up their grog rations. Abolitionism was another of his passions. In 1849 the Navy gave him command of USS *Perry*, and for two years he energetically pursued illegal slave traders off the African coast. He was a member of the 1855 Efficiency Board, and the following year went to East Asia as commander of the sloop USS *Portsmouth*. There he led three hundred sailors in storming the Chinese barrier forts outside Canton. When the war began, he had been assistant commandant at the Brooklyn Navy Yard for three years.[7]

Foote hardly fit the stereotypical image of a humorless crusader. He had been willful and headstrong as a young man, but religion and age had tempered and mellowed his personality, giving him patience, tolerance, and insight. Despite his controversial causes, he was popular in the officer corps. People respected his obvious sincerity and piety, and enjoyed his company. Indeed, Foote was an affable, hearty, and pleasant man full of funny stories and good cheer. Although his friend Du Pont, with unconscious irony, once called him a flatterer, others valued him for his ability to speak frankly without giving offense. There was nothing impressive about him physically. He was of average height, bearded but with a clean-shaven upper lip, and had a determined and fiery look in his eyes. Beneath his amiable exterior, however, Foote possessed ample supplies of tenacity and courage that enabled him to persevere in the face of adversity.[8]

Foote's appointment left John Rodgers at loose ends. Welles gave him the option of remaining out west to assist Foote or of reporting to the Navy Department for new duty. Rodgers deeply resented Foote's ascension over his head and felt that the Navy Department had treated him badly. He had nothing but respect for Foote personally and professionally, but he believed that Foote's elevation indicated that Welles disapproved of his Herculean, albeit unorthodox, efforts to build a naval force on the western rivers. To defend himself, he wrote to Welles: "When the plant thus watered and cultivated gives its first prematurely ripe fruit, the crop is turned over to another with cold words. . . . The appreciation of the Department must be very dear to every right-thinking officer. I do not pretend to be indifferent to it, and I only perform an official duty in endeavoring to defend myself against what I feel is an implied lack of worthiness."[9] Although Welles was normally not very tolerant of such impertinence, he made an exception in Rodgers' case. For one thing, he understood the difficulties under which Rodgers had labored. Moreover, Foote wrote to him to vouch for Rodgers, who had done well under difficult conditions, and Welles respected his old friend's opinions. Instead of the rebuke he would have ordinarily issued, Welles soothed Rodgers' hurt feelings with a kind letter of appreciation for his efforts. In the end, Rodgers agreed to stay in St. Louis temporarily to help Frémont outfit some vessels and Foote to square some accounts, but thereafter he returned east for a new assignment. Frank Du Pont knew talent when he saw it, so he secured Rodgers' services for his Port Royal expedition.[10]

Rodgers' days serving in the West were numbered, but Foote's frustrations were just getting started. Once he reached St. Louis, it did not take him long to recognize the enormity of the task he faced. In fact, his problems seemed downright overwhelming. He lacked sufficient money, manpower, and equipment to outfit the vessels that the Army and the Navy were purchasing or building. He had no naval yard, so he had to rely on the Army for the paraphernalia he required. Unfortunately, army officers had little sympathy for or understanding of the Navy's needs. This was especially true of Maj. Gen. Henry Halleck, Frémont's replacement as head of the Department of the West. Halleck, nicknamed "Old Brains" for his supposed military acumen, seemed to think that he knew more about naval affairs than Foote, which did nothing to promote a good relationship between the two men. To make things worse, Foote lacked the necessary rank to get things done. In the Army's eyes, he was at about the same level as a lieutenant colonel, and they were a dime a dozen. Foote complained that one lowly regimental commander even confiscated a gunboat without informing him. Foote asked Welles to either make him a flag officer or to dispatch someone senior to him to assume command of the flotilla with that rank. Foote felt so strongly about the matter that he told Welles he would rather be transferred to one of the Atlantic squadrons than continue under the present circumstances. Finally, Foote was unhappy with some of the officers the Navy Department was sending to captain his warships. He believed that these men were too old for the rigors of riverine warfare, and it hurt morale when they supplanted more energetic junior officers already with the flotilla.[11]

Welles and Fox did what they could to help Foote. The Navy Department transferred five hundred sailors who had been guarding Washington to man the flotilla, and the Army kicked in another thousand. On 13 November Welles appointed Foote a flag officer. Although the rank remained a courtesy title until Congress made it official in December, it was impressive-looking enough to put Foote at the same level as an army major general. The gunboats, though way behind schedule, were slowly arriving, and Foote began sending them on expeditions along the Ohio, Mississippi, and Tennessee rivers. Moreover, some of his officers were proving their worth. Foote's fleet captain, Cdr. Alexander Pennock, began constructing a naval yard at Mound City, Illinois. Foote was also impressed with young and aggressive Lt. Seth Ledyard "Sam" Phelps. Despite this trickle of good news, the pressures of command took a toll on Foote. He suffered from frequent and debilitating headaches, and even fell prey to self-pity. On 11 January 1862, he wrote to a sympathetic Fox, "I feel of course very sadly as I perceive . . . that I have not accomplished what has been expected of me. But could any officer here, in my circumstances, have accomplished more?"[12]

Happily for Foote, things were about to look up. The Lincoln administration was pressuring Halleck to do something to advance the war effort and justify the enormous investment in resources it had made in his theater. Old Brains was rarely capable of conjuring up any ideas of his own, so he was receptive to a plan Foote and Brig. Gen. Ulysses Grant, the commander of the District of Cairo, developed to attack Confederate Fort Henry and Fort Donelson. Located on the Tennessee and Cumberland rivers, respectively, the two forts controlled the gateway into central Tennessee. Once Halleck gave his approval, Foote and Grant moved fast. On 1 February Foote gathered seven warships, four of which were armor-clad gunboats dubbed "turtles," to escort transports carrying Grant's 17,000 soldiers upstream to Fort Henry. A warm spell had melted the ice on the Tennessee's tributaries, filling the rapid-flowing river with flotsam and slowing the warships and transports to a crawl. Even so, the flotilla reached the fort on 4 February, and Grant disembarked his soldiers. Despite its strategic importance, Fort Henry was miserably located on low ground that rapidly filled with water from the recent thaw. In fact, the rebels had already evacuated most of their troops to nearby Fort Donelson to make their stand there, leaving behind a skeleton force to delay the bluecoats. Foote did not know this, but, after a long morning of prayer, he was confident of God's help, so he issued orders for an all-out naval assault the next day, 6 February, before the soldiers came up. Now that he was certain God was in his corner, Foote visited the gunboat USS *Essex* to put its crew through drill and make sure that they also knew to place their trust in the Lord.[13]

The morning dawned sunny and warm, and a gentle breeze blew the smoke and fog away from the flotilla. Foote opened his attack around noon, with his four ironclads up front and the three wooden gunboats behind them, hugging the western bank. Foote's flotilla had fifty-four cannon at its disposal, considerably more than the nine available to the rebels upstream. Foote, imperturbable in combat, pushed his ironclads within six hundred yards of Fort Henry and blasted away with increasing accuracy. Whatever the heavy odds against them, the Confederates put up a stiff fight and pounded the Union warships repeatedly. Foote's flagship, the turtle USS *Cincinnati,* was hit thirty-one times, though her iron plating prevented serious damage. *Essex* was less fortunate, despite Foote's admonitions the previous day. A rebel cannonball punctured her center boiler, blowing superheated steam into the pilothouse. Thirty-two men were scalded, including her skipper, Cdr. William "Dirty Bill" Porter—David's estranged brother—and the vessel drifted downstream out of control. In the end, however, Foote's warships carried the day. By 4:00 PM Union fire had disabled five of Fort Henry's guns, and there were barely enough uninjured men to man the rest. Having bought sufficient time for the rest of the garrison to escape, the Confederates finally ran up the white flag. Their commander, Brig. Gen. Lloyd Tilghman, told Foote, "I am very glad to surrender to [so] *brave* an officer." Foote responded, "You should have blown my boats out of the water before I would have surrendered to *you*."[14] Whether he meant this as a compliment did not really matter, at least as far as the battle's outcome was concerned. The Confederates had suffered ten fatalities, in addition to ninety-four men taken prisoner. Foote's losses amounted to two killed and forty wounded, though some of the scalded sailors died later. No sooner had Union soldiers taken possession of the fort than Foote dispatched three gunboats under Phelps upriver to do as much damage as they could. While Phelps' vessels ranged all the way to Muscle Shoals in northern Alabama, Foote took his remaining warships back to Cairo, Illinois, to repair their damage and prepare for an assault on Fort Donelson.

Back at Cairo, Foote got his men busy hammering out the dents, clearing the debris, and patching the holes in their warships. Although preoccupied with a myriad of details, he still found time to thank God for his recent success. On Sunday, 9 February, Foote and some of his sailors attended the local Presbyterian church. After the congregation was assembled, someone whispered to Foote that the preacher was sick and unable to officiate. As undaunted here as at Fort Henry, three short days earlier, Foote volunteered to fill in. Taking the pulpit, he used his recent experience up the Tennessee as the basis for a short, plain, extemporaneous, and well-received sermon derived from Jesus' comment in John 14:1: "Let not your heart be troubled; you believe in God, believe also in me."[15]

As events proved, Foote's faith was about to be sorely tested. On 12 February he left Cairo with six warships to assist Grant in his offensive against Fort Donelson. Unlike Fort Henry, Fort Donelson was well situated on a high bluff on the Cumberland River's south bank. The defending rebels had only ten cannon targeting the river, but they could deliver a plunging fire on any advancing naval force. Foote planned on running past the enemy guns in an effort to get behind the fort so he could bombard its more vulnerable open side. He would have liked to bring in more warships, but, with Grant's army already on the move, he did not have enough time. Foote launched his attack on the afternoon of 14 February. His four ironclads—USS *Louisville*, the flagship USS *St. Louis*, USS *Pittsburg*, and USS *Carondelet*—advanced abreast, with two wooden gunboats bringing up the rear. Up until now, Union warships had invariably won once they brought their firepower to bear. Today, however, the Confederates showed what they were capable of when properly deployed and prepared. The rebels waited until the progressing Union flotilla was only four hundred yards away before they opened fire. *Louisville* was hit fifty-nine times before she dropped downriver, her steering gear demolished. On *St. Louis,* Foote had already been slightly wounded in the arm when he climbed into the pilothouse for a better view. Just then, an enemy cannonball plowed through the pilothouse, killing the pilot and wounding Foote in the left foot. *St. Louis* staggered out of line and drifted downstream, followed shortly afterward by *Pittsburg. Carondelet* made it to within two hundred yards of Fort Donelson before she too tumbled back, pummeled by thirty-five cannonballs at point-blank range. Foote's assault was a dismal failure, costing his flotilla fifty-four casualties and considerable damage to his ironclads.

In the end, Foote's failed attack did not matter much to the Union war effort. After a confusing battle on 15 February, Fort Donelson's 13,000 defenders surrendered to Grant the next day. The Union triumph shattered the Confederate defensive line, forcing the rebels to evacuate Nashville and abandon most of central and western Tennessee. In the warm glow generated by victories at Fort Henry and Fort Donelson, most people forgot, or anyhow overlooked, Foote's repulse. Indeed, when he returned to Cairo with his battered ironclads, congratulations for his success at Fort Henry were still pouring in. These included a heartfelt letter from Welles, whose gratitude was greater than most because he understood the tremendous difficulties under which Foote had labored. Despite all this acclaim, Foote felt little jubilation. For one thing, he was hobbling around on crutches on account of his wounded left foot. The injury initially appeared minor, but it festered instead of healing. In fact, it troubled him so much that he considered giving up his command until he realized, more with resignation than pride, that he was the only naval officer on the scene capable of holding his ragged flotilla together. In addition, he still had to deal with the usual logistical and personnel problems; victory did not change that. Finally, his experience at Fort Donelson revealed the dangers lurking behind every Confederate fortification, and seemed like an unwelcome harbinger of things to come. This was not the kind of conflict Foote had envisioned in his serene prewar days. Referring to the peculiarities of riverine warfare, Foote wrote to his wife: "It has added two years to my age constitutionally and is quite enough to break any man down. I would at this moment give all I am worth could I have been in the Atlantic, a captain of a good steam frigate, instead of out here under a pressure, which would crush most men, and how I have stood I can only account for that God has been my helper thus far. All is confusion and I am almost crazy and in despair."[16]

While Foote lamented his woes, the war continued. To block a Union advance down the Mississippi River, the rebels had fortified Island Number Ten—so named because it was the tenth island south of Cairo, on the Kentucky-Tennessee border—and the nearby high ground. The Confederates stuffed their defenses with 6,000 troops and seventy-five guns, including sixteen on a floating battery called the Pelican dry dock. Swamps surrounded the area, so the only way to approach the rebels was head-on. Maj. Gen. John Pope was prepared to do just that with his 25,000-man force, but he recognized that he could not make much progress without the help of Foote's flotilla. Halleck asked Foote to take his warships downstream as soon as possible to help Pope, even though Foote protested that he had not yet had time to repair all the damage his vessels had sustained outside Fort Donelson. Still, Foote wanted to cooperate, so on 14 March he left Cairo with seven warships, including his flagship, the powerful new ironclad USS *Benton*. He also brought along eleven hexagonal-shaped scows, each of which contained one 13-inch mortar bolted to its deck. The flotilla steamed southward, pushed forward by the fast-moving current that carried with it trees, fences, houses, and all sorts of imaginable and unimaginable debris, and reached Island Number Ten on 17 March.

Those officers expecting a repetition of Foote's aggressive tactics at Fort Henry and Fort Donelson were disappointed. Instead of again boldly throwing his warships against the rebel defenses, Foote settled down for a long-range siege. He brought up his mortar scows, anchored his vessels at the outermost range of the Confederate batteries, and commenced a methodical bombardment that did as little damage to the dug-in rebels as their fire did to him; that is, none at all. This was not like Foote, but he had his reasons. He had no confidence in Pope, so he did not believe he could rely on the Army. Moreover, unlike at Fort Henry and Fort Donelson, an injured Union warship drifting downstream would float toward the rebels, rendering it vulnerable to destruction or capture. Had similar circumstances prevailed at Fort Donelson, Foote would probably have lost all four of his ironclads. In addition, his wounded foot continued to bother him, limiting his mobility. Finally, Foote received awful news from home. One afternoon, while chatting with James Eads on the deck of *Benton*, Foote opened a letter, perused its contents briefly, and said calmly, "I must ask you to excuse me for a few minutes while I go down to my cabin. This letter brings me the news of the death of my son, about thirteen years old, who I had hoped would live to be the stay and support of his mother."[17] He returned fifteen minutes later, perfectly composed. On the inside, however, Foote was devastated. He wrote to his wife, "I am very much exhausted. My feelings are so much lacerated by the shock of dear William's death that I am unfit for anything but I must stand up to my duty. If it were not for you and the dear children, oh how I should welcome death at God's earliest pleasure."[18]

While Foote's flotilla frittered away the days fruitlessly bombarding rebel positions, Pope undertook his own efforts to end the siege. He put his men to work digging a canal on the Missouri side of the river that bypassed Island Number Ten. When it was completed, Pope hoped to float shallow-draft transports down the canal, load them with soldiers, and cross the Mississippi River south of the island. Once on the eastern shore, Pope could then sever the Confederate supply and communication line to Island Number Ten. To carry out his plan, Pope needed naval protection for his transports while they ferried his troops across the Mississippi. The only way to get the warships there, however, was to run them past Island Number Ten's batteries, but Foote and all but one of his ship captains believed that Confederate guns would blow out of the water any warship foolish enough to make the attempt. In the ensuing days, though, Pope exerted increasing pressure on Foote, and eventually Foote gave way. At a 29 March meeting with his captains on board *Benton*, Foote reluctantly approved Cdr. Henry Walke's offer to take his vessel, the ironclad *Carondelet*, past Island Number Ten. Walke prepared carefully for a swift and silent run downstream. He placed loose timber on the deck, wound chains around the pilothouse to protect the ship from rebel cannon fire, armed his men, and readied hoses that sprayed scalding hot water to repel any boarders. He also diverted steam through the paddle wheelhouse to cut down on noise. At 10:00 PM on 4 April, *Carondelet* slipped her cables and steamed downriver through the moonless and rainy darkness. Although the Confederates spotted the warship and opened fire, she survived the gauntlet and tied up at Union-held New Madrid around midnight. Two days later the ironclad *Pittsburg* repeated the trip with equal success. Now that he had the necessary naval protection for his transports, Pope got his troops across the Mississippi, seized the town of Tiptonville, and isolated Island Number Ten. Its five thousand rebel defenders surrendered on 8 April, thus opening the way for the Union juggernaut to continue downstream.

Foote garnered another round of public applause after Island Number Ten fell. In fact, he was the naval hero of the hour. Down at Port Royal, Du Pont, not usually the most fulsome of men, wrote to Fox, "Ain't Foote a hero! He is leaving us all out of sight."[19] Welles was equally effusive. In another congratulatory telegram, he thanked not only Foote for his success, but also "that Being" who protected him. Much though Foote appreciated these and other kind words, he was not about to rest on his laurels when the rebels were off balance and demoralized. The great commercial center of Memphis, Tennessee, was a mere 115 winding miles downriver. The only obstacle between it and Foote's force was Fort Pillow, 40 miles north of the city. Despite its name, Fort Pillow was actually a series of fortifications containing forty cannon situated on steep bluffs on the Mississippi's east bank. Foote's flotilla, spearheaded by seven ironclads, weighed anchor on 12 April and reached Fort Pillow the next day. Along the way the vessels passed a level, monotonous, and peaceful countryside, and islands full of willow and poplar trees. No sooner had Pope's soldiers arrived in their transports, though, than Halleck yanked the rug right out from under the operation. In the panicky aftermath of Ulysses Grant's brush with disaster at the Battle of Shiloh on 6–7 April, Old Brains ordered almost all of Pope's troops eastward to Pittsburg Landing to join the large army he was organizing there for a march on the railroad center of Corinth, Mississippi. With hardly any troops left, Foote was reduced to using his mortar scows to chuck shells ineffectually at Fort Pillow.[20]

Without the adrenaline generated by active campaigning to keep him going, Foote's physical condition further deteriorated. In addition to his infected foot, he now had contracted dysentery and a fever. He was in such terrible pain that he could barely get around, so he spent most of his time in bed in his cabin. Three surgeons examined him and told him he was unfit for duty and should leave the flotilla to recuperate. On 15 April Foote relayed the surgeons' conclusions to Welles and warned him that he might have to step down. Welles, however, was reluctant to detach him; Foote was a proven winner, and as such was hard to replace. By way of compromise, Welles gave Foote permission to go north and rest, but he would retain command of the flotilla and would resume his responsibilities as soon as he was well again. As for his temporary replacement, Foote noted that his favorite ship captain in the flotilla, Lt. Sam Phelps, was perfectly capable of taking over, but his appointment would alienate more senior officers. Instead, Foote repeatedly recommended his good friend Charles Davis for the post. Welles agreed, and on 22 April issued the necessary orders. Although Welles' official instructions were vague, Fox warned Davis in a private letter that he would in fact be taking over the Western Flotilla for now.[21]

Davis arrived off Fort Pillow on the evening of 8 May. Next morning he boarded *Benton* to breakfast with Foote. The two men broke down and cried when they met—Foote out of relief from the stress of responsibility; Davis out of shock at his old friend's appearance. Indeed, Foote was thin, weak, haggard, and obviously depressed. Despite his illness, Foote explained to Davis that he planned to return to the flotilla after a couple weeks of rest. Foote was popular among his officers and men, so *Benton*'s deck was crowded when he departed. The sailors huzzahed and applauded as Phelps assisted him across the deck. Foote stopped and, with tears in his eyes, complimented the men on their efforts, explained why he was leaving, expressed regret that he could not stay, and urged them to continue to do their duty. Once on board the side-wheeler USS *De Soto,* Foote took a seat on the guards and held a palm leaf over his face so *Benton*'s crew would not see him crying. The men gave him three cheers as *De Soto* backed away for her trip north, so Foote stood up and yelled, "God bless you all, my brave companions!"[22]

As things turned out, Foote never again saw his brave companions. When he arrived at his brother's house in Cleveland, Ohio, to recuperate, he still believed that he would return to the flotilla in a few weeks' time, especially after his fever broke and his dysentery abated. Unfortunately, his injured foot refused to heal. The doctors he consulted concluded that his recovery would require months of rest, not weeks. On 13 June Foote reluctantly informed Welles of his condition. Clearly he could not continue as flotilla commander, so Welles removed him four days after receiving his letter. He did so in a kind message expressing his appreciation of and gratitude for Foote's work out west, and he promised him a new command as soon as he got well. In addition, Welles appointed Foote one of the first four active-list rear admirals after Congress authorized the rank's creation in July. While he built up his strength, Foote kept a close watch on the Western Flotilla's activities along the Mississippi. He admitted to Davis that he envied him, but, generous as ever, he offered his successor nothing but support and encouragement. In July Welles put Foote in charge of the newly created Bureau of Equipment and Recruiting. When Foote arrived at the Navy Department building the following month, he was still on crutches. Welles took one look at him and told him to take more time off, though his excuse was that Foote's office was not yet ready. After a squabble with Welles about the appointment of bureau clerks, Foote got down to work in November. Although he treated Welles with a formality and deference that made Welles uncomfortable, the navy secretary was still very happy to have his boyhood friend whom he so admired working with him.[23]

Charles Davis' Eventful Summer

Within twenty-four hours of assuming command of the Western Flotilla, Davis received a rude introduction to the bare-knuckle world of riverine warfare. On the morning of 10 May, one of his mortar scows was three miles downstream from the Union flotilla's anchorage at Plum Run Bend, engaged in the routine duty of chucking shells every half hour at Fort Pillow. The ironclad *Cincinnati* was standing guard nearby, tied up to some trees on the shore, steam down, her crew holystoning her deck. At 7:00 AM sailors spotted the eight gunboats of the Confederate River Defense Fleet steaming upriver at full speed. Caught napping, *Cincinnati* hurriedly fired her furnaces, slipped her cables, and prepared to fight. Unfortunately, the rebel gunboats had the twin advantages of speed and surprise. *General Bragg* rammed *Cincinnati* and ripped a twelve-foot hole in her. The two vessels stuck together until a salvo from *Cincinnati* blew them apart. Then *General Sterling Price* hit *Cincinnati*'s fantail, followed by a blow from *General Sumter* to the port quarter. When Cdr. Roger Stembel went on deck for a better look, a rebel sharpshooter shot him in the mouth. His crew refused to surrender, though, even as the proud ironclad settled on the river bottom, her pilothouse sticking above the water for the sailors to cling onto. By now the rest of the Union flotilla at Plum Run Bend had heard the racket, and its warships were coming to the rescue as fast as they could get their steam up. *Carondelet* and *Mound City* were the first to arrive. The Confederate vessel *General Van Dorn* plowed into *Mound City* practically head-on, and, after breaking free, the Union ironclad staggered to shallow water and sank in the mud. *Benton* and *Pittsburg* now appeared, and their heavy ordnance began to exact a toll on the lightly armed rebel gunboats. Having done more damage than they had any right to expect, the Confederates called it a day and scurried back downstream to Memphis. Their casualties amounted to two killed and one wounded. Union losses were five injured—as well as the loss of two of the hitherto seemingly invulnerable ironclads.

Davis tried to put the best face on the defeat, claiming inaccurately to have damaged two of the rebel gunboats and sunk a third in the course of repelling the ferocious Confederate assault. In fact, the setback was temporary; within a month both *Cincinnati* and *Mound City* had been raised and repaired. Although the Navy Department initially accepted Davis' explanation of the Battle of Plum Run Bend, the truth eventually came out, so that on 18 May Fox noted, "Davis is overmatched and *things* are not well managed."[24] It was not that Davis intentionally lied to his superiors, but rather that he believed it was easiest to interpret events in the most positive light. He was a distinguished, charming, and amiable man with pursed lips, a flowing mustache that diverted attention from his weak chin, and a receding hairline. He did not like to make waves, so he gave his honest opinion only when directly asked. Unlike many of his contemporaries, he displayed a rare sensitivity toward the war's victims, once sending a boat upriver to look after the refugees scrambling to escape the violence. He especially lamented the children traumatized by the conflict, noting that the sight of them removed all the beauty from life. On the other hand, he viewed the war itself with a professional equanimity. His old friend Du Pont eventually concluded that Davis was too lazy and unorganized to make an effective squadron commander. Welles, however, was closer to the mark when he later identified Davis as more of an intellectual than a fighter. Indeed, he possessed extensive scientific and literary knowledge—in fact, he committed large parts of Shakespeare and Virgil to memory during his time afloat—and was probably the closest thing the Navy had to a legitimate scholar. As such, he had acquired a cosmopolitan outlook rare even in the Navy. While Davis was fully capable of coolly analyzing problems and developing reasonable solutions for them, he lacked the toughness and drive to attain those solutions regardless of cost.[25]

Actually, Davis' prewar career, like his personality, tilted toward the cerebral. He was born in Boston in 1807, the son of a man who was for thirty-three years Massachusetts' solicitor general. Although he was the youngest of thirteen children, his parents made sure he received a good education. He attended the Boston Latin School, and then Harvard for two years, before accepting a midshipman's warrant in 1823. He spent most of his early years in the Navy on routine duty, but in between cruises he studied mathematics at Harvard, eventually receiving his bachelor's degree in 1841. From that point on, his career was more scholarly than naval. He worked for the Coast Survey charting the New England shoreline, wrote and published scientific articles, and helped establish the Nautical Almanac Office, of which he was superintendent throughout much of the 1850s. When the war began, Paulding called him to Washington to help him in detailing officers, which brought him to Welles' attention. Du Pont recruited him as his chief of staff for the Port Royal expedition, but on 10 February 1862 the Navy Department brought him back to Washington.

When Davis returned to Washington in March, he discovered that there were no orders for him, just lots of speculation among those outside the command loop as to his future. He initially had little to do, and worried that Welles would place him on some dead-end board for a long time. On the other hand, he was encouraged by rumors that he might get to lead the new Bureau of Navigation. Welles and Fox, however, had their reasons for keeping Davis on ice. Capt. William McKean commanded the East Gulf Squadron, and it was common knowledge that his uncertain health could force him to surrender his post. When and if that happened, Welles and Fox planned to give his squadron to Davis. On 13 April, however, McKean wrote Welles that he felt hale enough to stick it out for now. By the time Welles received McKean's letter, though, he and Fox were already contemplating sending Davis out west to help Foote, who had first warned them on 27 March that his Fort Donelson wound was so serious that he might have to step down. On 22 April Welles finally acted and instructed Davis to go to Tennessee, and Fox elaborated in a private letter that he would actually be assuming temporary control of the Western Flotilla. Welles had good reasons for choosing Davis. First of all, Fox, Foote, and Du Pont all recommended him, and at that time Welles respected their opinions more than almost anyone else's. Welles also knew through personal acquaintance that he was of unquestioned loyalty. Finally, Port Royal had given Davis battle experience that most naval officers as of yet lacked. Davis was still a commander, albeit a senior one, ranking seventh on the list, but not many captains were clamoring for duty on the unfamiliar Mississippi anyway. Davis accepted his new assignment with his customary grace, telling Fox that he would do his best, but adding that he doubted anyone could adequately fill the shoes of his old friend Foote.[26]

Davis was not the only one on his way to Plum Run Bend that late spring. So was Col. Charles Ellet, a prewar civil engineer who had persuaded the new secretary of war, Edwin Stanton, to authorize the construction of a fleet of ram warships to help clear the Mississippi of rebels. Ellet bought a number of old steamships, braced their hulls and bows with heavy oak timbers, installed powerful engines, and thus created updated versions of the rams that had once terrorized the Mediterranean Sea. His new Mississippi Ram Fleet contained neither iron plating nor cannon at first, but instead relied on the vessels themselves for their offensive punch. On 25 May Ellet brought his eight rams—each of which was commanded by a member of his extended family—down to Plum Run Bend to cooperate with the Western Flotilla. Davis had no authority over Ellet's vessels, which was just as well because he scarcely knew what to make of them. As aggressive as he was innovative, Ellet proposed to run his rams past Fort Pillow's batteries. Davis did not think much of this plan, though it was certainly more aggressive than the ineffectual shelling of the fort that the Navy had been engaged in for the past six weeks. Fortunately, on 29 May Halleck finally occupied Corinth, Mississippi, after an agonizingly slow march. Doing so rendered Fort Pillow untenable, so the rebels evacuated the place the following day. The river road to Memphis was now open.[27]

On 5 June Davis' and Ellet's warships arrived off Memphis. As at Fort Pillow, rebel troops had already evacuated the city in the wake of Corinth's fall, leaving it defenseless on land. On the other hand, the eight gunboats comprising the Confederate River Defense Fleet remained in place. If they could do nothing to prevent the Union Army from seizing Memphis, they might at least inflict enough damage on the Union Navy to delay its advance downriver. Even this limited goal, however, might not be obtainable; the rebel gunboats mounted only twenty-eight cannon to oppose the sixty-eight studding Davis' ironclads. Moreover, Davis was determined to erase the stain of his defeat at Plum Run Bend by destroying the rebel warships that instigated it. At first light on 6 June, Davis and Ellet launched an all-out assault on the River Defense Fleet.

The two commanders chose not to coordinate their efforts, which might have proven fatal against a more powerful opponent. But not in this case. As the citizens of Memphis watched from the bluffs overlooking the river, Ellet's faster rams sped past Davis' five lumbering ironclads and made a bee-line for the rebel vessels. Ellet's flagship, *Queen of the West,* crashed into *General Lovell* amidships, forcing the wounded rebel warship to the Arkansas bank. The Confederate vessel *Beauregard* took advantage of *Queen of the West*'s preoccupation with *General Lovell* to ram her, and the Union ship followed her recent victim to the other shore. Another of Ellet's rams, *Monarch,* darted between two advancing rebel gunboats so quickly that they collided in her wake. By now Davis' ironclads had come up, and in short order they systematically demolished the remaining rebel gunboats. By the time the smoke cleared, all but one of the River Defense Fleet's warships had been captured, destroyed, or sunk, and the cheers from the bluffs had turned to groans. Ellet and Davis both sent representatives to Memphis to walk through the stunned and silent crowds to accept the city's surrender, and Union soldiers arrived a few hours later to secure the place. Ellet was the only Union casualty in the one-sided fight. A Confederate sharpshooter shot him in the leg, and although the injury did not initially seem serious, infection set in, and he died several weeks later.[28]

Although Memphis' fall was a foregone conclusion by the time Davis' and Ellet's vessels appeared, their victory was still important because it eliminated the only remaining organized Confederate naval force on the Mississippi. For Davis, it also restored the credibility he had lost at Plum Run Bend. Indeed, Welles not only congratulated him and his crews for their triumph, but he also made him a flag officer. It was quite an accomplishment for a mere commander, so it was hardly surprising that Davis wrote to his wife, "I hope I shall be able to fight through the summer and retain my flag, which, by the way, I am so lucky in getting. How many above me are without one!"[29]

As with much of life, however, there was bad news with the good, and it came in doses both small and large. By now the weather had turned steaming hot, and the lack of wind made the Mississippi feel like a sauna. Davis' small cabin in *Benton* was located between the Marines and the pantry, and he complained that it was as noisy as it was smelly. Visitors arrived almost daily to see the new flag officer, but Davis had no wine to serve them. All he had was Mississippi River water that was "more dirty than that which runs down the gutter on Beacon Street in a summer shower."[30] Davis could accept these inconveniences as part and parcel of naval life. No one, however, was prepared for the events on 16 June. The War Department had asked the Navy to open the White River to Union shipping in order to help supply Maj. Gen. Samuel Curtis' army in Arkansas. Welles passed the request on to Davis, who on 13 June dispatched three ironclads to lead an expedition up the river. Three days later, as the Union force approached St. Charles, the rebels ambushed the warships. A Confederate cannonball crashed through recently repaired *Mound City*'s casemate, destroyed her boiler, and sank the ship. The hot steam escaping from the boiler contributed to the high casualty count. One hundred twenty-five men were killed, another twenty-five were dreadfully scalded, and only twenty-five survived unhurt. Although the Army easily took St. Charles after this disaster, the river's declining water levels forced an end to the operation. Davis was unhappy with this hollow victory that cost him one of his warships, and horrified at the sight of the disfigured burn victims.[31]

On 28 June Davis opened a letter from Flag Officer David Farragut, commander of the West Gulf Squadron. Farragut and his fleet, Davis learned, were anchored 250 miles downstream near Vicksburg, Mississippi, the last major Confederate stronghold on the Mississippi River. Just three days earlier, Welles had written Davis urging cooperation between him and Farragut in opening the Mississippi to Union shipping. Thus prodded, Davis did not waste any time; he quickly got his vessels under way and headed downriver to meet Farragut. Three days later, at breakfast, Davis' warships steamed around a bend and spotted Farragut's fleet above the rebel citadel. Davis' and Farragut's men cheered themselves hoarse as Davis' column of vessels passed down the line of West Gulf Squadron vessels toward Farragut's flagship, the steam sloop USS *Hartford*. Davis was rowed to *Hartford*, and the two old friends greeted each other warmly. Davis noted that he and Farragut had last seen each other at Port Royal the previous March, and he never would have imagined that they would meet again as commanders of large fleets in the middle of the North American continent. Officers renewed old acquaintances, looked each other over, and swapped war stories. The contrast between the two fleets could not have been more striking. Davis' weird conglomeration of vessels was dominated by the squat, boxlike, mastless, armor-plated ironclads. Farragut's force, on the other hand, contained majestic, graceful, and rigged wooden vessels that towered over the surrounding countryside. Almost none of them would have guessed that it would take them more than a year to fulfill the mission that Welles had set for them.[32]

William Mervine, William McKean, and the Mississippi Delta

Among the Confederacy's few major cities, New Orleans was in a class of its own. With a population of 170,000, it contained more than four times as many people as its closest competitor, Charleston. In fact, it had more inhabitants than the other five largest Confederate cities combined. Its cosmopolitan makeup—39 percent of its populace was foreign born—differentiated it as well. Its economic significance also put it in its own league. New Orleans was one of the world's great commercial centers. Bankers, clerks, accountants, warehouse operators, and businessmen of all shades and stripes facilitated the transshipment of goods from the American heartland to the rest of the world. During the 1850s three thousand steamers a year visited the city, providing plenty of business for the machine shops, ironworks, and dry dock that sprouted up along the riverbank. Many of these steamers were engaged in the increasingly important cotton trade. In 1860–61 New Orleans transshipped 2.2 million bales of cotton worth $110 million. Considering its obvious importance, it was inevitable that the city would draw the Strategy Board's attention. Du Pont and his colleagues, though, were baffled by New Orleans' peculiar geography, which made blockading it extremely difficult. New Orleans was not on the Gulf Coast, but actually lay more than a hundred miles upriver because the Mississippi created an uninhabitable marshy delta and divided into four fingerlike distributaries as it flowed into the Gulf of Mexico. To complicate things further, various lakes bracketed the city and provided additional avenues for blockade-runners to operate. It was obvious that sealing the city off from the outside world would require a major undertaking.[33]

To lead the blockade along the Gulf Coast, Welles on 6 May 1861 appointed Capt. William Mervine. He did so reluctantly, and for reasons he later regretted. At the beginning of the war, Paulding and others encouraged Welles to rely on seniority in selecting squadron commanders, and Welles, still new to the Navy Department and its ways, reluctantly agreed. Welles initially wanted to give the post to Capt. Francis Gregory. Gregory was not only seventh on the list of loyal captains, but Welles also believed that he was a courageous and indomitable old sailor. Paulding dissuaded him from doing so, however, probably because he knew Gregory's health was uncertain. Instead, he recommended Mervine, who was eleventh on the naval register, one behind Silas Stringham. In fact, Mervine and Stringham had much in common. Both had joined the Navy as midshipmen in 1809, but Mervine spent most of the War of 1812 on the Great Lakes. Afterward he saw the usual service afloat in the West Indies, in the Mediterranean, and along the African coast. He capped his prewar career as commander of the Pacific Squadron from 1855 to 1857. Mervine, like Stringham, was an upright, brave, proud, and patriotic man who performed well under routine conditions. Unfortunately, again like Stringham, he was also overly cautious, stuffy, and detail oriented. Fox exaggerated when he later called Mervine stupid, but Mervine proved unwilling or unable to wage the kind of aggressive war Welles wanted.[34]

Welles' dissatisfaction with Mervine began almost immediately. Repairs on Mervine's designated flagship, the steam frigate USS *Colorado,* in Boston required more time than anticipated. Instead of taking another vessel to the Gulf to assume his new duties, Mervine waited around for *Colorado.* As the days passed, Paulding grew so frustrated with Mervine's dilatoriness that he contemplated asking Welles for the command himself. In the end, Welles peremptorily ordered Mervine to leave for his new station without *Colorado.* Welles' unhappiness grew worse once Mervine arrived at Key West, Florida, on 8 June to take charge of his new squadron. On 30 June the rebel raider *Sumter* slipped past the Union blockade of New Orleans and proceeded to seize more than a half-dozen Union merchant ships over the next few months. Welles felt that Mervine could prevent a repetition of the *Sumter's* escape by fortifying someplace along the Mississippi River downstream from New Orleans, but Mervine did not do so. Finally, although Mervine stated that he hoped to occupy strategically located Ship Island, halfway between New Orleans and Mobile, he made no effort to undertake the necessary work. Welles' growing anger toward Mervine was stoked by Cdr. David Porter, who had absconded with *Powhatan* just before the Confederates bombarded Fort Sumter. Porter and *Powhatan* were now part of the Gulf Squadron, but this did not prevent Porter from denouncing his commanding officer's lethargy in letters to Fox and Davis that found their way into Welles' hands. Welles acknowledged that Mervine was operating with limited resources, but he believed that more initiative on Mervine's part could overcome some of those deficiencies. On 23 August Welles warned Mervine, "There is great uneasiness in the public mind, as well as anxiety in the Department, on the apparent inactivity of your squadron."[35]

In general, Welles' charges against Mervine were similar to those he leveled against Stringham that same summer. Like Stringham, Mervine attempted to defend himself—with as little success, as things turned out. On 9 September he wrote Welles that he was deeply grieved by the navy secretary's recent allegations. It was, he said, the first time in his fifty-three-year career that anyone had ever accused him of neglecting his duty. Mervine noted that his primary responsibility was implementing the blockade, and he lacked the necessary vessels to engage in supplementary operations such as occupying Ship Island. He had fewer warships at his disposal now than when he first arrived on the Gulf Coast, and many of those were in poor repair. Indeed, Welles' own complaints about the blockade's ineffectiveness were evidence that he did not have the resources for other activities. Getting down to specifics, Mervine argued that he could not bring his big warships within two miles of Ship Island and its eight hundred defenders because they drew too much water. Finally, Mervine guessed correctly that one of his younger officers was denouncing him behind his back, and he regretted that Welles would lend credence to such unsubstantiated and irregular reports that violated naval regulations. He concluded by stating that although he was an old man, he was still prepared to do his duty at all costs.[36]

Mervine did not know it, but Welles had relieved him of his command three days before he penned his missive. In fact, the navy secretary had been contemplating such action since mid-August. He acted because he did not believe that Mervine would ever accomplish anything along the Gulf Coast. Three years later, recalling his reasoning in his diary, Welles scrawled that Mervine was "good for nothing as an officer for such duties the time required."[37] Nor was Welles alone in his thinking; those he consulted—Fox, Paulding, Du Pont, and Foote—agreed with his logic. On 21 September a devastated Mervine transferred his flag to USS *Rhode Island* in a ceremony full of pomp. When Mervine arrived at New York City, he wrote to Welles to ask for a court of inquiry or a court-martial because he feared that some would interpret his relief as evidence of misconduct or neglect of duty on his part. Although such judicial proceedings were relatively common in the prewar Navy, Welles rejected his request. Welles stated that the Navy Department had the right to select whomever it wanted to promote the public welfare, regardless of personal interests.

No one, he continued, doubted Mervine's honor and faithfulness, so a court of inquiry or a court-martial would just waste valuable time and energy. Summing up his thinking, Welles concluded, "In selecting you to command the Gulf Squadron no partiality was exercised, nor has any prejudice induced a change. Only the true and best interests of the country have been consulted in what has been done, and the Department must judge men as well as means best adapted to carry into effect its views."[38] Mervine was placed on the retired list the following year, served on various boards during the war, and died in Utica, New York, in 1868.[39]

Welles consulted many people and carefully scrutinized the naval register to find Mervine's replacement. He finally narrowed his choice to two captains, Charles Bell and William McKean. He asked Foote for his opinion, and, after much hesitation, Foote recommended McKean. Fox and Du Pont also gave their approval. Welles liked McKean not only because of his undoubted loyalty, but also because of the speed with which he returned from Japan in his warship, the screw frigate USS *Niagara,* during the secession crisis. Moreover, he was already in the Gulf, having been sent there on 4 May to temporarily command until Mervine superseded him, so he was familiar with the geography and problems he would face. Welles hoped that his reputation for good judgment, energy, discretion, and character would translate into success in his new assignment. While it was true that McKean was down on the register—he ranked forty-second out of the sixty-five loyal prewar captains—Welles' experiences with Mervine and Stringham helped convince him that he needed to combat, not encourage, the Navy's obsession with seniority.[40]

McKean was born in Pennsylvania in 1800, the grandson of Thomas McKean, one of the signers of the Declaration of Independence. He joined the Navy as a midshipman in 1814, and spent much of his career on routine patrols. He campaigned against pirates in the West Indies in the early 1820s, and later saw duty in the Mediterranean and off the South American coast. He was in charge of the naval asylum in Philadelphia, where midshipmen were then instructed, from 1843 to 1844. Later he and his cousin, Frank Buchanan, served on the board that recommended Annapolis as the site for the new naval academy. He commanded the USS *Dale* during the Mexican War until invalidated home, and was returning the Japanese embassy to Japan when the war broke out. McKean was a deeply religious man whom observers characterized as the prototypical officer and gentleman. His rectitude, however, had its limits; he was also the father of twelve children. Unfortunately, McKean was also hampered by unspecified ill health that did as much as anything else to undermine his efficiency.[41]

In placing McKean in charge of the Gulf Squadron, Welles stressed that enforcing the blockade was his primary task, especially along the Mississippi delta. Welles acknowledged the difficulties McKean faced, but he expected him to show more energy and initiative than his predecessor. McKean's 22 September response, therefore, did not encourage the navy secretary. McKean wrote that although he would of course do his best to live up to Welles' expectations, he could not thoroughly seal off the Gulf Coast until the Navy Department supplied him with light draft steamers to track down the small vessels the rebels used to run the blockade. Happily for McKean and his relationship with Welles, however, he was able to take credit for a couple of successes initiated by others. On 16 September Union naval forces seized recently abandoned Ship Island, securing the potential base in the region that Welles wanted. Several days later, a Union expedition led by Capt. John Pope—nicknamed "Honest John" to differentiate him from his less-than-candid army counterpart—crossed the bar, ascended Pass à l'Outre, and occupied Head of Passes, a two-mile-long deepwater anchorage located where the Mississippi diverged into four distributaries on its way to the Gulf of Mexico. Once ensconced there, Union warships could bottle up the big river and prevent Confederate blockade-runners from entering or leaving New Orleans by that route.

No sooner had McKean welcomed his unearned blessings, though, than he received disturbing news from Head of Passes. Before dawn on 12 October, the rebel armored ram *Manassas* swept downriver from New Orleans to attack Pope's anchored warships. In the ensuing predawn melee, the screw sloop USS *Richmond* was damaged and ran aground before *Manassas* headed back upriver. Worse yet, Cdr. Robert Handy misread one of Pope's signals, ordered his grounded ship, the sailing sloop USS *Vincennes,* abandoned, and set a fuse to her powder magazine. Fortunately, the vessel did not explode, so after a long wait an embarrassed Handy reboarded and refloated her. When he learned of the mayhem, McKean abandoned his plan to bombard Pensacola and steamed immediately to Head of Passes. There he relieved Handy and accepted a mortified Pope's resignation.[42]

After the debacle at Head of Passes, events along the Gulf Coast settled into a mundane routine that did little to advance the Union war effort.

One officer, Capt. Theodorus Bailey, wrote from Head of Passes, "I am afraid that we are destined to remain on this blockade for the remainder of the war, with nothing in view but a light house, a hamlet of houses called Pilot Town and some maid islands and lumps—not a vestige of vegetation and no prospect of a battle for us."[43] McKean regretted that he was unable to accomplish more, but grumbled to Welles that his insufficient resources made it impossible to adequately blockade any of the ports in his charge. Although McKean had accomplished little more than Mervine, Welles refrained from relieving him. Welles recognized that there was justice in McKean's complaints, and in January 1862 wrote him that he was grateful for his achievements, limited though they were. Besides, Welles had already concluded that McKean, honorable though he no doubt was, lacked the hard-driving killer instinct he was looking for. Until he found an officer with those qualities, Welles was prepared to tolerate McKean's mediocrity.[44]

David Farragut and New Orleans

On 9 November 1861, recently promoted Cdr. David Porter docked at the Brooklyn Navy Yard in *Powhatan* after a seven-month cruise to and from the Gulf of Mexico. Except for undermining Mervine, Porter had accomplished next to nothing on his voyage. These meager results, combined with the irregular manner with which he attained his command, gave him good reason to worry about the reception he would receive from Welles. During his time in the Gulf, however, Porter had formulated a plan to seize New Orleans that he hoped would deflect the navy secretary's wrath. New Orleans was defended by two forts, Jackson and St. Philip, located on each side of the Mississippi River about seventy-five miles downstream from the city. The former contained seventy-five guns, the latter forty. Porter believed that a flotilla of schooners, commanded of course by him, each mounting a heavy 13-inch mortar, could reduce both forts in forty-eight hours and enable Union warships to steam past them unscathed to New Orleans. Three days after he reached Brooklyn, Porter arrived in Washington and walked over to the Navy Department building. There Faxon, Fox, and others gave him the cold shoulder, and one of the bureau chiefs, Capt. Joseph Smith, commented, "Well, you didn't run away after all!" Porter was nothing if not brash, though, so he parked himself outside Welles' office, determined to speak with the navy secretary.

When senators John Hale and James Grimes showed up to congratulate Welles on Du Pont's recent victory at Port Royal, Porter explained his idea to them. The two men were impressed enough to bring Porter in with them to see Welles. To Porter's relief, Welles received him cordially. Even so, Porter did not want to push his luck, so he quickly sketched out his plan again in as few words as possible to an attentive Welles.[45]

As it was, Welles and Fox had already given serious thought to attacking New Orleans, whose strategic significance was as obvious to them as to anyone else. Fox had commanded merchantmen plying the Gulf in his prewar days, so he was thoroughly familiar with the city's peculiar geography. He believed that a naval expedition could steam past Fort Jackson and Fort St. Philip in the middle of the night and place New Orleans under its guns the next day. Porter, though, insisted that no assault on the city could succeed without first battering the forts into submission, and his mortar vessels were the key to doing that. After Welles heard Porter out, he summoned Fox, and the three men walked across the street to the White House to see the president. Lincoln liked Porter's idea and added, "This should have been done sooner. The Mississippi is the backbone of the Rebellion; it is the key to the whole situation."[46] Although Fox, with his usual prejudices against the Army, wanted the operation to be a Navy-only affair, clearly troops would be necessary to hold and defend New Orleans after the Navy seized the place. The four men, joined by Seward, then traveled to George McClellan's headquarters. McClellan was then general in chief, so he had to authorize the use of any soldiers. Not surprisingly, McClellan believed that any campaign against New Orleans was beyond the Army's resources because it would require 50,000 men that he could not currently spare. When informed that the Army's role would be secondary, however, he agreed to contribute a total of 15,000 soldiers under Maj. Gen. Benjamin Butler. While Fox began organizing the warships the planned expedition required, Porter got to work purchasing, arming, and manning his 20-mortar schooners.[47]

The attack on New Orleans promised to be the biggest and riskiest naval operation of the war so far. Welles understood that his main role would be choosing the right officer to lead the expedition, so he was even more thorough and careful than usual when he scrutinized the naval register and weighed the strengths and weaknesses of each captain. As usual, he disregarded the infirm and otherwise employed, which considerably narrowed his options. Scanning the register, his eyes settled on Capt.

David Farragut's name. Farragut was thirty-fourth on the list of loyal captains. As a bureau chief during the Mexican War, Welles had listened to Farragut outline a never-implemented plan to seize Fort San Juan d'Ulloa off Veracruz, and had been favorably impressed with Farragut's earnestness and modesty. When the Civil War began, Farragut, like so many loyal Southern-born naval officers, found himself sitting on the sidelines as a member of the retirement board because of Welles' doubts about such men. Welles, however, had heard that Farragut had abandoned Norfolk and all his property when Virginia seceded from the Union, which was the kind of sacrifice the navy secretary appreciated.

With this information in hand, Welles made some discreet inquiries. Fox, Paulding, and Porter all vouched for him. In fact, Paulding had written to Fox a month earlier, "Don't be unmindful of Farragut. He is a valuable officer whenever you can use him."[48] Since Porter was slated to play a crucial role in the upcoming campaign, his approval was equally important. Others who Welles respected—bureau chief Joseph Smith, Andrew Foote, Cdr. John Dahlgren, and Capt. William Shubrick—voiced their support. The consensus was that Farragut was a capable, energetic, faithful, and impetuous officer with good sense and sound habits. At the same time, though, many of these same officers expressed doubts that Farragut had the proper discretion and temperament to command a large naval force. Welles threw these and other opinions into his mental hopper as he continued to turn the issue over in his head. He noticed that Farragut was not part of any of the Navy's innumerable cliques, and had pulled no strings since the war began to advance his career. Although his prewar record was not stellar, it was solid. Finally, and perhaps most significantly, Welles was realizing more and more that great leaders required independence of mind, self-reliance, and an instinctive desire to seize the initiative whenever possible. Whatever the other positive qualities Stringham, Mervine, Goldsborough, and McKean may have had, they did not possess that fire in their bellies that Welles was seeking. His gut told him that Farragut might.[49]

However much Welles trusted his instincts, he was not about to rely on them completely when the stakes were so high. Instead, he sought more information. He dispatched Porter to New York to sound out Farragut about his willingness to lead a big and dangerous expedition, but without providing the name of the target. Porter did so and reported back what he had said repeatedly from the beginning: Farragut was the right officer. By mid-December, preparations were so far advanced that it was now imperative for Welles to designate a commander. On 15 December he directed Farragut to report to Washington. When he arrived, Fox and his wife invited him over to dinner with Fox's brother-in-law, Postmaster General Montgomery Blair. At the meal, Fox was so taken with Farragut's enthusiasm and good nature that he dropped all pretense and fully explained the operation the Navy Department wanted him to lead. Farragut of course knew he was a candidate for a major command, and he was willing and eager to get the appointment. Welles was equally impressed in subsequent meetings with Farragut. Farragut liked Fox's original idea to run past Fort Jackson and Fort St. Philip, and, in refreshing contrast to some of Welles' other officers, said he could fulfill his mission with even fewer vessels than the Navy planned to give him. As for Porter's mortar flotilla, Farragut agreed to take it with him, but he feared that a prolonged bombardment of the forts might merely tip off the enemy to his intention to attack New Orleans. Welles thought he had found his man, so on 9 January 1862 he issued orders placing Farragut in charge of the naval forces slated to assault the city.[50]

Farragut's appointment required some organizational changes in the Gulf Squadron. The previous July, Mervine had warned Welles that it was probably a good idea to divide the Gulf Squadron in two to ease the administrative burdens involved. Welles had agreed, but now he saw that doing so would serve more than simple administrative convenience. Assigning Farragut to the newly designated West Gulf Squadron might deceive rebels interested in his activities into believing that he was merely going to the Gulf of Mexico to assume his new blockading duties, not to launch an all-out attack on New Orleans. In fact, Welles' 9 January orders to Farragut did not mention New Orleans at all. As for McKean, Welles put him in charge of the less-important East Gulf Squadron—crowding him offstage, as it were, to make way for the new star in the navy secretary's latest production.[51]

As Welles had noted, Farragut's career had been respectable but hardly spectacular. He was born in 1801 near Knoxville, Tennessee. His mother died when he was seven, and he never saw his father after he turned nine. His family moved to New Orleans in 1807, and there Farragut's father earned the gratitude of the Porters by rescuing a family member from Lake Pontchartrain and nursing him through his final illness. David Porter's father and namesake took young Farragut in, made sure that he got a rudimentary education, and in 1810 secured for him a midshipman's warrant. David Porter Sr. brought Farragut with him on USS *Essex* during his famous voyage during the War of 1812. Although Farragut was barely a teenager, he fought bravely in *Essex*'s last bloody battle off the Chilean coast in 1814, during which he was captured and later paroled. After the war, Farragut served on a number of ships of the line in the Mediterranean, usually as the captain's aide. When possible, he took advantage of opportunities to further his education, eventually learning Italian, Spanish, French, and Arabic. He combated pirates in the West Indies in the early 1820s, but then, after such a promising start, his career slowed down. Despite—or, perhaps, because of—his connections with the powerful but controversial Porters, Farragut made little effort to use his connections to curry favor with the Navy Department. He spent his time ashore in Norfolk with his first wife and, after she died in 1840, his second. Although he captained the sloop USS *Saratoga* at the end of the Mexican War, he saw little action. His major professional accomplishment in the 1850s was his construction and command of the new naval yard at Mare Island in California. He was at home in Norfolk when the Confederates fired on Fort Sumter, and he spoke out strongly against secession. When his neighbors told him he could not live there with such sentiments, he said, "Well, then, I can live somewhere else."[52] He fled the city with his wife and son, settled in a cottage in Hastings-on-Hudson, New York, and reported himself ready for duty.

One person described Farragut as a man "of medium height, stoutly built, with a finely proportioned head and smoothly shaven face, with an expression combining overflowing kindness with iron will and invincible determination, and with eyes that in repose were full of sweetness and light, but, in emergency, could flash fire and fury."[53] He had been well liked in the prewar Navy, and with good reason. He was modest and affable, with a twinkle in his eye and a cheerful manner. While certainly no scholar, he possessed enough education to talk intelligently on a wide variety of subjects, making him a welcome officer in the wardroom. He discovered Christianity late in life, but combined it with an earthiness all his own. Once on the Mississippi he said grace in his cabin before dinner, and concluded by noting, "It's hot as hell here!" Like Du Pont, he could inspire intense loyalty among his subordinates and mold them into an effective team. Although he was often frank and brusque with them, he softened his hard edge with kindness and a self-effacing sense of humor

that won them over. One day, for example, he looked up from his paperwork and said to a group of officers, "Now, how in the devil do you spell
'Appalachicola'? Some of these educated young fellows from Annapolis
must know." Almost everyone commented on his energy. In fact, Farragut
told one person that he turned a handspring on every birthday just to
make sure he was not yet too old. He possessed the physical courage and
coolness under fire that characterized the Civil War officer corps, even
though he had seen little combat since his childhood. He was not much
for paperwork and red tape, often writing out his own orders on his knee
and using his coat pocket as his filing cabinet. Most of these were welcome
and well-known attributes, but there was more to Farragut that did not
become clear until the war began, and that Welles was one of the few to
suspect. Farragut possessed a knack for formulating simple and clear plans,
preparing them thoroughly, and implementing them ruthlessly and deci-
sively. He saw setbacks as personal affronts requiring rectification as soon
as possible, and was always impatient for victory. Welles may have poorly
chosen some of his earlier squadron commanders, but in Farragut he
uncovered a diamond in the rough.[54]

After Farragut learned of his new command and mission, he hurried to
assemble the force Fox had allotted to him, which totaled eighteen warships mounting 243 cannon. As he prepared his fleet, Welles wrote out his
final orders. He directed Farragut to seize New Orleans and then steam up
the Mississippi to affect a junction with Foote's flotilla. He also expected
Farragut to maintain the blockade, and reduce Mobile, Alabama. Despite
these daunting and somewhat contradictory instructions, Farragut blandly
responded that he would do his best to obey them. On 2 February
Farragut and his flagship, the steam sloop *Hartford*, left Hampton Roads
and arrived at Ship Island eighteen days later. Once there, Farragut undertook the difficult task of familiarizing himself with the local geography,
marrying the warships he brought with him with those he inherited from
McKean, and getting his forces in position for an assault on New Orleans.

As he did so, though, Welles began suffering from buyer's remorse.
Farragut's appointment had not met the kind of universal acclaim that followed Du Pont's selection. About the only thing that most congressmen
knew of Farragut was that he was a Southerner by birth, by marriage—
twice, in fact—and, until the war began, by residence. This was more than
enough to raise a few congressional eyebrows. Although Welles had not
sought input from Lincoln and the cabinet about his choice for West Gulf

Squadron leader, Seward offered his unsolicited opinion anyway. The secretary of state was not sure that Farragut was up to the job, and asked Welles why he had not simply assigned the position to Du Pont, who was a proven winner. Moreover, Farragut's own actions gave Welles cause for concern. Even before he left Hampton Roads, Farragut started making demands on the Navy Department. While Farragut believed he had sufficient resources to take New Orleans, he needed light draft steamers to scour the shallow Gulf waters for blockade-runners. Welles, however, did not initially recognize the distinction; he feared that Farragut wanted these vessels for the New Orleans assault, and he knew it would take a long time to provide them. Nor did he understand Farragut's allusions to an attack on Fort Livingston at remote Barataria Bay, or why blockade-runners were still slipping out of New Orleans. To Welles, Farragut was sounding a lot like Mervine or McKean.[55]

To make matters worse, David Porter did his best to stoke Welles' and Fox's doubts about Farragut. With his usual energy, Porter had purchased or constructed twenty-one small sailing vessels, mounted a 13-inch mortar and a couple of cannon on each, and recruited and trained seven hundred men to crew and operate them. With his flagship, the side-wheel steamer USS *Harriet Lane,* he brought his mortar flotilla to Ship Island on 13 March. Porter was a critical, unscrupulous, and ambitious man willing to use almost any means fair or foul to advance his career, even at the expense of others. He was tight with Fox, and exploited his unofficial correspondence with the assistant secretary to systematically undermine Farragut. He claimed he was doing so for the good of the service and the country, but in all likelihood his primary goal was to supplant his superior. In his letters to Fox, he mixed tepid praise for Farragut with damning indictments of his leadership. On 28 March, for example, he wrote, "Men of [Farragut's] age in a seafaring life are not fit for the command of important enterprises, they lack the vigor of youth. . . . He talks very much at random at times, and rather underrates the difficulties before him, without fairly comprehending them. I know what they are and appreciate them, and as he is impressible I hope to make him appreciate them also."[56] Porter criticized Farragut's dilatoriness in getting his vessels to the Mississippi delta, his lackluster administrative abilities, his informal relationship with his captains, and his improper understanding of the military situation. By way of eliminating any possible competitors among Farragut's subordinates, Porter denounced many of them as well. He told

Fox that he was unsure of the solution to the Farragut problem, but the fairly obvious subtext to all his messages was that he would do a much better job than his adopted brother. Unfortunately for Porter, although Welles and Fox were concerned, they chose not to take counsel of their fears just yet.[57]

The truth was that Farragut was nowhere near as clueless and befuddled as Porter intimated to Fox. He was well aware of the stakes involved not only for his country, but also for himself personally, writing, "Success is the only thing listened to in this war, and I know that I must sink or swim by that rule."[58] He had good reasons for making some of the decisions for which Porter criticized him. Most prominently, he did not want to concentrate his forces at the Mississippi River until Porter's mortar flotilla arrived because doing so too soon would weaken the blockade of other ports. While he waited, he took stock of the military state of affairs. Although Benjamin Butler was already there with 15,000 soldiers, he was content to follow the Navy's lead. Farragut was willing to take it, but he needed to know what he had to work with. He asked his ship captains for information about their vessels' tonnage, number of guns and their caliber, amount of coal on hand and consumed, rounds of ammunition available, crew size, kinds of mechanics, and so on. He also evaluated his chief subordinates. Du Pont may have skimmed off the best officers for his Port Royal expedition, but there were some good ones left over for Farragut. In fact, three of those present in the Gulf—four, once Porter arrived—later became squadron commanders: Capt. Theodorus Bailey, Farragut's second in command and skipper of *Colorado;* Cdr. Henry Bell, Farragut's chief of staff; and Cdr. Samuel "Phillips" Lee, captain of the screw sloop USS *Oneida.* Farragut would rely heavily on these men in the coming weeks and months.[59]

Farragut's first obstacle was not the rebels, but rather Mother Nature. In the year since the war began, the Mississippi's four fingerlike channels into the Gulf of Mexico had silted up, making it very difficult for deeper-draft ships to steam over the sandbars into the river. On 18 March Porter got his mortar fleet across the bar and into Pass à l'Outre, the delta's easternmost finger. The fleet's bigger vessels, however, drew too much water there, so Farragut sent them to Southwest Pass. There were two more feet of water over its bar, which provided enough room for many, but not all, of the warships to cross without much assistance. It took eleven days, though, for the sloop USS *Pensacola* and the side-wheeler USS *Mississippi* to get over, and they did so only after their crews stripped them bare at Ship Island and tugs dragged them through the bar's muddy bottom. Farragut put Porter in charge of the effort, and Porter told Fox it was the hardest work of his entire life.[60] As for *Colorado,* Farragut told Bailey, "*It must be done. . . .* You are my second in command in this *great undertaking,* and you will I know do all in your power to carry out the wishes of the Dep[artmen]t, and in the shortest possible time."[61] Despite Herculean efforts, it proved impossible. Farragut finally gave up, turned *Colorado* into a floating depot, and transferred her best guns and sailors to other ships. This put a somewhat downbeat coda on things, but Farragut could take comfort in the fact that by 8 April he had the rest of his fleet assembled in the river at Head of Passes. Now he could turn his attention to the rebels waiting for him upstream in their forts and warships.

On the morning of 18 April, Porter's mortar flotilla, its vessels camouflaged with tree branches tied to their masts to hide them from counterbattery fire, opened up on Fort Jackson and Fort St. Philip. In ten hours that day they lobbed nearly three thousand shells at the rebel positions, smothering them with smoke and haze. Porter may have believed that his flotilla could subdue the Confederate forts, but Farragut did not, even though he was willing to give Porter his chance. While Porter's mortar vessels blasted away, Farragut continued his preparations to run past the forts. In doing so, he displayed considerable imagination and thoroughness. Farragut had his sailors strip their warships of all nonessential materials, drill in firefighting techniques, smear mud and tar on the hulls of their vessels to obscure them at night, paint huge numbers on their smokestacks for quick identification, hang garlands of chains over the sides to protect their engines from cannonballs, whitewash the gun decks to see better in the dark, and pack the boiler rooms full of bags of ash or sand to shield the coal heavers from splinters. Readying the warships, however, was easier than readying the men. Many officers and sailors believed that any run past the Confederate forts would be suicidal. Even if some vessels survived the gauntlet, they said, rebel warships would destroy them soon afterward. To combat such demoralization, Farragut traveled from warship to warship to inspire the crews. When he learned that the officers in one vessel were unhappy, he said to a journalist with battle experience, "I hear that they are as blue as indigo in that wardroom over there. Go and cheer them up. Tell them some stories of the fights you've been in and come out of alive. It will stir their blood and do them good."[62] He held daily councils of war on *Hartford* to fine-tune his plan and reassure jittery officers. Throughout it all, Farragut remained optimistic and cheerful, as if Union victory was a foregone conclusion.[63]

Porter thought so too, at least at first, before his mortar vessels opened fire. For two days his flotilla pounded Fort Jackson and Fort St. Philip, but, except for exhausting his men and irreparably damaging the hearing of some, he accomplished little. Rebel artillerymen continued to respond, and in fact one cannonball wiped out nearly an entire gun crew on *Oneida*. Farragut was not surprised by Porter's lack of success, so on the morning of 20 April he signaled his subordinates to report to *Hartford* for a meeting. There he noted that the fleet was fast expending its ammunition, so he wanted to move now while he still had something with which to fight. Farragut explained his latest plan to run past the forts, assigned each warship her place, and revealed the most recent intelligence. Some officers still believed that the mission was suicidal, but others, such as Phillips Lee, had concluded that it was the only solution to the current military equation. Porter asked for more time, however, and Farragut agreed, in part because a couple of his vessels needed last-minute repairs. That night Henry Bell took two gunboats upstream and cut the chain boom across the Mississippi so the warships could steam upriver unhindered. He did so under heavy fire and after considerable effort, and Farragut was greatly relieved that his friend made it back alive. On the morning of 22 April, Farragut and Porter got into a heated argument about whether to give the mortars still more time to work their magic. Finally Farragut said, "Look here, David, we'll demonstrate the practical value of mortar work. Mr. Osbon, get two small flags, a white one and a red one, and go to the mizzen topmasthead and watch where the mortar shells fall. If inside the fort, wave the red flag. If outside, wave the white one." Osbon did as Farragut asked, and waved the white flag considerably more often than the red one. After a while, Farragut ordered him back down and said, "There, David. There's the score. I guess we'll go up the river tonight."[64] By that time Porter's mortar boats had fired more than 16,000 shells. As things turned out, heavy winds forced Farragut to postpone the assault for twenty-four hours, but he was ready to move on the night of 23–24 April.[65]

For his dash past the rebel forts, Farragut divided his seventeen vessels into three groups. Theodorus Bailey had transferred from *Colorado* to the little gunboat USS *Cayuga*—quite a comedown, to be sure, but one he was willing to make in order to get into the upcoming fight—and Farragut put him in charge of the Red Division, which would lead the charge. Farragut assumed direct command of the Blue Division, consisting of *Hartford* and the biggest warships. Finally, Bell brought up the rear in his Red and White Division, containing the small sloop USS *Iroquois* and five gunboats. As for Porter, his mortar flotilla would remain behind to provide fire support with its remaining ammunition. The fleet got under way at 2:00 AM, but it was slow going. Only eight vessels were beyond the chain boom Bell had severed three nights earlier when the rebels opened up on them. Someone compared the ensuing racket to the simultaneous rattle and roar of all the earthquakes and thunder on earth. Rebel warships entered the fray, and in the darkness and pandemonium it was impossible to distinguish friend from foe. No one knew what was going on beyond his own vessel. One participant wrote later, "Death and destruction seemed everywhere. Men's faces were covered with powder black and daubed with blood. They had become like a lot of demons in a wild inferno."[66] Farragut had stationed himself high on the mizzen rigging in a mostly futile effort to see and control the action. A fire raft hit *Hartford,* and its flames began licking the side of the Union flagship, threatening to burn her up. B. S. Osbon, a journalist who had helped Farragut disprove Porter's claims about his mortar boats a couple of days earlier, knelt to unscrew the fuse tops from several 20-inch shells. Farragut saw him and yelled, "Come, Mr. Osbon, this is no time for prayer!" Osbon responded, "Flag Officer, if you'll wait a second you'll get the quickest answer to prayer you ever heard of." He then rolled the shells over the side, and the resulting explosions blew the fire raft away and extinguished the flames.[67]

In the bewildering melee the Union warships, with their heavier ordnance, destroyed most of their rebel counterparts, including the ironclad *Manassas,* and pushed upriver past the forts. Once safely above, they gathered off Quarantine Station to count noses. Only one Union vessel, the screw steamer USS *Varuna,* had been sunk, and three small gunboats had turned back. Union casualties amounted to 36 killed and 149 wounded. As darkness turned into dawn and it became clear that the fleet had survived, and survived pretty much intact, elation spread from ship to ship. Officers rowed over to *Hartford* to report, swap stories, backslap each other—and to pay homage to their flag officer, who had insisted from the start that their mission would succeed. Farragut received their congratulations with the modesty of a man secure in the knowledge that his accomplishment would speak for itself. [68]

Steaming past Fort Jackson and Fort St. Philip was a harrowing and exciting experience for everyone involved, but it was also merely preliminary to greater ends that Farragut quickly set about attaining. That afternoon, Farragut ordered his fleet upriver toward New Orleans, only seventy-five miles away. The Chalmette batteries were the sole military obstacle between the Union fleet and its goal, and they opened up on Bailey's *Cayuga* as she approached Slaughterhouse Bend next morning. Fire from *Hartford* and *Pensacola,* however, scattered the rebels and put the advance back on track. At 1:00 PM on 25 April, in the middle of a pouring rainstorm, Farragut's fleet reached undefended New Orleans like phantoms out of the mist. As second in command, Bailey claimed the right to receive the city's surrender, so a shore party rowed him and Lt. George Perkins to a wharf under a flag of truce. Unfortunately for Bailey, there were no city officials there to greet them, just a large, wet, and belligerent mob. Undaunted, Bailey and Perkins walked unarmed through the baying crowd to city hall, doing their best to ignore the cheering for Jefferson Davis and booing for Abraham Lincoln. There the mayor and city council claimed that they lacked the authority to surrender because New Orleans was under martial law. When the local Confederate military commander, Maj. Gen. Mansfield Lovell, arrived, though, he informed Bailey that he was evacuating his soldiers from the city, so he could not surrender it either. This was hardly the dignified and solemn capitulation Bailey had envisioned. Somewhat nonplussed, he and Perkins slipped out of the building through a side door, hopped into a closed carriage a sympathetic citizen placed at their disposal, and returned to the wharf. Farragut was equally perplexed, but he finally sent Bell and the Marines into the city to occupy the important municipal buildings. Meanwhile, downriver, the garrisons at Fort Jackson and Fort St. Philip mutinied and forced their officers to submit to Porter on 28 April when they learned that Farragut's fleet had New Orleans under its guns. Two days later, Butler's troops showed up to secure the city.[69]

On 26 April Welles was at his desk, scribbling a letter to a friend, when he learned that Farragut's fleet was safely past Fort Jackson and Fort St. Philip. He interrupted the flow of his correspondence only to comment, "We have, just at this moment as I am writing, received news that our fleet has passed up the river and is off New Orleans. I have, for weeks been waiting this result with inexerable [*sic*] anxiety."[70] Indeed, the strategic importance of New Orleans' fall could scarcely be exaggerated. In one fell swoop, Farragut deprived the Confederates of their largest city and all the resources it contained, and placed a good chunk of Louisiana under Union control. It gave Union forces a logistical base in the region that they could use to advance up the Mississippi River, opening up a new front with which the rebels had to contend. And since the victory came at such small cost, this meant Union forces remained intact and ready for further operations. Coming as it did during a spate of other Union triumphs—Fort Henry and Fort Donelson, Shiloh, Island Number Ten, Roanoke, and Pea Ridge—it helped dampen Confederate morale during their gloomy spring. For naval officers, seizing New Orleans plugged one of the blockade's most porous parts, meaning they had one fewer port to guard. Along with *Carondelet*'s run past Island Number Ten, it provided further evidence that steam-powered warships could race past well-defended enemy fortifications, making them seem more vulnerable than ever before.

Considering both human nature and the campaign's significance, it was not surprising that claiming and distributing credit for conquering New Orleans absorbed almost as much energy as attaining the victory itself. Farragut was a humble man to begin with, and he seemed more interested in giving others, including especially God, the recognition he felt they deserved. Besides, there was no need for him to toot his own horn when so many others were more than happy to do it for him. Welles was elated for all sorts of reasons, not the least of which was satisfaction in seeing his selection of Farragut justified. In a congratulatory letter to Farragut, he expressed his increasingly important philosophy: "Your example and its successful result, though attended with some sacrifice of life and loss of ships, inculcate the fact that the first duty of a commander in war is to take great risks for the accomplishment of great ends."[71] Fox was equally fulsome in his praise, but he emphasized the happy fact that the Navy won the battle without much help from the Army. Farragut's performance inspired officers throughout the service, including combat-hardened men such as Andrew Foote and Percival Drayton. Farragut's achievement, though, placed David Porter in an awkward position. Before the run past the forts, Porter had repeatedly denounced Farragut as unfit for his job, but now he had to face the fact that Farragut had been right all along. He tried his best to backtrack in a disingenuous 10 May letter to Fox, stating, "It looks as if I was trying to make capital which I am not in the habit of doing—let Farragut have all the credit he can get. I wrote my report hurriedly and did not notice the impropriety of the remarks until after it had gone, and it was too late to correct it—tho [*sic*] Farragut has been pleased to consider me an *outsider,* and has not deigned to invite me to his public councils, I don't want to do anything that may look like pique." Even now, however, Porter could not help but attempt to hone in on Farragut's limelight by concluding, "Privately he has been confidential enough, [but] had he not been he would now be blockading the mouth of the Mississippi."[72] Porter might gnash his teeth, but Congress and the Union public were jubilant. Although it took time, on 11 July Congress voted Farragut its thanks, which enabled Lincoln a few weeks later to nominate him one of the first four active-list rear admirals.[73]

Farragut may have been the brightest star in the West Gulf Squadron's constellation, but there was plenty of room for lesser lights. As squadron commander and naval hero, Farragut was in a position to bring to the Navy Department's attention meritorious officers whom Welles and Fox could later assign to important and prestigious posts. Although he was normally chary with effusive praise, Farragut still wanted to give credit where he believed it was due. He acknowledged Porter's gallantry in managing his mortar flotilla so Welles could place a check in the positive side of the thick mental ledger he was keeping on him. Farragut gave considerably more public and private recognition, however, to Henry Bell and Theodorus Bailey for their roles in leading their divisions past the Confederate forts. Bailey had been in the Gulf since the previous July, and had performed his duty even though he suffered from hydrocele, the accumulation of fluid in the scrotum. Despite this painful condition, he insisted on remaining until New Orleans was in Union hands. Now, out of concern for his health and as a reward for a job well done, Farragut ordered him and *Cayuga* home with the official dispatches of his victory and the Louisiana state flag that Bell had pulled down from city hall as a souvenir. Later Farragut referred to Bailey as a "gallant old fellow," and regretted that Congress did not vote him its thanks for his actions at New Orleans.[74]

Phillips Lee also garnered a disproportionate amount of praise for his role in the campaign. It was not from Farragut, who gave Lee the same recognition he gave his other ship captains, but from others who had or got the Navy Department's ear. Bailey, for example, was grateful for Lee's help in protecting *Cayuga* during its run past Fort Jackson and Fort St. Philip, and informed Lee's brother-in-law, Postmaster General Montgomery Blair, of this when he arrived back north with Farragut's dispatches. Porter was not normally one to commend a fellow officer and competitor, but he did in Lee's case. Porter had learned that Lee's wife was denigrating his contribution to the campaign, stating that his mortars hid behind the trees while Lee did all the work. No doubt Porter did not want to antagonize the powerful Blairs, so he asked Fox to tell Lee's wife, "If no one else gives Sam Phillips credit I will for one, for I never saw a ship more beautifully fought or managed. . . . I admire Lee very much for his cool calm bravery, the highest quality an officer can possess, and he is properly estimated by the younger officers, who after all are the best judges."[75] For all these reasons, Lee's stock, like Bailey's and Bell's, was rising.[76]

Vicksburg Unbowed

New Orleans' occupation was important, but there was still much for Farragut to do. As things turned out, exploiting the victory was as problematic as achieving it in the first place. The immediate difficulty was that Welles' 20 January marching orders to Farragut were not explicit. Once New Orleans was in Union hands, Welles expected Farragut to both advance upriver to affect a junction with the Western Flotilla and reduce Mobile on the Gulf Coast. Farragut was a saltwater sailor with little taste for further fighting on the constricted Mississippi. He had no idea exactly where Foote's flotilla was, and he doubted that his big warships could operate safely any farther north than Natchez, Mississippi. Therefore, he decided to use the discretion he believed Welles had given him to concentrate on sealing off Mobile, leaving Phillips Lee and the gunboats to watch the river. He sent Porter's mortars to Mobile and prepared to join him with most of his remaining force. Before he could do so, however, he learned that the rebels had a large number of warships at Memphis—their River Defense Fleet, which Davis and Ellet destroyed a month later. Farragut was loath to leave behind any unfinished business on the Mississippi, so he decided to go upriver and see if the rumors were true. He told a dubious Porter that he did not expect to be gone for long, and left New Orleans on 8 May to join Lee at points north.[77]

Although Henry Bell was already feeling the first twinges of the dysentery that would eventually befall almost everyone on the Mississippi River, he could not help but comment in his diary on the exotic sights that unfolded before the Union fleet steaming through the Confederate heartland. He noticed a red dilapidated Catholic church surrounded by well-kept gravestones, sugarcane lining the riverbanks, slaves toiling hard in the fields, high levees, distant mansions, giant gloomy oaks, sweeping magnolias, and orange and pecan trees. The fleet passed recently occupied Baton Rouge, Louisiana, 140 miles upstream from New Orleans, and continued northward. At nightfall on 12 May, though, *Hartford* ran hard aground on a mudflat and remained stuck for two days. Farragut's crews toiled in the oppressive Mississippi heat to strip the vessel of everything—cannon, shot, shells, coal, etc.—before a gunboat managed to work her free.[78]

Upstream, Phillips Lee's frustrations equaled Farragut's. After New Orleans' fall, Farragut had sent Lee up the Mississippi with some gunboats, and Butler tacked on fourteen hundred soldiers under Brig. Gen. Thomas Williams. Lee's problem was Vicksburg, Mississippi, which was located on a high bluff at a bend in the river. Lee reached Vicksburg on 18 May and demanded the city's surrender. Unlike New Orleans and Baton Rouge, Vicksburg's geography was well suited for defense. The local garrison commander therefore replied that "Mississippians don't know, and refuse to learn, how to surrender to an enemy."[79] Somewhat at a loss as to how to proceed now that the rebels had called his bluff, Lee ended up doing nothing. Downstream at Natchez, Farragut grew increasingly impatient with Lee's dilatoriness. He finally went upriver on the gunboat USS *Kennebec* to have a firsthand look at the situation, arriving near Vicksburg on 22 May. Eager to chastise the rebels for their insolent response to Lee's surrender summons, Farragut focused his anger on a pair of enemy gunboats the Confederates had deployed under Vicksburg's batteries. He wanted Lee and his gunboats to destroy them, but Lee and a majority of his captains insisted that an attack in the face of Confederate cannon was suicidal. In fact, Lee stated adamantly that if someone wanted to take his ship, *Oneida,* and try, he was welcome to her. Thoroughly exasperated by his subordinates' defeatism, Farragut superseded Lee with the recently arrived Cdr. James Palmer, now skippering *Iroquois.* When Lee complained to Farragut about his decision, Farragut somewhat disingenuously explained that Lee himself had said that his force was too small for the job, so he was simply augmenting it. In doing so, he continued, he had to give the command to the senior officer, who was Palmer. Lee recognized the demotion for what it was and began pulling strings to get out from under Farragut and into a more conducive post.[80]

While Lee fumed, Farragut brought his remaining vessels up to Vicksburg and prepared to assail the city. He quickly discovered, though, that Lee's frustrations were understandable because there seemed no way to get at the target. Farragut was confident that he could steam past Vicksburg's batteries as easily as he had steamed past those at Fort Jackson and Fort St. Philip the previous month, but he did not think that would accomplish anything; doing so would merely sever his supply and communication lines to New Orleans. Seeking advice, on 25 May he called a council of war with his captains and General Williams on *Hartford*. Instead of providing answers, however, his subordinates merely reminded Farragut of the problems the fleet faced. Coal vessels coming up from Natchez required armed escort because rebel guerrillas infested the riverbanks at every rapid and bluff. The river itself was falling, threatening to strand the deeper-draft warships until winter rains raised the water level. Farragut's vessels so sorely needed repairs that he doubted they were in any condition to fight the ironclad the rebels were supposedly constructing upriver. Williams pointed out that his soldiers were sick, heavily outnumbered, and unable to find a suitable place in the vicinity to disembark. After articulating these difficulties, Williams and most of the captains—including Bell and Lee, but not Palmer—urged Farragut to give up any planned assault on Vicksburg. Farragut was a pugnacious man who disliked backing away from a fight, but, as he later candidly recalled, "I was very sick at the time, and yielded to their advice, which I think was good; but I doubt if I would have taken it had I been well."[81] Leaving the gunboats to keep an eye on Vicksburg, he fell back to New Orleans with the remainder of his fleet.[82]

Farragut may have considered his 20 January orders discretionary and vague, but the Navy Department certainly did not. As far as Welles and Fox were concerned, seizing the Mississippi River had top priority, and they felt that Farragut's instructions were crystal clear on that point. Indeed, they saw a naval advance up the Mississippi as the key to Union strategy out west. They had promised Foote and Halleck that Farragut would strike the rebels in Mississippi from behind, forcing them to either retreat or fight unprepared. Moreover, Foote might need Farragut's warships to deal with the Confederate River Defense Fleet. On 8 May *Cayuga* docked at Hampton Roads. Theodorus Bailey disembarked and hurried to Washington to deliver the official dispatches Farragut had entrusted to him. Welles was out of town when Bailey arrived the next morning, but he talked with Fox. The assistant secretary asked him how many vessels Farragut had sent upriver, and Bailey responded that as far as he knew, none. Fox declared that this was impossible because Farragut had preemptory instructions to open up the Mississippi. Bailey did not know anything about that, and could only lamely respond that perhaps Farragut had forgotten his orders. Increasingly worried, on 12 May Fox sent Farragut an unofficial reminder that the Mississippi River was more important than Mobile. Four days later, however, Fox read newspaper reports that Farragut was actually retreating downriver to New Orleans. By now Fox was so thoroughly alarmed that he was losing sleep. Moreover, the president shared his concern. Fox telegraphed to Hampton Roads and had a fast ship sent to New Orleans with explicit orders for Farragut to push northward toward Memphis. As Fox noted, "Mobile, Pensacola, and, in fact, the whole coast sinks into insignificance compared with this."[83]

Fox's unhappiness with Farragut gave David Porter another chance to apply his talent for discord. Farragut had sent Porter and his mortar flotilla to Mobile to bombard the city's outer defenses preparatory to the full-scale assault he hoped to launch later. Porter achieved little against the city, but his activities persuaded the Confederates to abandon nearby Pensacola on 10 May, thus providing the West Gulf Squadron with a sheltered harbor far superior to its exposed base at Ship Island. A week later, Fox wrote to Porter at Ship Island to complain about Farragut's decision not to proceed immediately up the Mississippi River. Porter knew an opportunity when he saw one, so he responded with a scathing letter a week later. He predicted that Farragut would get stranded upriver for the rest of the season, requiring the Navy Department to outfit a new squadron under a new commander—the implication being that he was the officer for such a job. Porter noted that Farragut was an amiable man manipulated by incompetent subordinates. Specifically, he denounced two of his chief competitors, Palmer and especially Bell. Porter acknowledged Bell's honesty and frankness, but added: "He is pig-headed and slow and has a bad influence on Farragut, wanting to cover himself with glory gained by other people's energy and intelligence. He is universally disliked for putting himself in positions he is not entitled to fill, and for not attending the duties of Fleet Captain which he was sent out here to perform. I am partly doing his duties here now, and you will have a nice mess of it if the public property in and out of these parts is left to the fostering care of Bell in his capacity of Fleet Captain."[84] In subsequent letters, he extended his range, criticizing Williams, Butler, and McKean as inept. Porter admitted that he was grumbling, but stated that his caviling was motivated by a hatred of the blunders he saw around him. In fact, Porter was so demoralized that he even asked Fox to be relieved of his command, though nothing ever came of his request.[85]

Not surprisingly, Farragut was displeased to receive the Navy Department's emphatic instructions directing him back up the Mississippi. He counted himself lucky to escape with his forces intact the first time, and he dreaded the idea of another trip upriver under conditions as bad as or worse than those he had previously faced. He was hard put to understand why Welles and Fox could not simply command Davis, with his warships specially designed for riverine work, to descend the Mississippi and come to him. Farragut wrote, "They will keep us in this river until the vessels break down, and all the little reputation we have made has evaporated. The Government appear[s] to think that we can do anything. They expect me to navigate the Mississippi nine hundred miles in the face of batteries, ironclad rams, etc." Despite his pique, Farragut knew that he had to obey, and added in resignation, "Well I will do my duty to the best of my ability, and let the rest take care of itself."[86] His subordinates, including the backstabbing Porter, were unanimous in their opposition to another foray upstream, complaining that the fleet lacked the coal, warships, troops, and water to fulfill its mission. Nevertheless, orders were orders, so Farragut hurriedly readied his vessels. He persuaded Butler to furnish Williams with thirty-four hundred soldiers to accompany the expedition, and sent for Porter's mortar flotilla because only its guns had sufficient elevation to reach Vicksburg's highest batteries. The fleet left New Orleans on 8 June. In addition to the familiar problems, fever and dysentery broke out among the sailors. Hartford reached Vicksburg on 25 June, and that night one of the Ellets—Charles Rivers—made contact with Farragut and informed him of Davis' whereabouts.[87]

Farragut believed that fulfilling the Navy Department's wishes required him to effect a junction with Davis' fleet. Reaching it, however, meant taking his warships past Vicksburg's batteries. To that end, on 27 June he called a meeting of his captains on Hartford to explain his plan. Basically,

Porter's mortar flotilla would provide covering fire while the warships steamed upriver, taking time to smother enemy batteries along the way. Weighing anchor at 2:30 AM, Farragut's vessels ran the gauntlet of fire. During the dash Palmer's *Iroquois* slowed alongside *Hartford*. Seizing a nearby trumpet, Farragut bellowed into the darkness, "Captain Palmer, what do you mean by disobeying my orders?" Palmer responded, "I thought, Admiral, you had more fire than you could stand, and I came down to draw part of it off."[88] It was the kind of gesture that appealed to Farragut, who made sure to praise Palmer in his dispatch to Welles summarizing the engagement. The bulk of the fleet made it upriver, losing about forty-five casualties in the process.

Despite his success, Farragut felt little jubilation. As he explained to Welles the morning afterward, the victory accomplished nothing more than lengthening his already vulnerable supply and communication lines. Moreover, until the Army provided at least 12,000 soldiers, Vicksburg would not fall, and Williams had nowhere near that many men. Finally, the Vicksburg run generated considerable controversy among Farragut's officers. Three vessels—including Capt. Thomas Craven's sloop, USS *Brooklyn*—turned back during the battle. According to Craven, Farragut's verbal instructions were that the warships should not continue upstream until they had reduced the Confederate batteries they faced, which he had been unable to do. He also blamed Porter, with whom he had a running dispute, for not providing sufficient fire support with his mortars. Farragut was normally easygoing about such misunderstandings, but in this case he issued a sharp reprimand to Craven, who replied that Farragut's charges were unfair and asked to be relieved. Farragut thereupon removed Craven from his command, sent him back north, and gave *Brooklyn* to Bell. Bell was hoping for new duty somewhere, anywhere, other than on the miserable Mississippi, so he was unhappy with his new assignment. Craven's relief became a divisive topic of conversation among Farragut's officers for a couple weeks.[89]

In the meantime, on 1 July Davis' Western Flotilla steamed down from Memphis and joined Farragut's fleet, bringing the two very different Union armadas together for the first time. Amid the celebrations and backslapping, though, there was an undercurrent of tension that went beyond the controversy Craven's removal generated. Some of Davis' officers, for example, were unimpressed with Farragut. Sam Phelps observed that Farragut was an impulsive man who considered purposeless activity better than doing nothing at all. As for Farragut's officers, some were disappointed that Davis brought no troops with him to storm Vicksburg. On 10 July Farragut wrote to Welles that although he did not want to tell the Navy Department what to do, he believed it was imperative that he should take his warships back to the Gulf Coast. Farragut noted that Davis had already destroyed all organized Confederate naval resistance on the Mississippi at the Battle of Memphis, so he had the situation well under control. Moreover, no matter how many warships the Navy deployed on the river, Vicksburg would remain in rebel hands until the Army made a commitment to taking it. Finally, Farragut stated that he needed to devote more attention to the blockade. In subsequent letters to the navy secretary over the next few days, Farragut warned that his sick list was lengthening and that in ten days there would not be enough water in the river for his warships to return to New Orleans, stranding them until autumn. Farragut speculated to Bell that the Navy Department was so concerned with McClellan's failed offensive toward Richmond that it was not paying attention to anything else. Indeed, Welles had already summoned Porter's mortar flotilla for service off the Virginia coast, even though it would take more than three weeks for the vessels to get there. Until he heard from Welles, all Farragut could do was fret and wait.[90]

For weeks Union naval officers had heard rumors that the Confederates were constructing an ironclad up the Yazoo River. Farragut discounted these reports, telling Welles that, even if they were true, he and Davis were ready for it. On 14 July a deserter informed Farragut that not only did the rebel ironclad exist, but she was coming down the Yazoo the next day. In response, Farragut and Davis decided to dispatch three warships—the side-wheeler USS *Tyler, Carondelet,* and *Queen of the West*—up the Yazoo to investigate. Next morning the three Union vessels steaming upstream encountered the Confederate behemoth ironclad *Arkansas* coming downstream looking for a fight. Seeing that their cannonballs had no effect on *Arkansas,* the Union warships commenced a noisy fighting retreat down the Yazoo. Along the way, *Arkansas* sank *Carondelet,* a proud veteran of the run past Island Number Ten, in the shallow waters along the riverbank. In order for *Arkansas* to reach Vicksburg, she had to get past Farragut's and Davis' combined forces. Union crews had heard the loud artillery exchange between *Arkansas* and her quarry for more than an hour, but they assumed that *Tyler, Carondelet,* and *Queen of the West* were merely firing at rebels along the shore. Therefore, when *Arkansas* came into view, Union warships were caught by surprise, with their steam down and boilers cold to conserve coal. *Arkansas* had taken a beating during her running fight down the Yazoo, and her engines were rickety under the best of circumstances. Even so, at least she was moving, which was more than could be said for the Union vessels. As she steamed slowly by the Union fleets, Union warships fired everything they had at her, but *Arkansas* survived the twenty-minute gauntlet and reached the safety of Vicksburg's batteries.[91]

Not surprisingly, Farragut was infuriated by this fiasco. He admitted that *Arkansas'* run was a "bold thing" and accepted his share of responsibility for the failure to stop her, but he was not going to let the matter drop. No sooner had *Arkansas* disappeared from sight than Farragut boarded Davis' flagship, *Benton*, and proposed an immediate attack by their combined forces on the rebel ironclad. Although Lee and Palmer supported their chief, Davis and others thought a daylight assault in the face of Vicksburg's batteries would be suicidal. Instead, Davis blandly suggested that Farragut wait until nightfall to go downriver after *Arkansas* with his squadron, and Farragut reluctantly agreed. At 6:00 PM Farragut's vessels got under way and steamed past the Confederate citadel, firing at *Arkansas'* approximate location. Unfortunately, it was too dark to see anything clearly, so *Arkansas* escaped serious damage. Union losses were minimal too, but this was small consolation for a still-seething Farragut. Now south of Vicksburg, he continued to advocate *Arkansas'* destruction, regardless of the consequences. Davis later remembered, "He was like an excited, hot-headed boy."[92] On 22 July Farragut and Davis tried again. This time they sent *Essex, Queen of the West,* and the former Confederate steamer USS *General Sumter* to assail *Arkansas* at dusk, but the Confederate ironclad survived after a confused melee of a battle.[93]

As for Davis, he took a more detached view of *Arkansas'* escape. Although Davis liked and continued to get along with Farragut, he was nonplussed by his old friend's impetuous attitude, and grateful that Farragut had subordinates such as Bell who could make their boss see reason. As Davis viewed things, setbacks were part of war and should be accepted as such, not taken personally. Davis wanted to destroy *Arkansas* too, but not by jeopardizing everything he, Foote, and Farragut had accomplished over the past six months. Besides, Davis did not believe that *Arkansas* was much of a threat now that she was penned beneath Vicksburg's batteries. Time, he felt, would ultimately resolve the *Arkansas* conundrum in the Union's favor. He explained to Farragut, "I have watched eight rams for a month, and now find it no hard task to watch one. I think patience as great a virtue as boldness."[94] Farragut of course disagreed, and attributed their differences of opinion to logistics. Davis had the luxury of time because his supply lines were relatively secure. Farragut's, however, were extremely tenuous, so he could not take chances with anything that might threaten them. There was perhaps some truth to this interpretation, but differences in personality and temperament provide a better explanation. For Davis, war was an intellectual exercise, a mental problem to be rationally calculated and solved. Farragut, on the other hand, viewed war as a no-holds-barred street brawl in which perception and reputation mattered as much as reality. While Davis and Farragut both believed in honor, Farragut saw no honor in delay and defeat.[95]

While Farragut and Davis grappled with *Arkansas,* Welles pondered the military situation out west. As he saw things, Corinth's fall and the Army's unwillingness to provide sufficient troops to successfully assail Vicksburg precluded the need for such a large naval force on the Mississippi. Even before he received Farragut's 10 July plea, Welles directed Farragut to return to the Gulf Coast and leave the bulk of the Mississippi to Davis' care. Farragut received Welles' orders on 24 July, and immediately prepared to retreat to New Orleans. He did not like to leave unfinished business behind, but the river was falling, so he felt he had little choice. After dropping Williams' tired soldiers off at Baton Rouge, he reached New Orleans on 28 July. With Farragut and the Army gone, Davis saw little reason to keep his sick sailors on station off Vicksburg, so he pulled back to Helena, Arkansas, in late July with most of his warships. Davis was heartbroken to abandon hundreds of slave women and children who had flocked to the area in search of freedom and refuge. He took some with him, gave provisions to others, and encouraged as many as possible to return home. Meanwhile, Davis' predictions about *Arkansas* proved accurate. On 5 August the Confederates attacked Williams' garrison at Baton Rouge. *Arkansas* was supposed to support the rebel assault, but engine troubles forced her commander to scuttle her as the Union warships Davis left behind approached. Although Williams was killed in the engagement, his Union soldiers successfully defended the city.[96]

Despite this welcome postscript, the fact remained that the first Union attempt to seize Vicksburg and gain control of the entire Mississippi River had fizzled. By late July the campaign was over, except for the finger-pointing and recriminations. Welles was mortified by *Arkansas'* run through the Union fleets, calling it "the most disreputable naval affair of the war," which, considering such fiascoes as the demolition of the Norfolk Navy Yard and Pope's confused brouhaha at Head of Passes, was saying a lot. Welles held both Davis and Farragut responsible for *Arkansas'* escape, and told them that he now expected them to remain in the Vicksburg area until they had destroyed the Confederate ironclad. By the time they received Welles' latest orders, however, *Arkansas* was history and Davis and Farragut had already pulled up stakes.[97]

Welles laid most of the blame for the campaign's failure on the Army's inability to provide the necessary troops to storm Vicksburg, but he also noted that things might have turned out differently had Farragut obeyed his instructions and immediately ascended the river after New Orleans surrendered. He was also increasingly aware that Farragut's management of the West Gulf Squadron left much to be desired. In fact, Farragut was not much for paperwork. Officers complained that he did not submit regular returns or keep his ship captains properly informed. As a result, Farragut's standing in the Navy Department plummeted in the late summer. Rumors of his impending relief circulated throughout the Navy and reached the Gulf Coast. Farragut heard them and scribbled to Bell that he would not mind getting away from the Mississippi River. On the other hand, Benjamin Butler, commander of army forces in and around New Orleans, wrote to Fox in Farragut's defense. Farragut, said Butler, was among the most energetic, conscientious, and patriotic officers in the region, so removing him would be foolish. Fox replied expansively, "Whence did you surmise that Farragut was to be relieved? We never heard the rumor here. The hero of that unequalled dash, despising the great obstacles, gave us victory, glory and New Orleans, is not to be forgotten or removed except at his own pleasure, and probably not even then. If he is with you I beg that you will assure him that we never heard of any such rumor this way."[98] In all likelihood, Fox was taking his cue from Welles. Having, he believed, single-handedly plucked Farragut from obscurity, the navy secretary felt a certain proprietary interest in him. Besides, unlike complainers such as Stringham, Goldsborough, and Mervine, Farragut demonstrated the daring and initiative Welles sought in his commanders, even if he did not always succeed. A few months later, Welles weighed Farragut's attributes in his diary: "Farragut has prompt, energetic, excellent qualities, but no fondness for written details or self-laudation; does but one thing at a time, but does that strong and well; is better fitted to lead an expedition through danger and difficulty than to command an extensive blockade; is a good officer in a great emergency, will more willingly take great risks in order to obtain great results than any officer in high position in either Navy or Army, and, unlike most of them, prefers that others should tell the story of his well-doing rather than relate it himself."[99] Farragut was obviously not perfect, but he was willing to grasp rebel nettles regardless of the resources at his disposal, and Welles appreciated that.[100]

While Welles was inclined to blame Davis and Farragut equally for *Arkansas'* run through the Union fleets, Fox placed most of the onus on Davis, believing that he had been caught napping. His friend Porter agreed. Porter and his mortar flotilla reached Hampton Roads on 26 July, and when he learned of the *Arkansas* fiasco, he immediately wrote to Fox that this was evidence that Davis was not up to his job. As it was, Davis' appointment as Western Flotilla commander had been temporary from the start because Welles and Fox always expected him to eventually take over the new Bureau of Navigation when Congress approved legislation creating it. Therefore, it was a matter of when, not if, Davis left the Western Flotilla. Fox wanted Davis to remain out west until he had achieved a significant victory—be it *Arkansas'* destruction or Vicksburg's fall—that would persuade senators that Davis deserved promotion to rear admiral. Welles, for his part, assured Davis of his support, but he increasingly viewed him as another William McKean—an adequate placeholder until he found someone better. Welles noted in his diary: "Davis . . . is more of a scholar than sailor, has gentlemanly instincts and scholarly acquirements, is an intelligent but not an energetic, driving, fighting officer, such as is wanted for rough work on the Mississippi; is kind and affable, but has not the vim, dash—recklessness perhaps is the better word—of Porter."[101] In mid-September Davis fell ill with chills, fever, and dysentery, and he informed Fox that a doctor had recommended that he leave the fetid Mississippi. Davis also humorously added that he asked the doctor which disease he could expect to contract next. This convinced Welles that it was time to establish a more permanent command structure for the Western Flotilla, so on 22 September Fox told Davis that he would be relieved in mid-October. Davis replied with professional equanimity that he was content to serve wherever the Navy Department thought best, but, after his travails on the Mississippi, he could not help but add that he would meet his relief with pleasure.[102]

Like Goldsborough, Davis spent the rest of the war behind a desk, in charge of the Bureau of Navigation. It was an important job that suited his intellectual tastes, encompassing the Naval Observatory, the Nautical Almanac Office, the Naval Academy, and the Office of Detail. In February 1863 Congress voted Davis its thanks for his actions at Memphis, Fort Pillow, and elsewhere on the Mississippi. Shortly thereafter, Welles had Lincoln nominate Davis an active-list rear admiral, and the Senate gave its approval, making Davis one of only seven officers to attain that rank during the war. Welles' reasoning is unclear, especially since he later referred to Davis as a "feeble officer" and one "without a strong naval fighting record."[103] Fox, on the other hand, appreciated Davis' actions on the Mississippi and had promised him that the department would promote him if he destroyed *Arkansas* or attained a significant victory such as the conquest of Vicksburg. While Davis failed to achieve either goal, he had won the Battle of Memphis, which was worth something. In all likelihood, Fox persuaded Welles to recommend Davis' elevation to the president against the navy secretary's better judgment. Davis himself had no doubt that Fox was primarily responsible for his promotion; when he heard the news, he threw his arms around the assistant secretary out of gratitude and joy. After the conflict, Davis served as head of the Naval Observatory, commander of the Brazil Squadron, commandant of the Norfolk Navy Yard, and again at the Naval Observatory until his death in 1877. His son and namesake also rose to become a rear admiral, and of his three daughters, one married Henry Cabot Lodge—and became John Hay's mistress—and another Brooks Adams.[104]

Too Far from Texas

Farragut's responsibilities included not only the lower Mississippi River, but much of the Gulf Coast as well. He had always felt that sealing off the Gulf Coast was more important than seizing control of the Mississippi, and his frustrating experiences up the big river at Vicksburg had not changed his mind; if anything, they reinforced his saltwater inclinations. Once back in New Orleans, Farragut resurrected his plan to assault Mobile, which he believed was poorly defended and ripe for the plucking. Unfortunately, circumstances conspired to prevent him from taking action against the city. Butler lacked sufficient troops for such an operation, and Farragut saw no reason to assail Mobile if the Army was unable to secure and protect it after it fell, although he was willing to try if the

Navy Department insisted. This, however, was not the case. Welles informed Farragut that his number one priority was protecting New Orleans, and that he and Fox did not think that Farragut had enough warships to initiate anything as ambitious as an undertaking against Mobile. In early September Fox explained to Farragut that the first *Monitor*-class ironclads would be ready for service in October, and Welles was giving them to Du Pont for his planned campaign against Charleston. Once Charleston was in Union hands, though, the Navy Department would send the ironclads to the Gulf for Farragut's use against Mobile and the rebel ironclad rumored under construction there. Until then Farragut had to cool his heals and concentrate on strengthening the blockade.[105]

Waiting did not come easily to Farragut. Without the stimulant of action, he had to focus his attention on the innumerable bureaucratic conundrums that took up the vast majority of a typical squadron commander's time. Farragut's dislike of paperwork made dealing with these problems all the more frustrating. Now that the war had slowed down in the Gulf, officers tried to finagle leaves to return north temporarily or permanently. Although Farragut was normally lenient about these things, his rejection of most such requests led to bitterness, resentment, and hurt feelings among some officers. As far as Farragut was concerned, though, they all had a job to do that took precedence over their personal problems. As he explained to his family, "They all complain; but I tell them that it is of no use, we must do our duty, and then when that is completed we can *all* go home."[106] Fortunately, not all his subordinates were troublesome. He continued to rely heavily on his friend and mainstay, Henry Bell, and eventually put him in charge of the Mobile blockade. He was also increasingly impressed with James Palmer, who had so gallantly led *Iroquois* past Vicksburg's batteries and supported Farragut so stalwartly throughout the exasperating summer up the Mississippi. With time on his hands, Farragut worried about the poor condition of his vessels, their difficulties in remaining on station in bad weather, and, most of all, the blockade's ineffectiveness. He told Fox that it was almost impossible for his warships to stop fast blockade-runners darting past Union warships on moonless nights at twelve to fourteen knots. On the other hand, he took comfort and pride in his promotion to rear admiral, especially because it was based on his record and ability, not seniority.[107]

Farragut had his woes, but he also had the privileges accorded to one with his rank. His quarters on *Hartford* were comparatively spacious and comfortable, he enjoyed New Orleans and its diversions, he had the company of his recently arrived young son, and he celebrated his new celebrity. For the West Gulf Squadron sailors engaged in enforcing the blockade off Mobile or the Texas coast, on the other hand, life was what one officer called "living death." The chief antagonist was not the rebels, but rather soul-killing tedium. Unlike their comrades who won glory steaming past Confederate forts on the Mississippi River, those stationed in the Gulf went for weeks without seeing the enemy or anyone else. Every night their vessels masked their lights and crept close to shore in a mostly unsuccessful effort to capture blockade-runners, and at daybreak they dropped back out of artillery range. Under this stultifying routine, many sailors exhausted their reservoir of anecdotes, books, newspapers, and patience. One officer noted that he eventually spoke no more than duty required. Even the sustenance was boring. The development of steamers with crude refrigeration techniques made it possible to periodically deliver fresh meat and vegetables to blockaders every few weeks. These provisions, however, only lasted a few days—just long enough to ward off scurvy—so most of the time the sailors' diet consisted of salted meat, cheese, hard bread, bad butter, and inferior tea and coffee. Blockade duty was anything but glamorous.[108]

To make things worse for the blockaders, their tedium was never perfectly predictable. There was always an off chance that the rebels might put in an appearance, so Union crews had to remain constantly alert. This was made abundantly clear at Galveston on New Year's Day, 1863. Texas was the most remote part of the Confederate coast, and Farragut had few available resources for it. Nevertheless, on 4 October 1862 five warships led by Cdr. William Renshaw occupied Galveston harbor without much opposition. Unfortunately Butler and the Army could spare only a few companies of Massachusetts troops to garrison the town. Their grip did not extend beyond the waterfront docks, so the rest of Galveston remained in rebel hands. This did not matter much as long as Renshaw's warships were on hand to provide artillery support for the isolated soldiers. During the moonless predawn on 1 January, however, the Confederates launched a combined land and sea assault on Union positions. Four rickety rebel gunboats assailed Renshaw's warships in the harbor. The only things the Confederates had going for them were audacity and surprise, but that was more than enough to do the job. The rebels boarded *Harriet Lane,* killed her skipper, Cdr. Jonathan Wainwright—his grandson and namesake would conduct a bigger and more famous last stand at Bataan and

Corregidor in the Philippines during World War II—and captured the vessel. Renshaw's flagship, the side-wheeler USS *Westfield,* ran aground in the confusion. After a short truce, Renshaw tried to scuttle her to keep her out of enemy hands, but the charges exploded prematurely, killing him while destroying the ship. In the panicky aftermath of these events, Cdr. Richard Law pulled out of the harbor with the remaining Union warships, leaving the Massachusetts soldiers to surrender shortly afterward. The dawn of the New Year found Galveston back in Confederate hands, where it remained for the rest of the war.

On 3 January Farragut and Bell were eating dinner together at the home of a well-known New Orleans doctor when they received the news of Galveston's fall. Impulsive as ever, Farragut initially contemplated steaming to Galveston himself to exact the necessary retribution, but he quickly realized that he had too many responsibilities in New Orleans to do so. Instead, he accepted Bell's offer to go. Farragut had great confidence in Bell's discretion and judgment, and had relied heavily on his advice and help since assuming command of the West Gulf Squadron. Farragut told Bell to take *Brooklyn* and six gunboats to Galveston immediately to restore Union control. Many of *Brooklyn*'s crew were ashore in New Orleans, but Bell rounded them up, got under way next morning, and reached Galveston on 7 January. Although Bell bombarded the harbor on 10 January, he ultimately decided against an all-out assault on the place. As he saw things, the rebel defenses were too strong, and he lacked sufficient knowledge of the narrow channel to risk his vessels there. Besides, he had no troops with which to occupy the town even if it surrendered to him. He knew he would be censured for his cautiousness, but he believed he made the prudent decision. The best he could do was to reestablish the blockade.[109]

Unhappily for the Union, Galveston was merely one setback among several that January in the Gulf. On 11 January Bell spotted an unknown mast over the horizon, and dispatched the gunboat USS *Hatteras* to investigate. The vessel was none other than the Confederate raider *Alabama,* which promptly sank *Hatteras* after a short engagement. When Bell arrived on the scene the next day in *Brooklyn,* he discovered two of *Hatteras'* masts sticking out of her watery grave, but *Alabama* was nowhere to be found. On 15 January, another rebel raider, *Florida,* slipped through the Union blockade at Mobile to begin her career of destruction. Finally, six days later, two Confederate vessels loaded with soldiers ambushed and captured two Union gunboats blockading Sabine Pass, a few miles east of Galveston.

In the big scheme of things, these defeats in the Gulf played little role in the war's outcome. Even collectively they could hardly be compared in terms of blood spilled or strategic significance with the Union disaster at the Battle of Fredericksburg the previous month. Still, these debacles generated plenty of fallout in the Navy. The loss of ships and crews were certainly regrettable, but Welles and Fox were mostly concerned with the evidence of mismanagement that contributed to these setbacks. They wanted blame assigned and punishment imposed as examples to everyone else. Farragut was equally depressed and distraught, using words such as "shameful" and "disgraceful" to describe January's events. Fortunately for him, he escaped the Navy Department's ire, but others were not so lucky. Law was court-martialed and suspended from rank and duty for three years. Bell also came under scrutiny. Welles and Fox believed that he had not acted with sufficient aggressiveness in his aborted effort to retake Galveston. Moreover, Welles had received reports that Bell had entertained Confederate officers in his cabin on *Brooklyn* after they came on board to inquire about the status of the truce to which Renshaw had agreed. Welles disliked such fraternization with the enemy to begin with, and Bell's North Carolina birthright only exacerbated his suspicions. Finally, Welles was put off by Bell's attempt to secure more prize money by claiming that Farragut's subdivision of his squadron entitled him to it. Farragut agreed that Bell should have been more forceful at Galveston, and was convinced that Bell's delays gave the rebels the time they needed to strengthen their defenses. Even so, he still defended his right-hand man, explaining to Welles that Bell was prompt, uncomplaining, and "the Government does not possess a more patriotic, zealous, and untiring officer."[110] If Bell had entertained those rebel officers, he continued—and, indeed, he had—then Farragut was certain that he did so solely out of the courteousness that helped make him such a good officer. Welles respected Farragut's opinion, so he let the matter drop. And so, on these discordant notes, the Union Navy began a new year in the Gulf of Mexico.[111]

Conclusions

The Union Navy made significant strides toward victory along the Mississippi River and its tributaries in the first year and a half of the war.

After a slow start, naval forces played an important, and sometimes decisive, role in the Union conquests of Fort Henry and Fort Donelson, Island Number Ten, Memphis, and especially New Orleans. By the summer of 1862, Vicksburg was the only remaining Confederate fortress on the Mississippi. The Navy's failure to seize Vicksburg was partially due to Farragut's inability to understand Welles' instructions, but mostly because the Army did not invest sufficient resources in the campaign. This was neither the first nor the last time that interservice discord hindered the Union war effort. Moreover, the Navy Department's emphasis on the Mississippi River came at the expense of operations on the Gulf Coast. The Navy lacked resources to actively campaign along the Mississippi and simultaneously seal off strategically vital Gulf ports such as Mobile, so Welles and Fox made a conscious decision to focus the Navy's attention on the Mississippi, even though doing so weakened the Gulf Coast blockade. Although it is possible to argue in favor of the strategic priority of either region, Welles and Fox deserve credit for choosing one over the other and sticking with their decision.

Union matériel superiority no doubt contributed enormously to the Navy's victories along the Mississippi River, but so did solid leadership. Welles' record in choosing his flotilla and squadron leaders in the region was creditable, and reflected his evolving philosophy on leadership. His reliance on seniority in selecting Mervine as Gulf Squadron commander at the war's start proved his biggest disappointment. No doubt Mervine labored under significant limitations, but he demonstrated little of the imagination, resourcefulness, and initiative needed to overcome obstacles and achieve results. In his subsequent appointments, Welles' deliberate decision to downplay or disregard seniority paid big dividends. With the exception of the ailing McKean, all these men contributed to the Union war effort. John Rodgers, Foote, Davis, and Farragut all overcame through determination and force of personality many of the same problems that had befuddled Mervine. Although none of them were comfortable or familiar with waging warfare on the Mississippi River, they all made the necessary adjustments. After nearly two years of trial and error, Welles was showing that he had a pretty good idea of the qualities a successful naval commander required in this war.

CHAPTER 4

Hammers on Anvils

The Blairs, Phillips Lee, and Wilmington

IN THE MID-NINETEENTH CENTURY, the Blairs were one of the most important families in American national politics. The patriarch and dynasty founder, Francis Preston Blair, had been a member of Andrew Jackson's kitchen cabinet, a Western spokesman, and editor of the *Globe,* one of the country's most powerful newspapers. He broke with the Democratic Party over the Kansas-Nebraska Act, which he interpreted as a betrayal of the Missouri Compromise, and became one of the first Republicans. Although he had long ago retired from public service, he continued to exert his influence. His sons were equally active in politics. The eldest, Montgomery, was a prominent Maryland lawyer who had represented both Dred Scott and John Brown before Lincoln appointed him postmaster general. Another son, Frank, played a prominent role in keeping Missouri in the Union in the war's early chaotic days. He later became one of the more effective political generals, winning praise even from the exacting William Sherman. The Blairs not surprisingly were deeply involved in Lincoln's administration, and they paid particularly close attention to the Navy Department. Francis Blair and Welles were friends, and the navy secretary even occasionally socialized at Silver Spring, Blair's District of Columbia estate. In addition, Fox, whom Montgomery more or less foisted on Welles, was related to the Blairs by marriage; he and Montgomery both wed daughters of Levi Woodbury, a former secretary of

the navy. These were significant connections, but the real focal point of
the relationship between the Navy and the Blairs was Samuel Phillips Lee.

Although Phillips Lee was born into one of Virginia's most distinguished families—his grandfather was Revolutionary War–era statesman Richard Henry Lee—he had a tragic childhood. His mother died when he was four, and his heartbroken father apparently suffered a nervous breakdown and took no part in his son's upbringing. Lee received a midshipman's warrant in 1825 and subsequently served in the West Indies, the Mediterranean, and the Pacific. In 1838 he volunteered to participate in Charles Wilkes' Exploring Expedition, and Wilkes assigned him to USS *Peacock* as first lieutenant. Unfortunately, Lee and Wilkes quarreled, and Wilkes sent Lee home in disgrace in February 1839. Later that year, while on his way to the Pensacola Navy Yard, Lee got into a dispute with a pilot on an Ohio River boat and shot him in the head. Happily for everyone, the pilot survived, and the local prosecutor declined to press charges.

Considering these occurrences, it was hardly unexpected that Francis Blair took a dim view of Lee's interest in his only daughter, Elizabeth. Their tumultuous on-again, off-again courtship lasted four years, during which her father did everything he could do to break up the young couple. In the end, though, true love prevailed; Blair grudgingly assented to Phillips and Elizabeth's marriage in 1843, and gradually warmed to his new son-in-law. Despite his inclusion into this powerful family, Lee had less success than he anticipated in advancing his career. He spent much of the 1840s and early 1850s in the Coast Survey, helping Matthew Maury's groundbreaking oceanographic work. Along the way he managed to get into the Mexican War and took part in the occupation of Tabasco, where he first established an ambivalent relationship with Fox. Lee was given command of the sloop of war USS *Vandalia* and ordered to East Asia just before Lincoln's election. As the country fell apart, Elizabeth hoped that the Navy Department would transfer her husband to the Pacific Coast so they could take stock of their options when and if war broke out, but Lee's commitment to the Union was never in doubt. Instead of steaming directly to East Asia, Lee dallied on his way across the Atlantic, waiting for news, before he finally decided to disregard his orders and bring his vessel back home to help crush the rebellion. When he returned, Fox chided him for disobeying his instructions, but Welles, while noting that Lee's actions were certainly irregular, made no effort to punish him. After a long wait, the Navy Department gave him the screw sloop *Oneida,* which he took to the Gulf as part of Farragut's squadron.[1]

Lee was forty-nine years old when the Civil War began, with a gray beard, curly hair, and bushy eyebrows hidden behind an ever-present pipe. He was so efficient, meticulous, and detail oriented that his nickname in the U.S. Navy was "Old Triplicate." Little escaped his penetrating gaze. He worked hard, and as a strict disciplinarian expected others to do the same. Although usually sensible and conscientious, he retained some of his youthful touchiness. He was capable of kindness to his subordinates—one young officer fondly remembered smoking pipes with Lee on deck—but unlike Du Pont and Farragut, he lacked the ability to inspire and shape them into a loyal team. Another officer recalled one of Lee's inspections: "He visited the ship[,] looked all around, asked some questions, and left without a smile."[2] Welles eventually complained that Lee did not possess sufficient drive and determination to be a great commander. In all fairness, however, Welles never really gave Lee a decent chance to prove himself in a major undertaking. Finally, Lee was an ambitious man willing to pull whatever strings were available to him. There was nothing unusual about this in the Navy, but in Lee's case the strings at his disposal were so formidable and obvious that some officers resented him and attributed his professional advancement more to favoritism than to ability.[3]

In fact, Lee's coterie of personal lobbyists with access to the Navy Department's corridors was impressive. Chief among them was his father-in-law, Francis Preston Blair. Blair had spent decades developing relationships and trading favors with powerful men, and he was thoroughly familiar with the language of understandings reached over dinner and drinks. Lee's was among the most important accounts in his personal portfolio. Elizabeth lacked her father's experience and access, but she more than made up for it with the single-minded ferocity with which she promoted her husband's interests. In a town full of pushy and ambitious military wives, Elizabeth was in a league of her own.

Yet, despite these connections, Lee never gained the rank and responsibilities he and his family thought he deserved, even though he rose to squadron command. There were several reasons for this. For one thing, the Blairs were hardly united in their affection and support for Lee. Montgomery Blair hated Lee for several reasons, some of which went back to his opposition to Phillips and Elizabeth's marriage nearly twenty years earlier. Indeed, Montgomery refused to speak to or even mention Lee's name for four years, until one day in April 1862 when he told Elizabeth that the Navy Department had received glowing reports of Lee's activities on the Mississippi River. The fact that Montgomery and his wife shared a Washington duplex with Phillips and Elizabeth made this family feud even more awkward. Nor was Fox much help. To be sure, he was always full of genial and syrupy reassurances about Lee's prospects, but he never went out of his way to push for Lee's advancement. Elizabeth had her doubts about Fox's commitment to her husband's welfare from the war's start, and as time went on and he failed to do all that she felt he could to aid Lee's career, she grew increasingly hostile toward him, snidely referring to him as "Fatty" in her correspondence. Finally, although Francis Blair had as much interaction with Welles as any other private citizen who was not kinfolk, he was usually unable to exert much influence over the unimpressionable and no-nonsense navy secretary.[4]

After his quarrel with Farragut outside Vicksburg in May 1862, Lee's immediate objective was to gain a transfer out of the West Gulf Squadron to a more conducive post. Elizabeth and her father believed that the key to rescuing Lee was securing his promotion, so they directed their efforts to that end. In response to Congress's 16 July law overhauling the Navy's ranking system, Welles established a board to sift through the register and recommend promotions. Elizabeth and her father were confident that the board would look favorably on Lee because he was one of the acknowledged stars of the New Orleans campaign. Nevertheless, they lobbied Fox just to be safe. Fox insisted that he had no influence over the board, but added that he had no doubt that Lee would receive the promotion he deserved. On 7 August Elizabeth was delighted to receive Lee's new commission. Upon opening it, though, she was bitterly disappointed to learn that her husband had been elevated only to captain, not commodore. Moreover, it said nothing about when and if he would return to Washington. Her father was equally upset, insisting that the Navy Department's treatment of his son-in-law was a slap in the face. Just days later, however, Lee unexpectedly walked through the door to greet his overjoyed wife.[5]

As things turned out, Welles had plans for Lee that he kept from Elizabeth and her father. On 15 July Louis Goldsborough had asked to step down as North Atlantic Squadron commander. Welles was more than happy to comply with his request, but now he had to find a replacement. Experience had confirmed Welles' belief that character and performance counted for more in choosing squadron commanders than seniority. After a year and a half of war, the Navy had a number of battle-hardened officers from whom Welles could select. In fact, Welles had already started tapping this newly uncovered pool of talent by assigning Charles Davis and James Lardner to flotilla and squadron commands. Fox apparently preferred Stephen Rowan for Goldsborough's job. Rowan had performed well in the North Carolina sounds, was familiar with the North Atlantic Squadron, and was currently waiting around to take over *Powhatan* after her refitting. Welles, however, opted for Lee. Even after his promotion to captain, Lee was still a relatively junior officer, but Welles was not too concerned about that. By all accounts, Lee had fought gallantly during the New Orleans campaign, so hopefully he had that fire in his belly that Welles usually sought in his squadron commanders. Since Farragut had downplayed his disagreement with Lee outside Vicksburg in his official report, Welles could not include it in his mental calculations. In addition, Welles felt that Lee's well-known meticulousness would counteract Treasury Department efforts to encourage trade with the Southern states by watering down or circumventing the blockade rules. Welles later insisted that Blair's lobbying on Lee's behalf had no impact on his decisions, but in all likelihood placating this politically powerful family during that tense summer certainly crossed his mind. On 19 July, just four days after Goldsborough submitted his resignation, Welles ordered Lee home from the Mississippi River.

Lee arrived three weeks later, and, after a well-deserved rest, he took over the North Atlantic Squadron on 4 September. He later claimed somewhat disingenuously that he did so reluctantly because he felt it would be dishonorable to let a junior officer such as Rowan take it instead. He also worried illogically that his new post would hurt his career prospects and his chances of earning prize money. Welles also appointed him an acting rear admiral so he would outrank the other officers in his new squadron and would be equivalent to the army major generals who might otherwise try to overawe him. Welles may have denied that favoritism had anything to do with Lee's assignment, but many people had a hard time believing that. John Rodgers' wife, for example, was incredulous at the news. Her husband, on the other hand, respected Lee and viewed his elevation more philosophically, explaining to his wife, "Lee has deserved promotion as much as any one—and has had the influence to get it."[6]

Geographically, Lee had three main areas of responsibility as North Atlantic Squadron commander. The first, patrolling Virginia's eastern shore, required comparatively few resources. After McClellan's failed Peninsula campaign, the Army of the Potomac concentrated on unsuccessful overland offensives toward Richmond in 1862 and 1863 that required minimal naval support. Second, Lee had to safeguard the North Carolina sounds. Although Union forces secured the sounds in the spring of 1862, the Army still wanted naval support to help protect its scattered garrisons from occasional rebel attacks. Finally, and most important, Lee was accountable for blockading Wilmington, North Carolina. Wilmington contained a prewar population of only 9,500 people, but during the conflict it became perhaps the most significant city in the entire Confederacy because it served as a haven for blockade-runners bringing in the supplies, equipment, and weaponry the rebels needed to prosecute their war. Indeed, Wilmington was ideally suited for this role. It was located only 570 miles from Nassau and 670 miles from Bermuda, and its railroads provided easy access to the Confederate interior. Situated 20 miles up the Cape Fear River, it was immune from direct naval assault. The Cape Fear River possessed two navigable entrances divided by Smith Island and Frying Pan Shoals that together extended 25 miles into the ocean. This meant that Union warships really had to blockade two places separated by 50 miles. To complement these geographical advantages, the rebels possessed two major forts, Caswell and Fisher, which protected Wilmington. Fort Caswell, opposite Smith Island, was an old masonry structure that local militia seized when North Carolina seceded from the Union. The rebels had gradually built Fort Fisher, at Confederate Point, from scratch until it became perhaps the most fortified site in the Confederacy. Its forty heavy guns helped blockade-runners by keeping Union warships at a respectable distance. As the war continued and the Union Navy slowly captured other rebel ports, Wilmington's value grew. From November 1863 to October 1864, for example, the Confederate government alone imported through the town 1.5 million pounds of lead, 1.85 million pounds of saltpeter, 6.2 million pounds of meat, 408,000 pounds of coffee, 420,000 pairs of shoes, 292,000 blankets, and 136,832 small arms. In the first nine months of 1863, the rebels also exported 31,000 bales of cotton from Wilmington.[7] As the war continued, and the Union Navy's constricting blockade coils tightened, Wilmington's and the Confederacy's survival became increasingly one and the same.

Although the Navy initially underestimated Wilmington's importance, by the time Lee assumed command of the North Atlantic Squadron the city's relevance to the rebel war effort was becoming apparent. Enforcing the blockade of Wilmington required a meticulousness that was Lee's forte, and Welles gave him a free hand to deploy his warships as he saw fit. Lee's biggest accomplishment in this area was to institute a true blockading system. Instead of simply ordering his vessels to patrol outside Wilmington without much rhyme or reason, Lee tightly coordinated their activities. His blockade was always a work in progress with which he constantly tinkered, but it got more and more sophisticated as the conflict progressed. His first goal was acquiring more and better warships. Lee explained to Welles that without sufficient vessels he simply could not successfully blockade Wilmington. The Navy Department gradually increased his strength, and Lee strove to organize his growing resources. He recognized geographical realities by dividing the Wilmington blockade into two parts. He also spread his warships farther apart to cover more ocean, painted them a dull gray as camouflage, removed masts and yards from smaller vessels to increase their speed and decrease their visibility, and provided lookouts with white suits to avoid detection. He did all this in very detailed and explicit instructions to his officers. All this was helpful, but Lee's biggest tactical innovation was deploying his warships in concentric cordons outside Cape Fear River. The innermost, stationary vessels detected blockade-runners leaving Wilmington and signaled the second line of faster roving warships to intercept them. Beyond that was a third ring of the squadron's fastest vessels that searched for targets of opportunity. These changes improved the blockade's effectiveness, but it took time. In fact, by May 1863, after more than six months in command, Lee's squadron had captured only one steamer trying to run the Wilmington blockade, although it got better results later on.

No matter how well Lee organized his vessels, communications troubles and new steam technology made it very hard to stop most blockade-runners. Explaining the problems that Union warships faced, one naval officer wrote after the war, "It was really extraordinary how easily the blockade runners evaded our forces. Frequently in the afternoon, we would see them drop down the river over the bar and anchor near the lighthouse, but in the morning they would be gone, without our having even sighted them again."[8] Another wrote, "People generally know nothing about our 'blockade' except that vessels run past and get in to

Southern ports quite frequently. They do not know the difficulties of stopping vessels during dark nights, and of the greater difficulties of finding and seeing them. There are no men in the employ of the government that endure more hardship and exposure and severe labor than those on board our blockaders, and who get so little credit for it."[9]

The obvious answer to the Wilmington conundrum was for the Union to simply seize the place, as it had New Orleans. Goldsborough had considered assailing Fort Caswell when he commanded the North Atlantic Squadron, but nothing ever came of it. After Lee replaced him, the Navy Department resurrected the idea. As Fox explained to Lee, "Though popular clamor centers upon Charleston I consider Wilmington a more important point in a military and political point of view."[10] Lee began collecting data on Wilmington's defenses, and initiated discussions with Maj. Gen. John Foster, commander of the Department of North Carolina, for a joint campaign against the city. Lee and the Navy Department believed that an assault on heavily defended Wilmington would require several of the new and scarce *Monitor*-class ironclads just coming into service. Unfortunately, the original *Monitor* sank in a gale on its way to Beaufort, North Carolina, on 31 December 1862. Without a sufficient number of ironclads available, the Navy Department called off the attack. Fox tried to revive the plan in February, but Du Pont insisted that he needed all the ironclads he could get for his big offensive against Charleston. After the Charleston operation failed in April, Welles and Fox refocused their attention on Wilmington. The main problem now, however, was that Maj. Gen. Henry Halleck, current army general in chief, stated that he had no troops to spare for the North Carolina coast, so any assault on Wilmington would have to be an all-naval affair. Welles was willing to try anyway, but Lee was not. As far as Lee was concerned, Du Pont's unhappy experience in Charleston harbor demonstrated the futility of engaging rebel fortifications when army support was simply not forthcoming. The blockade, imperfect though it certainly was, would have to suffice for now.[11]

At the same time Lee was plotting strategy from his flagship, *Minnesota,* at Hampton Roads, he was also waging a proxy war in Washington that had less to do with the Confederacy than with his own personal agenda. Lee was only an acting rear admiral, so when he left the North Atlantic Squadron he would revert back to a captain. As an ambitious man, he wanted to make his temporary rear admiralty permanent. According to Congress' 16 July 1862 law that reformed the Navy's ranking system, Lee needed a congressional vote of thanks, Lincoln's appointment, and Senate approval to gain one of the nine authorized active-list rear admiral slots. Although Lee was of course not usually on location to lobby directly for his promotion, his wife and father-in-law were, and they pursued his advancement with a resolute single-mindedness that was a Blair hallmark. Actually, Lee had already come close to gaining Congress' official thanks. After the fall of New Orleans, Congress had considered voting its gratitude to all the ship captains who participated in Farragut's campaign, but eventually decided that doing so would elevate the undeserving with the deserving, so the congressmen instead thanked only Farragut and Porter. Elizabeth and her father thought that reviving this bill might be the way to go until Welles explained that its passage would create a slew of competitors for Lee—Bell and Bailey, most prominently—for the few available rear admiral positions. Once they understood this, Elizabeth and her father concentrated on persuading Congress to give Lee its thanks for his actions as North Atlantic Squadron commander. In late February 1863, Elizabeth drew up a list of sympathetic senators to target, and Blair went to the Capitol determined to corner every congressman he could find to plead Lee's case. He was optimistic about his chances when he left Silver Spring, especially since Fox had told him that securing Lee's promotion should not be that difficult. After several frustrating days, however, Blair returned to Silver Spring discouraged, tired, and unhappy because so many congressmen responded to his entreaties with double-talk and obliquity. In all likelihood, they were reluctant to act without guidance from the Navy Department.[12]

Having been stymied on Capitol Hill, Blair decided to try another tack. On 7 March he secured an interview with his old friend Welles to make his pitch for his son-in-law. The two men discussed Lee's fitness, claims of past service, and future opportunities. Welles was cordial and friendly, and he assured Blair that he also wanted to see Lee advanced. The problem, Welles explained for neither the first nor the last time, was that although Lee was performing creditably as North Atlantic Squadron commander, he had not yet done anything extraordinary, such as seizing Wilmington, to merit his elevation to permanent rear admiral. Until he did, it would not be fair for Welles to give him special treatment. Welles and Fox, in short, were taking advantage of Lee's overweening ambition to hold his promotion hostage until he paid Wilmington as ransom. If there was something cold-blooded and cynical about their actions, it was no more so than the behavior exhibited by Lee and his pushy surrogates. Welles and Fox, though, could at least claim that they were acting in their country's best interests, and it was understandable that both men were put off with Lee's incessant and blatant lobbying, however indirect. Back in mid-February, when Phillips was briefly in Washington, Fox and his wife invited the Lees over to dinner. Unsurprisingly, Phillips and Elizabeth took advantage of the opportunity to raise the issue of his promotion during the meal. Afterward an exasperated Fox said to his wife, "Lee had better be off to Wilmington where if successful, he would be entitled to one."[13] Lee's future, like the Confederacy's, was now tied to this one stubbornly held North Carolina port.[14]

The Tragedy of Frank Du Pont

Like any good chief of staff, Cdr. Raymond Rodgers was sensitive to his boss's temperament and needs. He worked closely with Frank Du Pont, and knew the long, grueling hours he put in every day as South Atlantic Squadron commander. On 3 September 1862 Rodgers wrote to Fox that it would be a good idea for the Navy Department to bring Du Pont home for a few days of rest. Rodgers explained that while Du Pont remained his cheerful and energetic self, he had been at sea for almost a year now, and that was bound to take a toll on even the hardiest of men. Besides, Rodgers knew that the Navy Department was contemplating an assault on Charleston, and he believed that Du Pont should consult with Welles and Fox personally about the proposed attack to make sure that everyone was in agreement on the proper strategy to pursue. Welles and Fox acquiesced, and a week later they ordered Du Pont back north. After spending a few days with his wife in Wilmington, Delaware, the two of them traveled to Washington on 1 October to stay with Fox at Montgomery Blair's house. Next day Du Pont met with Welles and Fox at the Navy Department building. Whatever Rodgers' concerns, Welles found Du Pont hale and hearty. Indeed, there were good reasons for Du Pont's robustness that went beyond a few well-earned vacation days. If he had been afloat longer than any other high-ranking naval officer, he had also been the most successful, and had a promotion to rear admiral to prove it. Under Du Pont's leadership, the South Atlantic Squadron had seized Port Royal and sealed off almost all of the South Carolina, Georgia, and eastern Florida coasts. Stringham, Goldsborough, Mervine, McKean, Foote, and Davis had all come and gone as squadron or flotilla commanders, but Du Pont remained a stable and dependable presence. Although Farragut had had his victories too, his record was counterbalanced and tainted by setbacks of a kind unfamiliar to Du Pont. Welles may have grumbled in his diary about Du Pont's cliquishness and sycophancy, but the important thing was that the navy secretary retained his confidence in his South Atlantic Squadron commander.[15]

As the cradle of secession, Charleston was a tempting naval target, but its occupation would bring the Union more than just psychological comfort. By late 1862 Charleston was one of only three major ports east of the Mississippi River—Wilmington and Mobile being the other two—still available to blockade-runners. Seizing the city would therefore ratchet up the blockade another notch and make it more difficult for the Confederates to obtain the supplies, equipment, and weapons they needed to prosecute the war. Although Fox had started making noises about assailing Charleston in the spring of 1862, not much happened until that fall. Geographically Charleston was a tough nut to crack. Its enclosed harbor was lined by batteries that could pulverize attacking Union wooden vessels to smithereens. The new ironclads, on the other hand, would be impervious to rebel cannonballs. Fox believed that they could steam straight through the gauntlet of fire to Charleston's docks and demand the city's surrender. As he later melodramatically explained to Du Pont, "I hope you will hold to the idea of carrying your flag supreme and superb, defiant and disdainful, silent amid the 200 guns until you arrive at the center of this wicked rebellion and there demand the surrender of the forts, or swift destruction. . . . The sublimity of such a silent attack is beyond words to describe."[16] Unfortunately, ironclad construction was so far behind schedule that Welles did not order the first five to join Du Pont's squadron until 6 January 1863. There was also talk of army cooperation, but that eventually fell through because of dissension and confusion among the generals whose participation was required. It would therefore have to be an all-naval show. This was fine with Fox, who felt that the Army regularly slighted the Navy's contributions to the war effort. In fact, the previous June, Fox had written Du Pont, "It may be impossible, but the crowning act of this war [seizing Charleston] ought to be by the navy. I feel that my duties are two fold: first, to beat our southern friends; second, to beat the Army."[17] During his October visit to Washington, Welles and Fox explained to Du Pont his mission. Du Pont listened without much comment, a silence that Welles and Fox interpreted

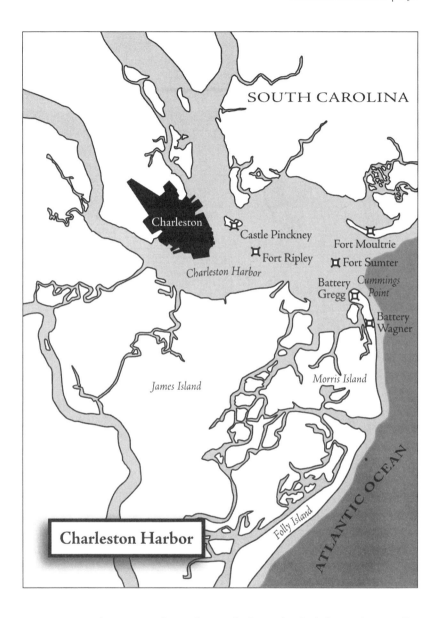

Charleston Harbor

as consent. A discussion of specifics might have clarified things, but Welles and Fox chose to leave the operation's details in Du Pont's capable hands.[18]

Du Pont had an excuse for his reticence. He was a good deal less enthusiastic than Welles and Fox about the proposed Charleston operation. He did not believe that his superiors really understood the difficulties an assault on Charleston entailed. Welles and Fox saw the South Carolina city as another New Orleans, and thought that Du Pont could steam past its defenses the same way Farragut had run by Fort Jackson and Fort St. Philip. In Du Pont's case, though, he would have the added advantage of using ironclad warships. Du Pont, on the other hand, recognized that Charleston harbor was really a cul-de-sac surrounded by enemy batteries that could trap an attacking fleet. As he had explained to his wife the previous June, "The truth is the harbor is a good deal like a porcupine's hide and quills turned outside in and sewed up at one end."[19] In addition, Du Pont was unhappy with continuing logistical problems in supplying his squadron, the difficulties in securing the army cooperation he felt was necessary for the operation's success, and the lack of detailed information on Charleston harbor's defenses. He also worried that a defeat might jeopardize the entire blockade, especially if one of the precious ironclads fell into enemy hands. But if the Navy Department expected him to rely on ironclads for his offensive, then he wanted as many of them as he could get.

Here too there were problems that extended beyond frustrating construction delays. After *Monitor*'s dramatic confrontation with *Merrimack*, many people such as Fox saw ironclads as the ultimate naval weapons. In reality, however, they had serious shortcomings. Du Pont suspected as much, but to get a better idea of their potential, he sent some of them down to the Georgia coast on several occasions to engage Fort McAllister outside Savannah. The results were troubling. While the ironclads proved themselves superb defensive platforms capable of withstanding the impact of dozens of cannonballs, their offensive capabilities were very limited. Each *Monitor*-class ironclad carried only two guns with an excruciatingly slow rate of fire. There were also serious questions about the durability of the new 15-inch Dahlgren cannon they sported. As for the vessels themselves, they were shoddily built with brittle iron, slow, and extremely difficult to steer and maneuver. Reading this information in his ironclad captains' reports did little to inspire or reassure Du Pont.

Although Du Pont had deepening doubts about the wisdom of assaulting Charleston with ironclads, he never expressed them clearly to Fox and especially Welles, who were in the best position to address his concerns. There were reasons for his diffidence. Du Pont was a career naval officer accustomed to obeying his orders, not questioning and debating them.

Besides, doing so would be an admission of failure, something that did not come easily to a man with a long record of professional achievement. If he stated flatly to Welles that the offensive could not succeed, the navy secretary might simply remove him from his prominent post and replace him with someone more enthusiastic, as he had done with naysayers Silas Stringham and William Mervine. Honesty, therefore, might cost Du Pont the Navy's most prestigious combat command. As a result, Du Pont resorted to allusions and insinuations to express his reservations. In his communications with Fox, he wrote elliptically of his misgivings, but always coupled them with assurances that he would do his best to overcome all obstacles and make the operation work. He told Fox that he smiled when he read the assistant secretary's suggestion that he steam silently and defiantly to Charleston's docks, but in a letter to his wife he guessed that Fox must have been drunk when he penned those outrageous words. Du Pont also forwarded to the Navy Department his ironclad captains' reports on the results of the Fort McAllister bombardments in the hope that Welles and Fox would draw the obvious conclusions and cancel or modify the operation. Finally, he conveyed his qualms to third parties such as Charles Davis so they could indirectly plead his case to Welles and Fox. In the meantime, Du Pont continued to wait and prepare, a process that took more than three months.[20]

Du Pont's anxiety about the Charleston operation was matched by that felt by Welles, Fox, and Lincoln, all of whom responded to the stress according to their individual personalities. As the weeks stretched into months and Du Pont continued to delay, Welles' emotions alternated between guarded optimism as to the ultimate outcome and growing frustration with the long wait. In his efforts to push Du Pont into action, Welles informed him that the country was impatient for victory, that he had stretched the Navy thin to supply him with ironclads needed elsewhere, and that closing down Charleston would compensate for the vessels the Navy had dispatched to search for the rebel raiders *Alabama* and *Florida*. He assured Du Pont of his support, pointing as evidence to the free hand he had given him to conduct the attack as he saw fit. At the same time, however, Welles was deeply concerned by signs of Du Pont's irresolution. To Welles, Du Pont seemed more interested in safeguarding his reputation than in risking it to gain victory. Indeed, Du Pont appeared to be wilting under the pressure of his responsibilities. Welles could not help but compare Du Pont's vacillation to that exhibited by George

McClellan during his unhappy days leading the Army of the Potomac. Welles lost sleep, but remained composed and calm in public, secure, he claimed, in the knowledge that he had done all he could to give Du Pont the tools he needed to win.

The president was equally discouraged by Du Pont's inactivity, and he too drew the McClellan analogy. Lincoln became so aggravated that he wanted Fox to go down the coast to talk with Du Pont, but Welles dissuaded him by noting that doing so would probably offend Du Pont, and that Fox had too much work to do in the Navy Department anyway. In early April Welles and Fox had to talk the president out of sending all the ironclads to the Mississippi River for Farragut to put to good use. Both Welles and the president took comfort in Fox's continued confidence in Du Pont's abilities. Fox was by nature an optimistic and energetic man, but the continued tension took a toll even on him. Elizabeth Blair Lee commented that he was becoming thin and haggard. Fox did his best to reassure, prod, and inspire Du Pont. He even tried to challenge and shame him by pointing out that Farragut had recently run *his* wooden vessels past Confederate batteries at Port Hudson on the Mississippi River. He also resorted to the most purple of prose to motivate Du Pont, writing, "Our flag staff stands surmounted by its gilded eagle waiting patiently for the downfall of Charleston to fling forth our beloved flag. May the Eternal in His Majesty and power watch over you and give you a naval victory according to your merits and righteousness in our cause."[21] As February and March came and went, Welles, Fox, and Lincoln continued to fume and fret.[22]

Lincoln had many reasons for his agitation and unhappiness. The Union was just emerging from a long, depressing, and defeat-laden winter. In Virginia in mid-December, the rebels shattered the Army of the Potomac's latest campaign against Richmond at the Battle of Fredericksburg. At the western end of the Union line, Ulysses Grant's two-pronged offensive toward Vicksburg collapsed with the destruction of his supply base at Holly Springs, Mississippi, and Maj. Gen. William Sherman's repulse at Chickasaw Bluffs. Although Maj. Gen. William Rosecrans' Army of the Cumberland fought off a ferocious Confederate assault at the Battle of Stones River near Murfreesboro, Tennessee, the victory did little to advance the Union cause. In fact, it took the Army of the Cumberland six months to recover from the experience. Nearly 30,000 bluecoats fell that winter in these engagements alone, but the Union had little to show for its heavy losses. As winter turned into spring, Lincoln did his best to prepare and cajole Union commanders into taking the field as soon as possible, but there was nothing easy about it. Lincoln's frustration with Du Pont was merely a microcosm of the entire Union war effort. A victory at Charleston, on the other hand, might be the harbinger of good things to come.

Down the Atlantic coastline at North Edisto, Du Pont put the finishing touches on his plan, such as it was. He hoped to steam past Charleston harbor's batteries and underwater obstacles with his nine ironclads in single file, bombard Fort Sumter's vulnerable northwest walls, and then head directly toward the city's docks. Once there, though, he and his ironclads would be cut off from the rest of the South Atlantic Squadron, whose wooden vessels could not survive a run past the gauntlet of rebel fire to reinforce and resupply the ironclads. Unless Charleston opted to surrender, Du Pont and his ironclads would have little choice but to return whence they had come empty-handed. Du Pont had recognized this dilemma from the start, and, having concluded that it was insurmountable, did little systematic planning to overcome it. Instead, he laid the groundwork to protect himself and his reputation from the backlash he suspected would follow a failed assault. He sent all of his official reports to his wife to give to Henry Winter Davis, his friend and political guardian angel, for safekeeping. He remained cheerful in public, but he often tempered his optimism with expressions of concern so people would know that he had doubted the wisdom of the operation. One visitor recalled: "On the whole the admiral felt that the chances were against us. Still he felt that very often difficulties grow less as you approach nearer to them— and his temperament was such that in doing his duty he was not affected by them. No day had been lost and everything had been done in his power to secure them against defeat. He felt that the public and the government had counted too much on the invincibility of iron clads. . . . It is fortunate that so able and good a man as Admiral Du Pont has charge. I have hardly ever in my life seen a man who impresses me so favorably."[23] In a letter to his wife the night before the attack, Du Pont lamented that he had not made his objections clearer to Welles, but noted that doing so would probably have resulted in his relief. Whatever his troubles, though, Frank Du Pont would do his duty and place his faith in God. Paraphrasing scripture, Du Pont wrote a friend, "*My* race is nearly run, whatever may happen."[24]

One thing that Du Pont did not have to worry about was the quality of the officers he would lead into action. Fox had assigned the Navy's best and brightest available officers to skipper the ironclads, men in whom Du Pont had considerable confidence. Many of them had served under Du Pont at Port Royal and elsewhere, and they had great affection and respect for their South Atlantic Squadron commander. Like Du Pont, though, they also harbored serious doubts about an assault on Charleston. One observer noted that from Du Pont down to the lowliest lieutenant, no one in the squadron thought much of the ironclads as offensive weapons. Percival Drayton, for example, told Du Pont with his usual frankness that even if the ironclads did manage to reduce Fort Sumter, there would be no place else for them to go. John Worden, who had commanded *Monitor* in her fight with *Merrimack* a year earlier at Hampton Roads, did not believe that the Navy could ever seize Charleston without army assistance, a conclusion that so depressed him that Du Pont made note of it. John Rodgers, on the other hand, was somewhat more optimistic, which comforted Du Pont because he considered Rodgers a genius. Du Pont also drew encouragement from Raymond Rodgers, who was still serving as the South Atlantic Squadron's chief of staff. Rodgers figured that assailing Charleston would be difficult, and he condemned the Navy Department for underrating its defenses, but he still thought the operation was doable. Whatever their individual inclinations, the ironclad captains were united in their conviction that the Navy Department was asking a lot of Du Pont, and they pledged to stand by him no matter what the outcome of the upcoming battle.[25]

Du Pont hoped to launch his attack on 6 April, but a heavy haze that blanketed Charleston harbor convinced him to postpone his assault for a day. Next morning the fog lifted and revealed clear skies, little wind, and smooth seas, optimal weather for the barely seaworthy ironclads. At about noon Du Pont and Raymond Rodgers crammed themselves into the pilothouse of USS *New Ironsides*, one of the two non-*Monitor*-class ironclads, and signaled the other eight warships to get under way. The vessels moved forward in single file, with John Rodgers' USS *Weehawken* in the van. To clear out any mines—or torpedoes, in contemporary parlance—*Weehawken* carried a large oaken structure dubbed a bootjack attached to her bow. Rodgers viewed the device's usefulness dubiously from the start, and his skepticism was confirmed when one of the bootjack's grapnels got caught in *Weehawken*'s anchor chain. Disentangling it required an hour, holding up the entire operation.

When the Union column finally entered the harbor, it quickly lost its cohesion. *New Ironsides* proved so unmanageable in the swift tides and shallow channel that Du Pont ordered her to anchor to avoid going aground, and she played little part in the battle. John Rodgers decided against leading the ironclads through a line of torpedo casks, and instead turned to engage Fort Sumter's northeastern face. The Confederates opened an extremely accurate and heavy fire on the ironclads. Raymond Rodgers later remembered, "The sight was one that no one who witnessed it will ever forget; sublime, infernal, it seemed as if the fires of hell were turned upon the Union fleet. The air seemed full of heavy shot, and as they flew they could be seen as plainly as a baseball in one of our games."[26] Du Pont later recalled that the sky was black with rebel cannonballs and shot. And John Rodgers, for his part, compared the rebel fire to a swarm of bees. Fortunately, the ironclads' armor offered Union sailors some comfort and protection. Even so, it was still an awful experience in the hot, dark, and cramped vessels. One participant compared the sound of the cannonballs hitting the ironclads to the cracking of giant walnuts. Sailors learned to stand on their tiptoes and not lean on walls to avoid concussions. The gun crews struggled to operate their Dahlgren cannon, but it took up to ten minutes to load them, an agonizingly slow process while in combat. After about forty minutes under intense Confederate fire, Du Pont asked Raymond Rodgers for the time. When Rodgers replied that it was crowding five o'clock, Du Pont decided to retire until the next day because it was getting too late to continue the contest. He signaled the ironclads to withdraw, and they steamed through the rebel bombardment back down the channel and anchored out of range.[27]

Du Pont's desire to renew the assault next morning evaporated upon seeing the condition of his ironclads. Raymond Rodgers recalled that USS *Keokuk,* the other non-*Monitor*-class ironclad, was riddled like a colander by rebel cannonballs. Her skipper, Cdr. Alexander Rhind, had been wounded in the foot. He hobbled across the deck to holler to Du Pont that his vessel was in no condition to fight. Once the ironclads were secured, all the captains boarded *New Ironsides* to meet with Du Pont in a large cabin. Each one reported his vessel's damage and contribution to the battle. Du Pont heard them out without much comment, and then retired for the night, leaving them to talk among themselves. Although he did not tell them then, he had already canceled plans to resume the battle in the morning. Du Pont's cursory analysis had convinced him that another such assault on Charleston's defenses would lead to the destruction of some or all of his precious ironclads. All nine of them had been disabled to one degree or another, and five of them had lost the use of at least one gun. The specifics, once the details were known, were grim. During the battle the rebels fired approximately 2,200 shots from their seventy-six cannon. They hit *Weehawken* 36 times; USS *Patapsco*, 47; USS *Catskill*, 20; USS *Nantucket*, 51; USS *Nahant*, 36; USS *Passaic*, 35; USS *Montauk*, 14; *New Ironsides*, more than 50; and *Keokuk* an astonishing 90 or more times. Indeed, *Keokuk* was in such bad shape that she rolled over and sank next morning. In all the Confederates had struck the ironclads about 400 times. In exchange, Union gun crews managed to fire only 154 shots that did almost no damage to Fort Sumter. While it was true that the vessels' armored plating kept casualties surprisingly low—Union losses amounted to one killed and twenty-one wounded—it was hard to see how another attack on the harbor's defenses, this time with structurally weakened warships, would accomplish more than the last one. A night's sleep did nothing to change Du Pont's mind. When he talked with Raymond Rodgers on deck next morning, Du Pont explained his rationale:

> I have given careful thought during the night to all the bearings of this matter, and have come to a positive determination from which I shall not swerve. I ask no one's opinion, for it could not change mine. I have decided not to renew the attack. During the few minutes we were under the heaviest fire of the batteries we engaged, half of our turret-ships were in part or wholly disabled. We have only encountered the outer line of defense, and if we force our way into the harbor we have not men to occupy any forts we may take, and we can have no communication with our force outside except by running the gauntlet. In the end we shall retire, leaving some of our ironclads in the hands of the enemy, to be refitted and turned against our blockade with deplorable effect. We have met with a sad repulse; I shall not turn it into a great disaster.[28]

Meeting with his ironclad captains again that morning, Du Pont discovered that they unanimously agreed with his decision and the logic behind it. If Welles and Fox wanted to take Charleston, they would have to find another way or another commander.[29]

Confusing rumors of the Union repulse at Charleston reached the Navy Department several days before Du Pont's official dispatches arrived.

Fox attributed these reports to warships probing the harbor's defenses, but Welles could not shake a nagging feeling that something had gone wrong. On 12 April Alexander Rhind hurried to Welles' house with official, albeit sketchy, word from Du Pont about the engagement. Du Pont had chosen Rhind to deliver the bad news not only because he had lost his command and possessions when *Keokuk* went down, but also because he wanted Welles to hear the details from someone who experienced the ordeal first-hand. Unfortunately for Du Pont, Rhind overplayed his hand. When Welles had assigned him to *Keokuk,* Rhind had been full of enthusiasm for the ironclads. Now, once again in Welles' presence, Rhind was the picture of dejection and despair, and had little good to say about the vessels he once championed. This was hardly the attitude Welles sought in his officers. Welles took Rhind over to the White House to meet Lincoln, and Fox and Massachusetts senator Charles Sumner later joined them. According to Rhind, Fox did most of the talking, and did his best to defend the ironclads.

Additional details that trickled in over the succeeding days did little to improve Welles' disposition. The navy secretary could not understand why the ironclads had stopped to bombard Fort Sumter instead of steaming directly past the underwater obstructions to Charleston's wharves. Moreover, the ironclads' armor was obviously effective because there was only one Union fatality in the battle, yet Du Pont gave up after little more than a half hour of fighting. Indeed, *New Ironsides* saw scarcely any action at all, which Welles found suspicious. Although these things upset Welles, he was most disturbed with what he interpreted as Du Pont's efforts to distance himself from the operation. Welles believed that he had given Du Pont not only sufficient authority, autonomy, and resources to successfully seize Charleston, but also plenty of opportunity to speak out against the assault if he saw fit. Du Pont had not done that; in fact, he had seemed committed to the campaign when he visited the Navy Department the previous October. Welles had had his doubts about Du Pont's resolution and determination for months because of his foot-dragging and demands for more ironclads. Even so, he always felt that Du Pont would come through in the end. Now, however, Du Pont was behaving as if the Navy Department had pressured him into attacking Charleston over his objections, against his better judgment, and contrary to his advice. The fact that Du Pont's officers backed him indicated that he was using his charisma to line up their support against the Navy

Department. The more Welles thought about it, the more convinced he became that perhaps Du Pont was not the right man to lead the South Atlantic Squadron.[30]

Down at Port Royal, Du Pont's disenchantment was growing too. He knew through unofficial channels that Welles and Fox were unhappy with him, and he resented the silence to which the Navy Department initially subjected him. On 15 April the *Baltimore American and Commercial Advertiser* ran an article by Charles Fulton that placed the onus for the Navy's defeat at Charleston on Du Pont. According to Fulton, the ironclads had not sustained serious damage, so Du Pont could have renewed the battle the next day if he had not lost his nerve. Fulton got much of his information from Alban Stimers, the Navy Department's chief engineer, who had witnessed the failed attack and was part of what Du Pont called the "ironclad interest." When Du Pont discovered Stimers' role, he had him arrested and preferred charges against him, but nothing ever came of it. Du Pont was also suspicious of Fox. According to Navy Department rules, Fulton was supposed to clear his article with Fox before the *Baltimore American* could publish it. Although Du Pont initially believed Fox's claim that he never saw the piece, he eventually concluded that Fox had in fact authorized the article's publication in an effort to shift the blame for the battle's outcome from the ironclads' deficiencies to Du Pont's timidity. The fact that the newspaper was tied to the Blairs lent further credence to Du Pont's conclusions. So did Welles' refusal to release to the public Du Pont's official reports on the battle. When Welles finally responded at length to Du Pont's dispatches in mid-May—he had waited at Fox's request in the hope of hearing good news from Du Pont that might break some of the tension—Du Pont was offended by the tone and tenor of his letter. Although Welles tried to be conciliatory, by then it was too late; Du Pont had already concluded that the Navy Department was conspiring to destroy his reputation and honor. Du Pont admitted to his wife that he was exhausted by months at sea, and this probably contributed to his poor judgment. He increasingly saw Welles as weak-willed, and Fox as an implacable enemy who was "a liar and scoundrel if ever there was one inside the Department."[31] No doubt Welles and Fox were disappointed with Du Pont's failure, and they erred in placing so much faith in the ironclads, but Du Pont's desire to protect his reputation above all else turned a series of misunderstandings into a tragedy that demolished his career.[32]

Henry Bell served as Farragut's chief of staff and right-hand man before becoming a squadron commander. (Naval Historical Foundation)

John Dahlgren's insatiable ambition succeeded in securing him squadron command, but also alienated fellow officers, obscured his other accomplishments, and contributed to his defeat outside Charleston. (Library of Congress)

Charles Davis fought competently at Port Royal and on the Mississippi River, but Welles kicked him upstairs because of his lack of aggressiveness. (Library of Congress)

Samuel Francis Du Pont (center) was Welles' and Fox's star squadron commander until he failed to seize Charleston. (U.S. Naval Institute Photo Archive)

David Farragut's victories at New Orleans and Mobile Bay made him the Union Navy's greatest commander. (U.S. Naval Institute Photo Archive)

David Farragut (r) and Percival Drayton. Drayton served as Farragut's chief of staff during the Mobile Bay operation, and helped compensate for Farragut's administrative shortcomings. (U.S. Naval Institute Photo Archive)

Andrew Foote was not only Welles' childhood friend and close confidant, but he also gave the Union some of its earliest and most important victories while commanding the Western Flotilla. (U.S. Naval Institute Photo Archive)

Gunboat USS *Fort Hindman*. Warships such as the *Fort Hindman* played a crucial role in securing the Mississippi River and its tributaries for the Union. (Library of Congress)

Gustavus Fox. Assistant Secretary of the Navy Gustavus Fox not only served as Welles' de facto chief of staff, but also provided the navy secretary with valuable intelligence about the officer corps through his informal correspondence network. (U.S. Naval Institute Photo Archive)

Samuel Phillips Lee. Although a thoroughly competent officer, Lee's heavy-handed string pulling eventually alienated Welles and contributed to the navy secretary's decision to shunt him to the war's periphery. (U.S. Naval Institute Photo Archive)

Military camp on Morris Island, South Carolina. The vessels in the background underscored the symbiotic relationship between the Union Army and Navy in their combined operations against Confederate coastal positions. (Library of Congress)

David Porter's meteoric rise through the Union Navy's hierarchy exemplified the changes that the Civil War brought to the Navy's command and organizational structures. (U.S. Naval Institute Photo Archive)

Henry Thatcher. Like most Union Navy squadron commanders, Thatcher labored in comparative obscurity to maintain the blockade and support Union Army operations. (Naval Historical Foundation)

Monitor-class ironclad USS *Catskill* in Charleston Harbor. Ironclads were among the most important new pieces of naval technology that Union Navy officers incorporated into their way of waging war. (Library of Congress)

View of Fort Sumter from a sandbar. Fort Sumter was one of the linchpins of the Confederate defense system protecting Charleston. (Library of Congress)

Gideon Welles. As secretary of the navy, Welles was responsible for the selection of Union Navy squadron commanders. (U.S. Naval Institute Photo Archive; artist: Matthew Wilson, 1883)

Charles Wilkes. The troublesome Wilkes gained fame by seizing the *Trent*, but eventually alienated Welles by exceeding his orders and constantly demanding additional resources. (U.S. Naval Institute Photo Archive)

Although Du Pont compared himself to a pilloried man pelted and abused by his tormentors, he had no intention of passively enduring the refuse and insults hurled at him. Instead, he hoped to use his formidable personal and political connections to fight these attacks on his character and reputation. Du Pont was a popular man with plenty of powerful friends in and out of the Navy whose support he could count on. On 18 April, for example, seven of the nine ironclad commanders, including the highly respected John Rodgers, signed a report authored by Percival Drayton endorsing Du Pont's version of the failed Charleston assault. Even Charles Steedman, one of the South Atlantic Squadron's occasional malcontents, came out on Du Pont's side. Outside of the squadron, influential senior naval officers such as Capt. William Shubrick and Capt. Cornelius Stribling wrote Du Pont encouraging letters. Du Pont could also rely on two bureau chiefs and friends, Charles Davis and Andrew Foote. Almost everyone commented to Du Pont that the vast majority of naval officers sympathized with him. Finally, Du Pont received considerable assistance from his friend Henry Winter Davis, an important Maryland politician. With all these allies committed to his cause, Du Pont believed he could generate sufficient public pressure on the Lincoln administration and Navy Department to bend Welles and Fox to his will.[33]

Yet in the end all this support for Du Pont did not amount to much. Welles and Fox initially refused to be drawn into a messy and public squabble with Du Pont that might hurt the Navy. They told those who met with them to plead Du Pont's case that they admired and sympathized with Du Pont, and that they had no desire to scapegoat him for his role in the failed assault on Charleston. Du Pont's friends thereupon tried to reconcile the two parties, but that was not what Du Pont wanted. The problem was that Du Pont's goals were never clear. No doubt he wished to preserve his honor, but what exactly did that mean? Did it require Welles' and Fox's resignation, or some sort of formal apology from them, or something else? In public at least, Welles and Fox seemed willing to go the extra mile to preserve the relationship. Du Pont's continuing hostility, on the other hand, made him appear more and more like the troublemaker and malcontent. Finally, it was increasingly evident that no one except Du Pont had much to gain from perpetuating the dispute, so there was no incentive for thoughtful and ambitious officers to put their careers on the line for a man who had little to offer them. In early June Percival Drayton gave Du Pont some astute advice. Drayton warned Du Pont that whenever an officer disagreed with the Navy Department, most people tended to fall in line behind the Navy Department regardless of their personal feelings because they wanted to see the war prosecuted effectively. A Du Pont victory, however defined, would not do this. In short, after emotions cooled, it was obvious that it was in everyone's best interests, except Du Pont's, for the controversy to go away so the Navy could get back to the pressing business of crushing the rebellion.[34]

Drayton discovered these truths after the Navy Department ordered him back north in late April. Drayton had commanded the ironclad *Passaic* during the attack on Charleston, and he firmly believed that the failed assault proved the impossibility of seizing the city through naval action alone. Drayton returned to Washington determined to defend Du Pont all the way up to the president if necessary. When he arrived on 7 May, however, he was surprised to learn that almost everyone now agreed with his and Du Pont's interpretation of events, including all the naval officers with whom he talked. Indeed, he felt somewhat foolish about the combative attitude with which he entered town. The only thing people criticized Du Pont for was his attempt to place the blame for the Charleston debacle on the Navy Department. Drayton met with Welles and Fox on 8 May, and spoke to them with his usual bluntness and honesty about the ironclads' shortcomings. He also admitted that Du Pont was overly sensitive, and attributed it to his physical and mental exhaustion from months at sea. Welles liked and respected Drayton, so he listened respectfully. Welles asked him to assure Du Pont that he was satisfied that he had done all he could under the circumstances, and that he still had the department's support. Welles also suggested through Drayton that Du Pont should not overreact to public condemnation of his efforts. Fox was equally kind, even though he was obviously hurt by charges that he was behind Fulton's newspaper article. On 12 May Drayton secured an interview with the president, who was also sympathetic and understanding toward Du Pont. In summing up his trip to Du Pont, Drayton urged him to follow Goldsborough's and Davis' advice to lay low until the controversy blew over, and he added that now might be the time to send Fox a friendly letter. As far as Drayton was concerned, he had done a lot to clear the air and mend fences.[35]

Henry Winter Davis had an experience similar to Drayton's when he visited the White House to lobby on Du Pont's behalf. As an outspoken opponent of Lincoln's selective suspension of habeas corpus, Davis would not have normally sought the president's help, but he made an exception for his old naval friend. He met with Lincoln alone on 2 May, and told him of Fulton's article and Fox's alleged connection to it. Lincoln responded that as far as he knew, Du Pont was one of Welles' and Fox's favorite officers, in whom they retained considerable confidence. He added that he had heard no one censure or speak badly of Du Pont. The president admitted that he had not understood Du Pont's long delay in assailing Charleston. Like most people, he had expected a gradual bombardment that lasted days or even weeks, so he was surprised that Du Pont called off the assault after only an hour. The main point of contention, Lincoln continued, was whether Du Pont could have steamed through the underwater obstructions directly to Charleston's wharves. When Davis pointed out that Du Pont had opposed a purely naval operation from the start, Lincoln stated that this was news to him. He remembered meeting with Du Pont the previous October, and Du Pont had then seemed in agreement with the Navy Department's strategy. Davis replied that it was probably a purely social call, or that Du Pont may have been waiting for the president to bring up the issue. The point, Davis emphasized, was that Fox in particular was manipulating the facts to suit his own purposes, and doing an injustice to Du Pont in the process. Lincoln admitted that Welles and Fox were obsessed with seizing Charleston, and he promised to read all of Du Pont's reports. Davis left the White House impressed with Lincoln's clarity and kindness, and he urged Du Pont to make any further appeals directly to the president. Like Drayton, Davis believed that he had done much to promote Du Pont's cause.[36]

Sadly for Du Pont, he failed to take advantage of the opportunities his friends developed to restore his deteriorating relationship with Welles and Fox. In fact, Du Pont was not interested in reconciliation, but rather he sought justice for the perceived wrongs that the Navy Department and its allies were inflicting on him. He believed, for instance, that Welles and Fox had manipulated Drayton for their own purposes. Du Pont appreciated Drayton's sincere efforts at mediation, but he thought that Drayton had actually hurt his campaign to reclaim his honor by misrepresenting Welles' and Fox's true intentions. As Du Pont interpreted things, his hard-line tactics were responsible for Welles' and Fox's pretense of civility, so he saw no reason to let up the pressure now that the Navy Department was caving in. In so thinking, Du Pont was deceiving himself. Although many people honestly hoped to patch up things between Du

Pont and his superiors, almost no one wanted to wage a proxy war on his behalf. This was especially true of the president, who was in the best position to punish Welles and Fox and vindicate Du Pont. Lincoln of course wished to end the disharmony in the Navy Department, but not at the expense of alienating Welles and Fox. Before the long-awaited Charleston assault, an impatient Lincoln had compared Du Pont to George McClellan, which was another way of saying that he saw Du Pont as an impediment to the war effort. Whatever their faults, Welles and Fox advocated the kind of aggressive war that Lincoln desired, even if it sometimes caused difficulties. Besides, Lincoln respected Welles and liked Fox, and he had no intention of disrupting his administration by removing either of them. If Du Pont was unwilling to meet the Navy Department halfway, then Lincoln would not intercede for him.[37]

In their conversations with Drayton, Henry Winter Davis, and others, Welles and Fox repeatedly expressed their support and respect for Du Pont. This fondness, however, declined with each increasingly defensive and strident letter that Du Pont posted to the Navy Department. By the end of the summer, Elizabeth Blair Lee noted that Fox could scarcely converse for five minutes without complaining bitterly about Du Pont. Welles, on the other hand, reserved most of his vitriol for his diary. To Welles and Fox, Du Pont was obviously distancing himself from responsibility for the failed Charleston assault and lining up support from the officer corps for a war with the Navy Department. Welles believed that Du Pont was more interested in protecting his reputation than in fighting against the rebels. Du Pont wanted the Navy Department to release his report on the battle, but Welles believed that doing so would merely alert the rebels to the ironclads' vulnerabilities and weaknesses. Moreover, Du Pont had no plans to attack Charleston again, and seemed content to let others bear the burden of prosecuting the conflict. Welles tried to understand Du Pont's actions, which he attributed to mental and physical exhaustion. As Du Pont's letters grew more and more discordant, Welles' hitherto overall positive opinion of him changed. He started to describe Du Pont as cliquish, proud, insincere, duplicitous, selfish, and scheming. Under these circumstances, there was no way Du Pont could continue as South Atlantic Squadron commander. Welles had initially contemplated removing Du Pont from his post in mid-April, and by mid-May he had made up his mind to do so. Fortunately for the navy secretary, Du Pont had already offered his resignation—sort of. On 16 April he had written

Welles: "I have to request that the Department will not hesitate to relieve me by any officer who, in its opinion, is more able to execute that service in which I have had the misfortune to fail—the capture of Charleston. No consideration for an individual officer, whatever his loyalty and length of service, should weigh an instant if the cause of his country can be advanced by his removal."[38] This was one Du Pont statement with which Welles fully agreed. All that remained was for Welles to determine Du Pont's successor.[39]

Du Pont had always prided himself on the conviviality of his meals on board *Wabash*, but maintaining a genial atmosphere was difficult when Cdr. Thomas Turner was present. Turner had skippered *New Ironsides* during the attack on Charleston and was one of Du Pont's most stalwart boosters. Even so, Turner's long-winded monologues tried Du Pont's patience. On 8 June, perhaps out of gratitude for his continued support, Du Pont invited Turner to *Wabash* for dinner. During the feast Du Pont saw Raymond Rodgers' expression change abruptly after reading a message someone handed him. Du Pont recognized immediately that something was up, and shortly thereafter Rodgers slipped him a note informing him that the Navy Department had finally removed him from his command. Du Pont did not believe that he had asked to be relieved, but he was hardly surprised; he knew Welles and Fox had been contemplating such action for some weeks. Next morning he received his official orders to return north when his replacement arrived. Although he told his wife that a great wave of relief came over him when he learned the news, this did not mean that his anger against the Navy Department had abated one bit. In ensuing letters Welles went through the motions of expressing his gratitude for Du Pont's services and accomplishments. Du Pont, though, was in no mood to reciprocate. A week before he turned over his command, Welles had chastised Du Pont for permitting the rebels to remove *Keokuk*'s guns from the sunken vessel under the Navy's nose. In response to this insulting communiqué, Du Pont concluded his official correspondence with the Navy Department as South Atlantic Squadron commander on 5 July by writing Welles: "Having indulged the hope that my command, covering a period of twenty-one months afloat, had not been without results, I was not prepared for a continuance of that censure from the Department which has characterized its letters to me since the monitors failed to take Charleston. I can only add now that to an officer of my temperament, whose sole aim has been to do his whole duty, and who has passed through forty-seven years of service without a word of reproof, these censures of the Navy Department would be keenly felt if I did not know they were wholly undeserved."[40] With that departing salvo, Du Pont hauled down his flag for the last time.[41]

Du Pont's bitterness remained even after he returned home to Wilmington in mid-July. Although friends such as William Shubrick and Charles Davis urged him to come to Washington to bury the hatchet with Welles and Fox, Du Pont believed it was too late for that and refused. It was just as well; the navy secretary was by then in no mood for reconciliation either, and he fully expected Du Pont's caviling to continue. Du Pont tried to maintain pressure on the Navy Department by sending his official reports on the failed Charleston assault to sympathetic officers and congressmen for their perusal and use. With Henry Winter Davis' help, on 22 October 1863 he penned a letter to Welles providing a detailed explanation of his role in the Charleston operation, in which he regretted that the Navy Department had no interest in the truth of those events. Thirteen days later, Welles responded with a scathing and acerbic broadside of his own that reiterated in detail his grievances against Du Pont. Welles' letter contained none of the friendly statements he had previously sprinkled in his dispatches to his former star admiral, but instead concluded that Du Pont's charges were unworthy of an American naval officer. Numerous officers who read Welles' diatribe deplored its sentiment, but they all urged Du Pont to cease his war of words because there was no profit in it. In fact, some of them even confessed to Welles that they believed Du Pont was acting unwisely. Du Pont finally took the advice, but this did not end things. Henry Winter Davis, now back in Congress, used Du Pont's experience at Charleston as a stick with which to publicly beat the Navy Department. In an effort to appease Congress and destroy Du Pont's reputation and standing with the officer corps once and for all, Welles eventually released all the relevant correspondence between Du Pont and the Navy Department. Welles' all-out assault was so ferocious that Du Pont expected the Navy Department to arrest and court-martial him. Instead, in June 1864 Du Pont received a curious proposal via Charles Davis. Welles' anger may have still glowered, but Fox, whose geniality was in such marked contrast to the navy secretary's, had apparently recovered some of his former affection for Du Pont. Fox asked Davis to see if Du Pont would be interested in commanding the small Pacific

Squadron. It was, Davis explained to Du Pont, an important position because of the troubled Union relationship with Mexico and because rebel raiders were in the region. While it is possible that Fox saw this as an opportunity to exile Du Pont to a remote part of the globe where he could cause the Navy Department no further trouble, Davis for one believed that the assistant secretary's motives were pure. Du Pont dismissed the offer, however, as too insignificant and did not reply. In the spring of 1865, as the war came to an end, Du Pont traveled to Philadelphia to serve on a promotion board. There he fell ill, probably with pneumonia, and died in a hotel room on 23 June with his wife at his bedside.[42]

In his diary Welles frequently complained that Du Pont had formed a clique within the South Atlantic Squadron that was more loyal to him than to the Navy as a whole. In fact, Du Pont had a knack for promoting personal loyalty, and his team contained some of the Navy's best officers. Once Du Pont had assumed an antagonistic attitude toward the Navy Department, Welles and Fox broke up his clique as part of their effort to undermine and discredit him. Although it is perhaps too harsh to state that Welles and Fox deliberately punished these Du Pont supporters, it is also probably true that some of them did not subsequently get the important positions they would have had they not so closely associated with Du Pont. Raymond Rodgers, for example, was one of Du Pont's most fervent supporters. He hoped to run the Naval Academy, but the Navy Department eventually gave him command of *Iroquois* and exiled him to the East Indies via Europe to hunt for rebel raiders. It was a necessary job, but one that took him far from the central and decisive part of the war and prevented him from helping Du Pont fight the Navy Department. Another Du Pont ally, Thomas Turner, found himself assigned to special duty in Washington and New York City for the remainder of the war. Percival Drayton's career, on the other hand, remained on track. Drayton's sympathy for Du Pont was less unequivocal than Turner's and Raymond Rodgers', and he was honest about Du Pont's shortcomings in his conversations with Welles and Fox. The Navy Department ordered him home in May, and then sent him to the Brooklyn Navy Yard to serve as ordnance chief there. Before the end of the year, Farragut asked him to become his chief of staff. At the end of the war, he succeeded Davis as head of the Navigation Bureau, but he died after a short illness in August 1865.[43]

As for John Rodgers, on 17 June 1863, while skipping *Weehawken*, he encountered the Confederate ironclad *Atlanta* in Wassaw Sound, Georgia. Superbly accurate firing from *Weehawken* forced the rebel vessel to surrender in just fifteen minutes. Welles and Fox were of course happy about this victory for all the ordinary reasons, but they also saw in it an opportunity to detach the well-respected Rodgers from the Du Pont camp. They issued a congratulatory public letter to Rodgers that ultimately led to a congressional vote of thanks. Welles eventually concluded that doing so played a role in wooing Rodgers away from Du Pont. This was not exactly true; Rodgers remained a Du Pont ally who distrusted Fox in particular, but he was realistic enough to recognize that fighting the Navy Department on Du Pont's behalf would not accomplish much. In June 1863 the Navy Department directed Rodgers to Boston to assume command of a new ironclad under construction, USS *Canonicus*. Unfortunately, illness prevented him from doing so. When he recovered, the Navy Department gave him another ironclad, USS *Dictator*. Rodgers spent the remainder of the war trying without success to prepare the warship for action. After the conflict he was one of the most respected naval officers in the service and held a variety of important positions. He served as commander of the Pacific and Asiatic squadrons, commandant of the Boston and Mare Island naval yards, superintendent of the Naval Observatory, chairman of the Lighthouse Board, and president of the United States Naval Institute. When he died in 1882, he was the senior active-list rear admiral.[44]

There was plenty of blame to go around for Du Pont's failure to take Charleston, beginning at the very top of the Navy Department. From the start Welles and Fox fell victim to several false analogies. They tried to take the lessons of New Orleans and Hampton Roads and apply them to Charleston. Welles and Fox wrongly assumed that Du Pont could steam into Charleston harbor without army support the same way that Farragut had steamed to New Orleans, only Du Pont would have the added advantage of ironclads. They failed to understand that ironclads, while terrific for defensive purposes, were poor offensive weapons due to their small batteries, slow rate of fire, low speed, questionable seaworthiness, and slapdash construction. Nor did they recognize that Charleston's geography was much different from New Orleans'. Once Farragut had passed Fort Jackson and Fort St. Philip, his way was clear to New Orleans. Charleston's harbor, on the other hand, was surrounded by dozens of rebel cannon and full of underwater obstacles that could easily trap an advancing Union fleet. Finally, Welles and Fox were unable to secure the necessary army cooperation. No doubt the Army had plenty of problems that made coordination difficult, but it is also true that Fox in particular had no desire to share the glory of occupying the cradle of secession with that other service and did not do enough to attain its help. Once Welles and Fox had formulated these basic presuppositions about Charleston, it tainted all other aspects of the operation.

But Du Pont was also culpable. Unlike Welles and Fox, he had a pretty good idea of the problems attacking Charleston entailed. Unfortunately, he failed to make his objections clear to his superiors, mostly because he feared he would lose his prominent command if he did. Instead, he hoped that if he let the facts—or anyhow his understanding of the facts—speak for themselves in his reports and memos, then Welles and Fox would eventually draw the proper conclusions and modify the operation. When Welles and Fox refused to see the light, he decided that the Navy could not on its own seize Charleston. Having made up his mind, he ceased to conduct the detailed planning a successful assault required. In particular, he did not reconnoiter the harbor's underwater obstacles as much as he should have, or explain exactly what his ironclads would do when and if they passed those obstacles. Defeat, in other words, became a self-fulfilling prophecy for Du Pont. While no one questioned his physical courage, he lacked the moral backbone to speak his mind when it counted.

Although it is possible to censure both Du Pont and the Navy Department for their actions leading up to the unsuccessful attack on Charleston, Du Pont was primarily responsible for the unhappy series of events that followed. Welles ultimately removed him from his command not so much because he failed to seize Charleston, but rather because of his poor attitude afterward. Du Pont could have soothed Welles' concerns by directing his efforts toward the immediate military situation in a frank and forthright manner, but he chose instead to concentrate on a campaign to protect his reputation and honor from detractors. The irony was that most people—both fellow naval officers and those in the press—sympathized with Du Pont. Du Pont, however, decided to focus on the minority who blamed him for the defeat, or who he believed blamed him for the defeat. He devoted an inordinate amount of attention to the *Baltimore American* article, and read the worst interpretation into every message

Welles sent him. Indeed, he even resented the fact that Welles did not immediately communicate with him after the abortive assault. Welles and Fox were certainly disappointed with Du Pont's setback, but they initially had no intention of destroying his career. Welles, for example, continued to salt his correspondence with Du Pont with compliments. Du Pont turned them into enemies by treating them as such. Moreover, once he went to war with the Navy Department, he never defined exactly what victory would entail. As at Charleston harbor, he plunged into battle without much of a plan, and therefore squandered the support he had. In the end he burned all his bridges, and with them his career.

John Dahlgren's Tragic Quest for Glory

Frank Du Pont and Phillips Lee were certainly not the only ambitious and self-serving naval officers with whom Welles and Fox had to contend. Captain John Dahlgren was just as troublesome, if not more so; whereas Lee's and Du Pont's influence within the executive branch usually stopped at the Navy Department, Dahlgren's extended all the way to the president himself, giving him a court of last appeal beyond Welles' control. Dahlgren was of Swedish extraction, born in Philadelphia in 1809. He was educated in a Quaker school and entered the Navy as a midshipman in 1826. His early career was the usual mix of sea duty, time ashore, and bickering with brother and superior officers. In 1834 the Navy Department assigned Dahlgren to the Coast Survey because of his proficiency in mathematics. He served there for four years, but the strain damaged his eyes, so he took a leave of absence until 1843 to recover. After a cruise in Cumberland in the Mediterranean, in 1847 he received orders to report for duty to the Bureau of Ordnance and Hydrography, where he made a name for himself. In the next fifteen years, during which his first wife died, he turned the Washington Navy Yard into a full-fledged ordnance establishment, with a boring machine, ball cutter and finisher, cap cutter, lathes, drill press, and so on. Before Dahlgren, ordnance was a craft plied by tradesmen who relied mostly on instinct and experience to make cannon. Dahlgren, however, turned ordnance into a science by implementing systematic experimentation, standardization, and keeping and publishing data. Using these methods, he pioneered advances in ranging, metallurgy, and shells. He was most famous for his development of the eight-ton bottle-shaped 9- and 11-inch smoothbore Dahlgren guns that graced American warships starting in the mid-1850s. The fact that he was able to overcome bureaucratic obstacles and persuade the Navy to adopt many of his inventions and ideas was even more impressive. After Frank Buchanan resigned as commandant of the Washington Navy Yard, Dahlgren took over the position, and on 18 July 1862 he became chief of the Bureau of Ordnance.[45]

There was much to admire about Dahlgren. He possessed a scientist's intelligence, work ethic, meticulousness, and persistency, which he combined with a practicality that appealed to Americans. John Hay, Lincoln's assistant secretary, called him "the greatest brain, so far, that the war had produced."[46] His forceful personality enabled him to get things done, making him a useful subordinate under many circumstances. Physically Dahlgren was on the thin side, with a large forehead, unkempt hair, and a heavy mustache that offset a hard look in his eyes. He lacked a sense of humor, but he was an attentive conversationalist and boon companion. On the other hand, there were some disturbing things about Dahlgren that prevented him from becoming one of the Union Navy's foremost commanders. He possessed little social insight, so he never understood how others might interpret and respond to his actions. He was, for example, perfectly willing to bypass his superiors to attain his objectives, without recognizing that doing so might cost him their future support. Dahlgren's Civil War career was marked by insatiable ambition and a thirst for glory. As Lee's and Porter's experiences indicated, Dahlgren was hardly unusual in having these characteristics. In Dahlgren's case, however, he combined them with a suspicion that his fellow officers did not appreciate him. Even though he was in many ways at the cutting edge of his profession, he got little respect from colleagues who believed in measuring an officer's worth by his time afloat. In fact, back in 1855, the Efficiency Board had considered dismissing him from the service because he was ashore too much. Only a strong and heartfelt speech on his behalf by his old friend Andrew Foote persuaded board members to recommend in his favor.[47]

When the war began, Dahlgren realized that no matter how well he performed in the Ordnance Bureau, it would not bring him the fame he craved, so he looked for other opportunities to gain the public's acclaim and his fellow officers' esteem. At the same time, though, he was always reluctant to take the risks a leader must to acquire greatness. Fortunately for Dahlgren, he was as accomplished a courtier as Du Pont. In order to get the Navy to adopt his mammoth gun, he had learned how to lobby congressmen and his superiors. In marketing his ideas, he also marketed himself. During his years in the capital, he had gained many important friends and allies in the government. It was not until the first days of the war, though, that Dahlgren secured his biggest and most important booster. After Fort Sumter fell and Washington was temporarily cut off from the rest of the Union, the president frequently visited the naval yard, and there he made Dahlgren's acquaintance. The two men hit it off immediately and became fast friends, despite their very different personalities. Dahlgren and Lincoln both shared an interest in practical science, and the president enjoyed listening to Fox and Dahlgren talk naval shop. Oftentimes Lincoln insisted on consulting "Dahl" about important naval matters, and he quickly became Dahlgren's biggest booster.[48]

Dahlgren was more than willing to use his relationship with the president to advance his career, and Lincoln was happy to oblige. Lincoln not only had Congress change the law so that Dahlgren could continue to run the Washington Navy Yard even though he was only a commander, but later pushed for Dahlgren's promotion to captain. Dahlgren, however, had higher goals than naval yard commandant or, later, bureau chief. Like Lee, he wanted to become a rear admiral, so in July 1862 he wrote to Lincoln to ask for an appointment to one of the five remaining slots once he— Dahlgren—had secured Congress' thanks. By way of justification for his bold request, Dahlgren pointed to Welles' last annual report, in which the navy secretary had stated that Dahlgren deserved as much credit for Union victories as the most successful squadron commanders because of his work with the Ordnance Bureau. Not surprisingly, Dahlgren's chumminess with the president irked Welles. The navy secretary had a pretty good understanding of people, so he recognized Dahlgren for what he was. Welles valued Dahlgren's positive attributes, but not his selfish ambition. Lincoln repeatedly told Welles that he would make Dahlgren a rear admiral if Welles gave the word. Welles resisted because he did not believe that Dahlgren had done enough to earn it, and he did not want to set a bad example by encouraging such favoritism. In February 1863, though, Congress finally voted Dahlgren its gratitude for his services in the Ordnance Bureau, and Lincoln nominated him a rear admiral over Welles' objections. When Dahlgren stopped by Welles' house to thank him for the honor, the navy secretary grudgingly told him to direct his appreciation to the president instead because it was his decision.[49]

Welles predicted quite accurately in his diary that the president's partiality toward Dahlgren would generate resentment in the officer corps toward him and obscure whatever contributions he made toward the war effort. While most officers freely acknowledged Dahlgren's merit, they simply did not believe that he had earned the rewards bestowed upon him. After all, he had never commanded a warship or been under fire. Fox, for example, opposed Dahlgren's appointment to rear admiral and bluntly told him so. Du Pont complained that Dahlgren's elevation was improper and demonstrated that duty ashore was worth more than duty afloat, which demoralized those officers serving in harm's way. Officers such as Drayton and Rowan, he argued, had served as well or better. Even Farragut, normally not one to insert himself into controversy, called Dahlgren's promotion "a great perversion of the law."[50] Still, there was nothing unique about Dahlgren's politicking; he was simply more brazen and successful than most. The jaded John Rodgers, for example, explained Dahlgren's activities to his wife, and concluded philosophically, "So goes the world."[51] Dahlgren, however, failed to recognize this bitterness, which would eventually cause him great difficulty.[52]

While Dahlgren welcomed his promotion to rear admiral, it did not quench his thirst for glory. Only a major combat command could do that, and he had been working hard to achieve that too. His brashest attempt to catapult himself into prominence occurred five and a half months before Lincoln appointed him a rear admiral. Dahlgren believed that combat and glory went hand in hand, so he needed to find some way to lead warships in action to put himself in the spotlight. He had tried to secure command of the James River Flotilla the previous July, but Welles and Fox had thwarted his efforts. On 1 October 1862, though, he set his sights considerably higher. In a letter to Welles, he asked to take charge of the planned attack on Charleston with the rank of acting rear admiral. Charleston was within Du Pont's bailiwick, but Dahlgren claimed disingenuously that Du Pont would probably enjoy a respite from his long and arduous duties. Du Pont happened to be in Washington at the time to consult with the Navy Department about the planned Charleston operation, so Dahlgren asked his friend Foote to approach him to see if he would go along with the scheme. Not surprisingly, Du Pont could hardly believe his ears. As Du Pont saw things, Dahlgren was a glory hound who had long ago chosen a particular line of work that offered little by way of celebrity, and he could not change now. With a line officer's usual prejudices toward staff, he explained to Foote, "He [Dahlgren] was licking cream while we were eating dirt and living on the *pay* of our rank."[53] Welles, on the other hand, was more diplomatic and accommodating. He explained that he could not deprive Du Pont, who had fought so well so far, of the honor of assailing Charleston. Nevertheless, he sympathized with Dahlgren's desire for action. Welles was willing to let Dahlgren take a temporary leave of absence from the Ordnance Bureau to skipper an ironclad participating in the proposed assault on Charleston. Moreover, Du Pont supported the idea. Dahlgren, however, replied that commanding a single warship would be a "painful humiliation," and renewed his request, along with specific reasons why it made sense. Welles wearily responded that he had no desire to humiliate Dahlgren, but was instead trying to do him a favor against his better judgment because he felt that Dahlgren was most useful to their common cause running the Ordnance Bureau. Besides, Welles could not help but adding, Dahlgren had received more special treatment than any other naval officer. Dahlgren did not see that Welles' offer would do much to advance his career, so he withdrew his request and bided his time.[54]

No sooner had Du Pont's assault failed than the War Department suddenly got its act together and agreed to cooperate with the Navy in an interservice effort to seize Charleston. Du Pont's unwillingness to undertake any new operations against the South Carolina city was one of many reasons why Welles removed him as South Atlantic Squadron commander. Welles took his time in doing so, though, because he was uncertain with whom to replace him. Charles Davis offered his services, but Welles had concluded that Davis' talents lay in the naval bureaucracy, not on a warship's quarterdeck, so he did not take his application seriously. Ideally Welles would have liked to give the position to Farragut, whose aggressiveness at New Orleans and elsewhere impressed the navy secretary. Moreover, he was popular both in the Navy and with the public. Farragut was already busily employed at the head of the West Gulf Squadron, though, and transferring him would cause as many problems as it would solve. Welles toyed with the idea of putting David Porter in charge of the squadron. Porter, like Farragut, possessed a pugnacious streak that appealed to Welles. Unfortunately, he was still a relatively junior officer whose elevation would offend his former superiors, and his reputation for bad-mouthing others did not endear him, so Welles passed him over. Mulling things over, Welles' mind focused on Francis Gregory, a rear admiral on the retired list. The Connecticut-born Gregory ranked sixth on the list of loyal captains when the war began, having entered the Navy as a midshipman back in 1809. He had spoken strongly in favor of relieving Fort Sumter in the tense days following Lincoln's inauguration, and Welles appreciated his determination. In fact, Welles had considered Gregory for the Gulf Squadron, but Paulding had persuaded him to assign it to William Mervine instead, a decision Welles later regretted. Gregory had so far spent the war supervising warship construction in private shipyards, a job that included overseeing ironclad production, so he was familiar with those vessels. Although Fox and Foote both endorsed Gregory, Welles ultimately decided against him because of his age—he was seventy-two years old—and his uncertain health.[55]

After all this mental back-and-forth, Welles finally settled on Rear Adm. Andrew Foote, the current chief of the Bureau of Equipment and Recruiting, as Du Pont's replacement. In many ways Foote was the ideal choice. He was battle tested and a proven winner, possessed plenty of rank, was well respected in the service, and was closer to Welles than any other naval officer. Welles' only major concern was that Foote greatly admired Du Pont and was reluctant to supplant his old friend. Du Pont had recently taken offense to one of Foote's memos, and Foote did not want anything to hinder his efforts to make amends. Even so, Foote recognized that he was an obvious candidate to replace Du Pont. He had recently applied for duty afloat because, he told Welles, the fresh air would improve his health, but no doubt he was also eager to get back in the thick of things. Although he had willingly accepted Welles' offer to run a bureau, Foote was at heart a man of action who lacked the temperament for an office job. On 27 and 29 May, Welles had long talks with Foote about his assuming command of the South Atlantic Squadron and assailing Charleston again. Foote's initial doubts that Union forces could seize the city receded after Welles brought in Brig. Gen. Quincy Gillmore, who was slated to lead the Army's contribution to the proposed interservice operation. Gillmore unrolled his maps and charts and explained his plans, which seemed sound enough at the time. Foote told Charles Davis he could have his office, and then traveled to New York City to begin preparations for the Navy's part in the new campaign. Welles wanted Foote to hurry in case the War Department changed its mind about attacking Charleston, so he was disconcerted when illness held up Foote's activities. On 9 June, however, a resolute and determined Foote was back in Washington. With great emotion, he asked Welles for a few more days' delay before he set sail for Charleston. His wife, he explained, was gravely ill and would not survive to see him return, so he wanted to go to New Haven to kiss her good-bye and make her final arrangements. He did not mention that he had an equally sick child at home too. Welles could only agree, and Foote rushed back north to his family for what was no doubt a heartrending farewell.[56]

There was unhappily one piece to the new Charleston puzzle that Welles could not quite make fit: John Dahlgren. Dahlgren desperately wanted to lead the next assault on the South Carolina city, and he recognized that Du Pont's discomfiture provided an opportunity to do so. On 21 April he reminded Fox that he still wanted duty afloat. Three weeks later he impressed Welles with an upbeat analysis of the Charleston imbroglio. Welles knew that appointing Dahlgren South Atlantic Squadron commander would please the president, who remained his persistent advocate. Welles disliked Dahlgren's relationship with Lincoln, and he undoubtedly realized that dispatching him all the way to Charleston would weaken the bond between the two men. As far as Welles was concerned, this would be all for the good. On the other hand, Dahlgren's recent promotion to rear admiral had generated considerable resentment among officers, and assigning the Navy's most prestigious squadron to him would add fuel to the fire. In fact, Thomas Turner, skipper of *New Ironsides* and a strong Du Pont ally, had already written the Navy Department that he would not serve under Dahlgren. The best solution, Welles finally decided, was to persuade Dahlgren to play a subordinate role under Foote. This would please Lincoln, get Dahlgren out of Washington and away from the president, give Dahlgren the opportunity to prove himself, and all without unduly provoking the officer corps. As Welles put it, "It would be the best arrangement I could make."[57] The trick, of course, was getting Dahlgren to go along with the scheme.[58]

Welles may have believed that he had formulated the ideal political resolution to the Dahlgren problem, but it was hardly a militarily sound plan. Dahlgren recognized this immediately when Fox broached the idea with him. Two rear admirals in one squadron, with unclear lines of authority between them, would surely cause confusion. Dahlgren felt that the best solution would be for him to lead the South Atlantic Squadron, with authority over both naval and army forces in the area. With his usual insensibility, Dahlgren told Fox that he doubted that anyone would find his proposal objectionable. Welles for one certainly did, and upon receiving Fox's report he immediately dismissed Dahlgren from his Charleston calculations. The navy secretary thought that Dahlgren had just thrown away the opportunity of a lifetime because he would never again get such a chance to prove himself worthy of a major combat assignment. To Welles, Dahlgren was placing his own pride and ambition ahead of his country's best interests. Foote, ever kindhearted and solicitous, offered to talk to his old friend, but he had no luck persuading Dahlgren to accept the offer either. On 2 June, while discussing other business, Welles told Dahlgren that he regretted that Dahlgren and Foote were unable to come to terms. Dahlgren explained his reservations, but, after Fox joined them, he noted that he would be willing to accompany Foote if he could command the ironclads only. Welles and Fox did not ask Dahlgren what motivated his change of heart, but instead at once grasped the opportunity before them. Foote had already left for New Haven to visit his wife, but Welles and Fox urged Dahlgren to meet Foote in New York City and see if he would be amenable to this compromise. Dahlgren rushed to the train station while Welles telegraphed Foote to set up a conference. Next day, Foote, Dahlgren, and Gillmore met in a New York City bookstore to discuss the new campaign against Charleston. Although Foote was fighting a blinding headache, he graciously agreed to give Dahlgren control over the ironclads. Having secured Foote's consent, Dahlgren hurried back to the Navy Department to break the news to Welles. The navy secretary, it seemed, had indeed made that best arrangement.[59]

Within days, however, Welles's carefully constructed personnel plan began to unravel. The problem was not Dahlgren, but rather Foote. On 16 June Welles learned from Paulding that Foote was seriously ill in a New York City hotel room, laid low by Bright's disease, a chronic inflammatory disease of the kidneys. Within days it was clear that his affliction was fatal. Although Welles dispatched the chief of the Bureau of Medicine and Surgery, Dr. William Whelan, to New York to attend to Foote, he could do little but make his patient more comfortable. Foote lingered for ten days, fading in and out of consciousness, during which time both Dahlgren and Percival Drayton visited him. When he saw Dahlgren, an incoherent Foote inquired, "Who will fight for Dahlgren? Dahlgren's boys." Foote also asked Drayton to make sure that Du Pont knew that he had nothing to do with his removal from command. Finally, on 26 June, he expired. His remains were taken to New Haven without fanfare and interred there. Welles was deeply distressed by his old friend's death for both personal and professional reasons. Foote had not only delivered the Union Navy some of its earliest and most important victories, but he had also given Welles good advice and served as a valuable sounding board. Shortly thereafter, Foote's wife followed him to the grave. Despite his fiery devotion to controversial causes, Foote was a popular man within the officer corps and among the general public, so his passing was widely lamented. Amid all the eulogies and commentaries, though, Foote probably would have appreciated Henry Bell's simple notation in his diary: "Admiral AH Foote, the Christian admiral, is dead."[60]

Foote's illness and subsequent death put Welles in a quandary. Army preparations for a joint assault on Charleston were proceeding rapidly, so Welles needed to find someone to replace Du Pont and organize the Navy's contribution to the operation as soon as possible. Dahlgren was not only readily available, but also familiar with the Army's plan. From this perspective he was Du Pont's logical successor. On 21 June, a Sunday, Welles summoned Dahlgren from church to his house to appoint him the South Atlantic Squadron commander. It seemed, intoned Welles, that destiny had fated Dahlgren for the position. During their interview Welles was blunt with Dahlgren. Welles explained that his decision was born of unique circumstances and entailed considerable risks for both Dahlgren and the Navy Department. Welles anticipated that some senior officers might refuse to serve under Dahlgren because his advancement had caused so much discontent. If so, said Welles, these officers could be transferred without prejudice. Recalling his experiences with Du Pont, Welles stressed that he expected Dahlgren to be completely open and honest in his dispatches with the Navy Department. Welles considered Dahlgren's assignment temporary, so he insisted over Dahlgren's strong objections that he continue as chief of the Ordnance Bureau. Despite this and other meetings over the next few days, Welles still had deep misgivings about his selection. While he got along with Dahlgren personally and believed he was an intelligent man who ran his bureau efficiently, he questioned his character. He worried about Dahlgren's overweening ambition and selfishness, had persistent doubts about his willingness to assume ultimate responsibility, and wondered if he possessed the stoutness of heart that characterized Farragut and Foote. Nevertheless, having cast his lot, Welles could do little more than await the results.[61]

Dahlgren left Brooklyn Navy Yard on 30 June on the small schooner USS *Augusta Dinsmore,* accompanied by Stephen Rowan, Thomas

Turner's designated replacement as skipper of *New Ironsides*. Five days later, on Independence Day, *Dinsmore* reached Port Royal. The Navy Department had initially kept Du Pont in the dark as to his successor after Foote's illness and then death precluded him from taking over. Scuttlebutt stated that Rowan would assume command, but Du Pont guessed that one of the available rear admirals—Goldsborough, Davis, or Dahlgren—would replace him, most likely Dahlgren. At 10:00 AM, Dahlgren was rowed over to *Wabash* to meet Du Pont. The two men exchanged cordial, even warm, greetings in a cabin full of officers curious to see their new chief. Du Pont may have been pleasant enough, but the response within the Navy's officer corps to "the foundry man's" ascension was less so. Tom Turner was merely the most public and vocal officer who resented Dahlgren's appointment. Others complained privately that Dahlgren's high-profile assignment was due primarily to his string pulling in Washington. James Palmer, for example, called Dahlgren's selection as South Atlantic Squadron leader an "outrage" and a "shame" because there were other officers without Dahlgren's connections who deserved the post more. John Rodgers observed that Dahlgren had never commanded a warship or heard a shot fired in anger. Discontent extended even into Congress. Iowa senator James Grimes noted that Dahlgren was little more than a conceited courtier. All this carping meant that there were plenty of people looking for opportunities to skewer Dahlgren for any mistakes he might make, giving him little room for error or forgiveness.[62]

The same afternoon he assumed command of the South Atlantic Squadron, Dahlgren went ashore at Hilton Head Island to confer with Quincy Gillmore about the upcoming campaign. Gillmore's basic idea was to deploy his ten thousand troops on Folly Island, about ten miles southeast of Charleston. From there, he would cross Light House Inlet and land on nearby Morris Island. Once ashore, the bluecoats would march to the island's northern tip and seize the Confederate batteries Gregg and Wagner at Cumming's Point. Gillmore planned to then emplace Union artillery there that would help the Navy pound Fort Sumter into submission, opening the way for Dahlgren's warships to enter the harbor and put Charleston under its guns. Dahlgren did not think much of the Army's part of this scheme because he felt that Gillmore's narrow approach would enable the rebels to stoutly resist him. On the other hand, he was confident that he could successfully carry out his assigned role. Dahlgren believed that if properly and carefully deployed, his ironclads could accurately pulverize Fort Sumter from close range and render the place untenable. As one person recalled, "He says he came here to *take* Charleston, and is confident he will take it."[63]

On the morning of 10 July, Union soldiers landed on Morris Island. The operation went smoothly, and by the end of the day almost the entire island was in Union hands. Unfortunately, the bluecoats were unable to thrust all the way to Battery Gregg and Battery Wagner, giving the rebels there time to brace themselves. Union assaults on the eleventh and the eighteenth failed miserably, so Gillmore settled down for a siege. In the ensuing weeks, he methodically pushed his men and guns closer and closer to the Confederate works in preparation for a final all-out attack. Offshore, Dahlgren's warships blasted the rebel fortifications with increasing accuracy and proficiency. Naval gunners, for instance, learned to ricochet shells off the water and into Confederate positions. As the summer wore on, Union artillery ashore and afloat gradually reduced Battery Wagner and Battery Gregg, and turned nearby Fort Sumter into a heap of rubble.

Although the Navy's role was secondary, this did not mean it was easy for anyone. Dahlgren rotated ironclads into action almost every day, which gradually took a toll on their effectiveness. The rebels hit them a total of 882 times, bending armor plates, shearing bolts and nuts, and jamming turrets. Ship bottoms became fouled with barnacles that reduced their speed. The constant wear and tear was equally hard on the crews. Heaving coal and loading guns in the cramped and sweltering ironclads exhausted the men. Fortunately, battle casualties were minimal, but they could still be gruesome. Dahlgren's chief of staff, Capt. George Rodgers, for example, was killed by a metal splinter after a rebel shot shattered the roof of *Catskill*'s pilothouse. Dahlgren survived the summer even though he spent a considerable amount of time in the ironclads under fire. On the other hand, unending stress and overwork led to debilitating gastritis. To make things worse, his son was severely wounded in Maryland after the Battle of Gettysburg, necessitating the amputation of his leg. Dahlgren worried about his boy as much as he worried about seizing Charleston. As the siege reached its climax, Dahlgren was scarcely able to get out of bed. On 28 August, he wrote in his diary, "My debility increases, so that today it is an exertion to sit in a chair. I feel like lying down. My head is light. I do not see well. How strange—no pain, but so feeble, it seems like gliding away to death. How easy it seems! Why not, to one whose race is run?"[64]

To complicate the situation, Dahlgren did not receive much help or support from some of his subordinates. This was hardly surprising; his advancement had generated substantial bitterness within the officer corps. Indeed, Welles had predicted as much when he appointed him to lead the South Atlantic Squadron. His biggest malcontent was Commo. Stephen Rowan, skipper of *New Ironsides,* whose sixteen cannon made her the most powerful of all the ironclads. Rowan had made a name for himself in February 1862 by destroying a rebel flotilla outside of Elizabeth City, North Carolina. He had a reputation as a ferocious and forceful officer, and in fact some felt that he deserved to command the South Atlantic Squadron more than Dahlgren. Under Dahlgren, however, Rowan's aggressiveness disappeared. Now that the Army and the Navy had supposedly pounded Fort Sumter into impotence, Dahlgren wanted to overcome the nearby underwater obstructions and push into the harbor toward Charleston. As the siege of Battery Wagner progressed, Dahlgren made several tentative attempts to do so, but nothing came of them. In the privacy of his diary, Dahlgren complained frequently about Rowan's timidity and negative attitude. In fact, Rowan repeatedly objected to any operation that put *New Ironsides* at risk. For whatever reason, his heart simply was not in the effort. On 2 September, for example, Rowan refused to steam past the underwater obstacles, claiming that it was too dangerous and futile. Although Dahlgren had warned Fox that Rowan might be a problem, he made no effort to remove him from his command. Instead, Dahlgren presented Rowan's obtrusiveness as one of many excuses for not taking his ironclads directly to Charleston's wharves. He also blamed difficult tides, continuing fire from Fort Sumter, the battered condition of his ironclads, and a lack of information on the underwater obstructions.[65]

On the night of 6–7 September, just before Gillmore ordered a full-scale attack, the Confederates evacuated Battery Wagner and Battery Gregg. With that, Gillmore believed that he had carried out his orders; Morris Island was in Union hands and Fort Sumter neutralized, so the way was now open for the Navy to steam to Charleston. Dahlgren, however, was not so sure because he was by no means convinced that Fort Sumter had been rendered impotent. After the rebels there rejected his surrender summons, Dahlgren decided to launch an amphibious assault on the fort with four hundred sailors and Marines. As things turned out, Gillmore was also planning to assail Fort Sumter with a couple of his regiments. Unfortunately, instead of coordinating their efforts, Gillmore and

Dahlgren squabbled over which service would command the operation. When they were unable to reach an agreement, Gillmore withdrew the Army's contribution, leaving Dahlgren to carry out the attack on his own on the night of 7–8 September. Several hours before the assault began, Dahlgren placed Cdr. Thomas Stevens in charge of the effort. An appalled Stevens tried to refuse because he did not think the operation could possibly succeed, but Dahlgren insisted that Confederate opposition would be minimal, so Stevens reluctantly agreed. At around midnight, the assault boats pushed away from the tug USS *Daffodil* toward Fort Sumter. Within minutes, everything went wrong. Not only were the boats unable to stay in formation, but the Confederates were fully alert. Rebel fire ripped into the bluejackets and Marines as soon as they disembarked, pinning them down in the rubble and debris outside the fort. After a half hour of this punishment, the officers opted to retreat. By the time the shooting stopped, Dahlgren had lost six men killed, fifteen wounded, and more than a hundred captured. Of equal importance, the fiasco also poisoned the relationship between Dahlgren and Gillmore, ruining any prospect for future interservice cooperation.

Despite this botched amphibious assault, Dahlgren continued his preparations to force his way into Charleston harbor. As summer turned to fall, Union batteries ashore and afloat repeatedly bombarded Fort Sumter. At the same time, Dahlgren gathered intelligence on those troublesome underwater obstacles. He believed that he needed all the reinforcements he could get; not only were day-to-day operations taking a toll on his vessels, but the Confederates and fate were steadily weakening his squadron. On 5 October, for example, a rebel torpedo boat damaged *New Ironsides* in a daring night raid. Two months later, *Weehawken* accidentally rolled over and sank at her anchorage. Dahlgren claimed that he was willing to risk an all-out offensive through the Confederate defense network, but only if Welles explicitly authorized one. Welles, however, thought that Dahlgren was in the best position to make the judgment, so he refused to do so. Instead, he suggested that Dahlgren poll his chief subordinates for their opinions. On 22 October Dahlgren met with his highest-ranking officers to contemplate the pros and cons of an attack. After six hours of discussion, they voted six to four against an assault, with the most senior officers providing the majority. Dahlgren agreed with the verdict, feeling that he required additional ironclads for a successful assault. Unhappily, ironclad construction was behind schedule as usual, and anyway Welles was not sure he wanted to risk those vessels in such a dangerous operation as Dahlgren proposed. These frustrations did little to improve Dahlgren's precarious health. In fact, on 10 October he was so sick that he asked Welles for a leave of absence to return north to consult a doctor, though he had recovered by the time he got an answer. On the other hand, there was little relief from the mental strain he was under. He complained in his diary that the Navy Department did not understand his situation, and that Welles and Fox made little effort to publicly defend him from newspapers that criticized his lack of progress. Although he regretted his failure to occupy Charleston, he tried to put the best face on things by emphasizing and exaggerating his accomplishments. In a letter to the president, for instance, he falsely noted that he had completely sealed Charleston off from the outside world and in the process tied down 20,000 to 25,000 rebel troops.[66]

The Navy's protracted failure at Charleston stood in stark contrast with the series of Union successes in the latter half of 1863. On Independence Day, Confederate soldiers began their long trek back to Virginia after losing the Battle of Gettysburg. That same day, Vicksburg and its 30,000 defenders surrendered to Ulysses Grant's besieging army. Two hundred winding miles downstream, Port Hudson, Louisiana, succumbed five days later, opening the entire Mississippi River to Union shipping and severing the Confederacy in two. Although the rebels defeated Maj. Gen. William Rosecrans' Army of the Cumberland at the Battle of Chickamauga on 19–20 September, two months later Union forces under Grant drove those same rebels into headlong retreat at the Battle of Chattanooga. Moreover, minor successes accompanied the major ones. Union soldiers repelled a rebel assault on Helena, Arkansas, the same day Grant's men marched into Vicksburg. In Ohio, Confederate general John Hunt Morgan was captured with the remnants of his force on 26 July after a three-week-long raid. Except along the eastern seaboard, Union fortunes were on the upswing.

After Du Pont's repulse in April, Welles had grumbled in his diary that Charleston really was not worth all the resources that the Navy was investing in its capture. Once Gillmore provided the often elusive ingredient of army cooperation, though, Welles recovered his enthusiasm for the Navy's continuing effort to seize the cradle of secession. He and Fox tried to be patient, but it was not easy as the weeks stretched into months and Charleston remained in rebel hands. Instead, they became increasingly frustrated and concerned with the lack of progress. There was, in fact, much to worry about, such as the deteriorating relationship between Dahlgren and Gillmore. On 3 October two of Gillmore's subordinates, Brig. Gen. Alfred Terry and Col. Joseph Hawley, appeared in Welles' office, most likely on their superior's behalf. The officers denounced Dahlgren as unfit for his command because he was too proud, hypersensitive, and unpopular. Although Welles disapproved of this irregular and out-of-channels complaint, he could hardly ignore its implications for interservice coordination outside Charleston. He admitted in his diary that there was probably some truth behind Terry's and Hawley's accusations; in fact, some of them dovetailed with his own misgivings about Dahlgren. He was, for instance, disturbed with the carelessness and haste that accompanied the failed assault on Fort Sumter. While he knew that Dahlgren labored under burdens foreign to most squadron commanders—critical and constant newspaper coverage, jealous and resentful subordinates, troublesome army officers—he also recognized that Dahlgren's character flaws created or exacerbated many of these difficulties. He was especially bothered by Dahlgren's unwillingness to accept responsibility for a full-scale assault into Charleston harbor. To Welles, Dahlgren and Du Pont were similar in that they both would rather endure unprofitable stalemate than risk their reputations on a setback. On the other hand, Welles believed that Dahlgren had worked hard, and his attitude toward the Navy Department was an improvement over Du Pont's. Besides, removing Dahlgren would not only upset the president, but it would also support Du Pont's contention that Charleston was unassailable. After weighing his options, Welles determined to stand by Dahlgren for now, and in his letters to Dahlgren he tried to be understanding and supportive.[67]

Dahlgren appreciated Welles' sympathy, but kind words did not bring him the victory and fame he sought. He believed that that objective required additional ironclads so he would have the strength to push past those underwater obstacles into Charleston harbor. After the New Year began, Dahlgren got permission from Welles to travel to Washington to discuss future operations. Leaving Rowan temporarily in charge of the squadron, he steamed north in the side-wheeler USS *Harvest Moon* and reached the capital on 2 March 1864. The voyage did him good; Welles for one noted that he appeared surprisingly hale. As soon as he arrived, Dahlgren made the rounds to renew old friendships and garner future support. One of his stops was of course the White House for an interview with the president. There he complained about the War Department and the constant, critical, and unfair newspaper attacks to which he was sub-jected. Lincoln, who knew something about hostile press coverage, heard him out and responded, "Well, you never heard *me* complain."[68] Welles and Fox were equally friendly, but not especially helpful. Dahlgren asked for the new ironclads just now coming into service, but Welles and Fox had decided instead to send them to the Gulf to help Farragut storm Mobile Bay. Without these reinforcements, Dahlgren did not believe he could take Charleston. The best he could do, therefore, was to continue to maintain the blockade and tie down rebel troops. Although he had told the president that these things were important accomplishments, they guaranteed little by way of glory. After he digested this bad news, Dahlgren asked Welles to relieve him of his command. Welles was aston-ished. He stated that he had been happy with Dahlgren's efforts and pre-ferred him to remain with the South Atlantic Squadron. On his way out of Welles' office, Dahlgren reiterated his request to Fox, whose response echoed Welles'. Moreover, Fox added, the newspapers would think that the department had forced him out. Dahlgren curtly and dishonestly replied that he did not care what the press thought. Welles and Fox, how-ever, did. If Dahlgren quit without conquering Charleston, it would give Du Pont and his allies additional ammunition in their war with the Navy Department by proving that seizing the South Carolina city was impossi-ble. Now that Welles and Fox had placed Charleston on the back burner, they had nothing to lose in keeping Dahlgren there, though of course they did not put it this way to Dahlgren. Dahlgren was disappointed with all this bad news, and in the following weeks he received little indication that his situation would change anytime soon. On 1 March Lincoln appointed newly promoted Lt. Gen. Ulysses Grant general in chief of the Union Army, and Grant got to work devising a grand strategic plan to win the war that did not include immediately assaulting Charleston. While Grant had no authority over the Navy, Welles wanted to be cooperative, so he was willing to provide ironclads to assist Grant's proposed offensive up the James River. This decision pushed Charleston even lower on the Navy's priority list. On 28 April, as he was preparing to return to his squadron, Dahlgren informed Welles that he could not accomplish much without additional ironclads that were obviously not forthcoming. Welles, though, urged Dahlgren to do the best he could with the ironclads at hand.[69]

In addition to Dahlgren's professional woes, he also faced a devastating personal loss that spring. After his son, Ulric, had recovered from his injury during the Gettysburg campaign, he volunteered to play an important role in Brig. Gen. Judson Kilpatrick's cavalry raid on Richmond. The raid went badly, and Ulric was killed in an ambush on 2 March 1864. Although initial reports indicated that Ulric had survived, on 8 March Phillips Lee sent definitive word that the boy was in fact dead. Over the course of the next few weeks, Dahlgren traveled down to Hampton Roads twice in an unsuccessful effort to secure his son's remains. Nothing could relieve the anguish he felt at Ulric's death. Indeed, Welles noted that Dahlgren was inconsolable during an 8 April meeting. The navy secretary had also lost a child since the war began, and he urged Dahlgren to use work as a curative for his grief. Perhaps to help him take the first steps, on 23 April Welles ordered Dahlgren to return to his squadron.[70]

By the time Dahlgren returned to Port Royal on 2 May, he had recovered some of his former élan. After all, he still had at his disposal a sizable force of ironclads with battle-tested crews and veteran skippers. When he arrived, however, he learned that Gillmore had gone north with 10,000 troops to participate in Maj. Gen. Benjamin Butler's Grant-designed campaign up the James River. Dahlgren was happy to see Gillmore go, but his departure left barely more than 14,000 bluecoats in the Charleston region, hardly enough to conduct offensive operations. Despite this unwelcome news, on 10 and 12 May Dahlgren convened his ironclad captains to discuss a naval-only assault on Fort Sumter. Unfortunately, no one had a clear understanding as to the objective of such an attack, beyond occupying the rubble-strewn island in Charleston's harbor. After prolonged conversations, they voted seven to two against another attempt on the island, the majority feeling that they lacked sufficient strength to neutralize it and push into the harbor. The ballot results reinforced Dahlgren's conviction that he was now commander of a military backwater, with little opportunity for glory or recognition. On 14 May he renewed his application for relief, arguing that he wanted to take a more active role in the war than was possible leading the South Atlantic Squadron. For two months he received no answer from the Navy Department, but he took comfort in rumors that his frustrating days off the South Carolina coast were numbered.[71]

Or so he thought. When Welles and Fox received Dahlgren's application, they were at a loss as to what to do with him. Fox believed that Dahlgren was physically and mentally exhausted due to overwork, hostile newspaper attacks, and a lack of support from subordinates such as

Rowan. At the same time, though, Fox observed smugly that Dahlgren had discovered that running a squadron was a lot more difficult than overseeing a naval yard. He advised Welles against bringing Dahlgren back to Washington, no doubt because doing so would enable him to rekindle his counterproductive relationship with the president, and instead recommended shipping him off to the Mississippi River. Welles did not like that idea; he had months earlier concluded that Dahlgren, while physically brave, lacked the moral courage to accept responsibility for hard decisions. Giving him another squadron command would not change that. On the other hand, turning down Dahlgren's application might provoke him into appealing to the president. Welles procrastinated for two months before penning a reply on 15 July. In it he acknowledged Dahlgren's request, but refused to give him a direct answer. Instead, Welles noted that a major command shake-up was in the works, and he would keep Dahlgren's wishes in mind when he made the necessary changes. As things turned out, although there was in fact a shuffling of positions that autumn, Dahlgren was not a part of those changes. Welles ultimately opted to retain Dahlgren where he was, but without explicitly telling him of his decision. By stringing him along with false hope, Welles kept Dahlgren in a gilded cage along the backwater Carolina coast where he would cause the Navy Department little trouble.[72]

Dahlgren remained South Atlantic Squadron commander for the rest of the war. As such, he had plenty to do even without the added responsibility of storming Charleston. Dahlgren kept busy maintaining the blockade, watching rebel warships in Charleston harbor, and supporting various minor army operations along the South Carolina, Georgia, and Florida coasts. It was important work, but it lacked the glory that had motivated Dahlgren to secure the squadron in the first place. Despite considerable evidence that the Navy Department was no longer interested in conquering Charleston, Dahlgren continued to search for ways to force his way into the harbor. He still possessed his ironclads, and his cordial relationship with Gillmore's replacement, Maj. Gen. John Foster, made army cooperation a possibility. Moreover, in mid-January 1865 Welles assured him that he would receive additional ironclads after Wilmington was in Union hands. In response to these positive harbingers, on 15 January 1865 Dahlgren held a meeting with his ironclad skippers on board *Harvest Moon* to discuss another assault on Charleston harbor. They saw little profit in doing so, and although Dahlgren was disgusted with their lack of enthusiasm, he as usual abided by their conservative recommendation. To make things worse, in early February Gillmore replaced an ailing Foster as local army commander, which eliminated any chance of interservice cooperation in the region.[73]

In the meantime the war passed Dahlgren by. In September 1864 Maj. Gen. William Sherman's armies seized Atlanta, Georgia, and then began their March to the Sea that culminated in the occupation of Savannah on 21 December. Dahlgren hoped that Sherman would target Charleston next, but the Union general had no intention of doing that. Instead, Sherman opted to cut his way through the Carolina interior, confident that Charleston would fall as soon as he severed the city's railroad lines. Sure enough, the rebels evacuated Charleston on 17 February after Sherman plunged into the middle of South Carolina, enabling Union troops to occupy it the next day. When Dahlgren heard the good news, he crossed the bar in *Harvest Moon,* picked his way through the numerous underwater obstacles, docked at the city wharves, and walked through the nearly empty city with his staff. No doubt he was pleased with the city's fall, but also well aware that it did nothing to help his career.

Two weeks later, on the morning of 1 March, *Harvest Moon* weighed anchor at Winyah Bay en route to Charleston. Dahlgren had just finished dressing, and was pacing about his cabin waiting for breakfast, pausing every now and then to squint at the shoreline. Suddenly, just before 8:00 AM, he felt a heavy shock and heard a crashing noise. The bulkhead separating his cabin from the wardroom shattered and fell toward him, and all the loose articles in the room tumbled to the floor. Dahlgren initially thought that *Harvest Moon*'s boiler had burst or her magazine had exploded. He put on his pea jacket and cap and rushed outside, passing frightened and shouting men struggling to lower the boats into the water. As he ascended a ladder to the upper deck, he saw that a rebel mine had blown a giant hole in the ship's hull and main deck. *Harvest Moon* was obviously sinking, and within five minutes she settled on the harbor bottom, her upper decks still above the water. A tug hurried over to evacuate Dahlgren and his staff, and all but one of the crew survived.[74]

Harvest Moon's unceremonious sinking was an appropriate ending to Dahlgren's frustrating and unfulfilling Civil War career. In his avid quest for fame and glory, Dahlgren had lobbied hard to lead the assault on Charleston. He had served diligently, courageously, and dutifully, but had failed to make a name for himself. He returned to Washington on 17

June, and officially gave up his squadron the following month. Two months later, he married Madeleine Vinton, a wealthy forty-year-old widow. Unhappily Dahlgren's postwar years were as trying for him as his time outside Charleston. He directed most of his professional energy toward protecting his reputation and protesting slights real and imagined. Welles initially placed him on various boards, but in 1866 he assigned him to command the South Pacific Squadron. Dahlgren went reluctantly, in part because his new wife was still recovering from giving birth to twins. When he returned two years later, Welles put him back in charge of the Ordnance Bureau. Dahlgren was not especially happy there either, and devoted more of his tenure toward defending the guns bearing his name than in developing new weapons such as rifled cannon. He stepped down in 1869 to become commandant of the Washington Navy Yard, and died of heart failure a year later, on 12 July 1870.[75]

Dahlgren's Civil War tragedy was that he placed personal aggrandizement above winning the war. As an ordnance pioneer, he could have made enormous, albeit unheralded, contributions to the Union war effort by remaining with the Ordnance Bureau, but he was more interested in celebrity. He pulled every string he could to secure his promotion to rear admiral and his South Atlantic Squadron command. In doing so, however, he alienated brother officers whose assistance he needed and planted serious doubts in Welles' mind as to his fitness for his new post. Unlike Du Pont, Dahlgren had army support for his campaign against Charleston, but it was not enough. As Welles initially suspected, and eventually concluded, Dahlgren, although brave under rebel fire, lacked moral courage. Like Du Pont, Dahlgren was never willing to put his reputation on the line by launching an all-out attack on Charleston. Instead, he blamed his failure to do so on many different elements, including his unenthusiastic subordinates. While it is true that a full-fledged assault in Charleston's harbor may have proved unsuccessful, his limited attacks accomplished very little. Unlike bold officers such as Farragut and Porter, Dahlgren saw impediments as excuses for inaction, not as obstacles to overcome. Dahlgren's innate caution may have prevented the Union from suffering a major naval defeat at Charleston, but it also deprived him of his one opportunity for the fame he so ardently sought.

CHAPTER 5

The Peripheries

The East Gulf Coast Squadron

EY WEST, FLORIDA, lies at the westernmost end of a string of islands looping toward Florida's southeastern tip. Its remote location and garrison of federal troops at Fort Taylor meant that it remained in Union hands throughout the war. In fact it became a stopping point for all Union vessels heading to and from the Gulf of Mexico. Despite beautiful placid waters covering miles of coral reefs, nineteenth-century Key West was anything but glamorous. It was located on a narrow and sandy spit of land baked by an unforgiving sun that kept temperatures in the eighties on hot summer nights. With a population of about twenty-six hundred, the town contained wide and dusty streets lined with squalid shops and buildings shaded by coconut and tamarind trees. Bored sailors amused themselves by trying to ride the small donkeys that inhabited the island. For officers, on the other hand, social life centered on Russell House, a fine hotel with a cool and spacious piazza that contained most of the conventional amenities. Before the conflict, Key West's economy revolved around salvaging wrecked ships. Once hostilities began, though, catering to Union forces became the town's economic staple. Its secessionist population declined as time went on, leaving behind an increasing number of Union-loyal black residents.[1]

Key West served not only as a way station for Union vessels, but also as headquarters for the East Gulf Blockading Squadron. On 20 January

1862, Welles divided the Gulf Coast Squadron in two to make way for Farragut's New Orleans expedition. The new East Gulf Squadron's territory extended from St. Andrew's Bay, about a hundred miles east of Pensacola, to halfway up Florida's Atlantic shoreline at Cape Canaveral, as well as Cuba and the Bahamas. Although the squadron was responsible for some large harbors—St. Marks, Tampa Bay, Charlotte Harbor, and Cedar Key, for example—the lack of railroads to the Confederate heartland restricted their usefulness to blockade-runners. As a result, the Navy's mission in these unpopulated and secluded waters was usually limited to hunting for small sailing ships and, as the war progressed, destroying the saltworks that the Confederacy's commissary relied on to preserve its meat products. Considering these comparatively minor objectives, it was hardly surprising that the Navy Department treated the East Gulf Squadron as a neglected stepchild.

Except for maintaining the blockade and avoiding international incidents, Welles had few plans for or expectations of the East Gulf Squadron when he first created it. His limited hopes stemmed at least in part from his lack of confidence in the squadron's first commander, Captain William McKean. Welles had chosen McKean the previous September to lead the Gulf Squadron, but had been disappointed with his subsequent lackluster performance. Rather than rile the officer corps by relieving him, Welles had shunted McKean off to the new East Gulf Squadron to make way for David Farragut, for whom the navy secretary had high aspirations. By the time McKean assumed control over the new squadron on 10 March, Welles had concluded that he was a mediocre officer burdened with poor health. Unfortunately, McKean did little to change Welles' low opinion of him during his short tenure with the East Gulf Squadron. In his defense, he labored under some of the same shortcomings that had hobbled him when he was responsible for the entire Gulf Coast. He lacked sufficient vessels to cover the vast Florida shoreline, especially the shallow-draft steamers that would enable him to closely guard some of the harbors. Worse yet, he had to send five of his precious steamships home because they were in such disrepair. As late as mid-May 1862, his squadron contained only nineteen warships, the majority of which were slow sailing vessels. Finally, David Porter, leader of the mortar flotilla during the New Orleans campaign, indulged his talent for discord by denouncing McKean to Fox. Porter believed that McKean was stupid and incompetent because he was not as ready as Porter felt he should have been when Farragut's

New Orleans expedition appeared in the Gulf. McKean was not, however, completely bereft of ideas. He believed that strong Union sentiment existed in western Florida, and he wanted to protect and promote such attitudes by occupying undefended Tampa, Cedar Keys, and Apalachicola, but the Army lacked the manpower to garrison these towns. As a result, the best he could do was to try to uphold his leaky blockade with the few warships at his disposal. His greatest success occurred in late May, when two of his vessels seized the rebel steamer *Swan* out of Mobile loaded with nine hundred bales of cotton and two hundred barrels of resin. It was a nice little victory that earned him kudos from Fox.[2]

McKean was a product of the pre–Civil War Navy, and was therefore very susceptible to slights both real and imagined. Not surprisingly, Welles' decision to shift him to the East Gulf Squadron displeased him because he recognized it for the demotion it was. All the important rebel ports in the Gulf—Pensacola, Mobile, and New Orleans—were now in Farragut's jurisdiction, giving McKean little opportunity to distinguish himself. Moreover, his health, never good, was failing under the strain of his extended sea service. After he informed Welles of these concerns, on 5 April the navy secretary gave him permission to surrender his post and return to Boston if he wanted. By the time McKean received Welles' instructions, however, his patriotism and vigor had reasserted themselves. He decided to set aside his personal sensitivities for the Union's sake and stick it out in the eastern Gulf for the time being. Within a month, though, his health finally and irrevocably gave way, and on 4 June he stepped down as East Gulf Squadron commander and returned to Boston eight days later. McKean had been in the Gulf since the war's start, so he was shocked by the changes that the country had undergone since the conflict began, especially the huge amount of money the Navy Department was spending on the war. He never held another important command, and died in Binghamton, New York, on 22 April 1865.[3]

Welles was neither surprised by nor unhappy with McKean's resignation. In fact, he had made plans for this contingency the previous February by directing Charles Davis to report to Washington from Du Pont's squadron. Davis was one of the stars of Du Pont's Port Royal expedition, giving him the kind of experience that appealed to Welles and made him a logical candidate to supplant McKean when it became necessary. By the time McKean finally submitted his request for relief, though, Welles had already sent Davis out west to fill in for Andrew Foote as the head of the

Western Flotilla. In his hasty search for a new replacement, Welles turned to Capt. James Lardner, current commander of *Susquehanna*. Lardner had fought valiantly at Port Royal by remaining with Du Pont's *Wabash* after other vessels had deviated from the attack plan. Du Pont appreciated Lardner's performance enough to praise him in his official report, and to complain later when the Navy Department transferred him and *Susquehanna* to the North Atlantic Squadron. There Lardner earned Goldsborough's respect for his role in keeping an eye on *Merrimack,* though the feeling was not mutual. As an aggressive officer who had served successfully in combat under two very different men, he was bound to attract Welles' attention. Moreover, he was readily available. On 19 May Welles ordered Lardner detached from *Susquehanna* and sent to Key West on the sloop USS *San Jacinto* as McKean's successor. Lardner was prompt and reached Key West on 4 June.[4]

James Lardner was born in Philadelphia in 1802. He entered the Navy as a midshipman at age eighteen, and subsequently served in many of the Navy's overseas stations, including the Pacific, European waters, off the African coast, and in the Mediterranean. The highlight of his early career was his participation in USS *Vincennes'* around-the-world voyage in the early 1830s. He found time to marry in 1832, but his wife died fourteen years later, leaving behind three children. Lardner remarried in 1853 and sired two more offspring. Unlike so many of his colleagues, he saw no combat in the Mexican War, but instead spent the conflict captaining a receiving ship in Philadelphia. By the time the Civil War began, he still possessed a tall and domineering presence, only he now accentuated it with a close-cropped head of white hair and a bristling white mustache. Welles had not met Lardner when he appointed him East Gulf Squadron commander, but he later described him as prudent, discreet, and overcautious. Away from authority, however, Lardner was anything but that. He had an eccentric streak that included collecting seashells, and he disliked the petty annoyances of naval life. Although he was normally courteous and kind, he also possessed a terrible temper. A strict disciplinarian, he hated anything smacking of insubordination or disrespect. He could curse like the saltiest of sailors, which won his men's respect and affection even when he directed his tirades at them. One day in the Caribbean, for example, his steward accidentally hit him in the back of the head while trying to swat a fly. Lardner interpreted this as the beginnings of mutiny, so he grabbed a carving knife and chased the terrified steward around the deck.[5]

Welles instructed Lardner to do little more than maintain the blockade and respect the territorial integrity of Spanish-controlled Cuba, but even these limited objectives taxed the East Gulf Squadron's meager resources. During his tenure Lardner usually had only about twenty warships under his command, less than half of which were steamers capable of catching most blockade-runners. Lardner, however, did not throw his hands up in despair. In addition to enforcing the blockade to the best of his ability, he also conducted the first naval raids along the Florida coast. In particular, he targeted the rebel saltworks at St. Andrew's Bay and St. Joseph's Bay in September 1862. Unlike McKean, Lardner did not even consider the possibility of cooperation with the Army, probably because he knew it would not be forthcoming. No sooner had Lardner initiated these early efforts to bring the war to Florida, though, than yellow fever broke out on *San Jacinto* and at Key West, requiring him to quarantine crews and curtail operations. Unfortunately, the Navy Department provided little assistance. Fox explained that the yellow fever, the current poor condition of so many of the Navy's warships, and the crisis atmosphere in Washington created by the Confederate invasion of Maryland made it impossible to provide Lardner with the reinforcements he wanted. The best he could do was to remove the Bahamas from Lardner's list of geographic responsibilities. By way of moral support, on 15 September Welles promoted Lardner to acting rear admiral. Lardner scarcely had time to savor his new rank before he succumbed to an attack of yellow fever that put him in bed for more than two weeks. He became so ill that he believed he had no choice but to surrender his post, and on 4 November Welles gave him permission to do so. By the time he learned of his relief, though, cooler weather had dampened mosquito activity and eradicated the seasonal yellow fever problem. Indeed, Lardner's case was just about the last one in the squadron. Lardner soon felt so much better that he regretted his decision to step down. Still, orders were orders, so in mid-December he boarded the supply steamer USS *Blackstone* and headed to New York City.[6]

After a year and a half of war, the Union Navy now had a cadre of high-ranking, battle-tested officers from which Welles could choose Lardner's replacement. One of the most obvious candidates was Stephen Rowan, who had made a name for himself by destroying a rebel flotilla in the North Carolina sounds the previous February. Rowan was currently awaiting orders in New York City, and as soon as he learned of Lardner's illness he immediately wrote to Fox to ask for the open position. Unfortunately for Rowan, the day before he penned his letter to the assistant secretary, Welles had assigned the East Gulf Squadron to another underutilized officer with an impressive combat record: Commo. Theodorus Bailey.[7]

The Navy Department had sent Bailey to the Gulf as skipper of the screw frigate *Colorado* right after the war began. There he served faithfully and well under Mervine, McKean, and finally his old friend Farragut, even though he suffered intensely from hydrocele, the accumulation of fluid in the scrotum. Despite his illness, Bailey fought valiantly during Farragut's New Orleans campaign. He led one of the divisions of warships past Fort Jackson and Fort St. Philip, and later defied an angry mob by walking unarmed through the streets of New Orleans to demand the city's surrender. Although Farragut strongly commended him for his actions, he also sent Bailey home because his health had broken down. On 30 May Welles appointed Bailey commandant of the naval yard in Sacketts Bay, New York. Within a few months, however, Bailey was ready to return to more active duty, though he worried that the Navy Department might send him to another quiescent post far from the war. On 15 October Bailey asked Fox for command of a "fighting squadron," and suggested that he could lead an ironclad assault on Charleston. Fox was no more inclined to replace Du Pont with Bailey than he was with Dahlgren, but he appreciated Bailey's attitude, as did Welles. In fact, Bailey had most of the characteristics Welles and Fox sought in a squadron commander. Bailey was experienced, courageous, spirited, and ready and willing to fight. Moreover, Farragut vouched for him, and Welles placed great stock in his opinions. On 4 November Welles promoted Bailey to acting rear admiral and ordered him to report to Key West as Lardner's successor as head of the East Gulf Squadron.[8]

The New York–born Bailey was fifty-seven years old when he assumed command of the East Gulf Squadron. As a child, the American victory over the British at the Battle of Plattsburgh on nearby Lake Champlain inspired him to enter the Navy as a midshipman in 1818. He thereafter served off the African coast, in the Pacific, and with Lardner on *Vincennes'* around-the-world cruise. During the Mexican War, he took the storeship USS *Lexington* and a detachment of soldiers that included William Sherman and Henry Halleck on a trip around the South American tip to California. In the mid-1850s he commanded USS *St. Mary's* in the South Pacific, and on his way home he helped protect American citizens during unrest in the Colombian province of Panama. Bailey was a solidly built man with bushy eyebrows and mustache, a heavy chin, and curly hair. Like his friend Farragut, he was full of courage, integrity, amiability, aggressiveness, and heart. Unlike Farragut, though, he kept careful records, tried to invest his prize money wisely, and was not above pulling strings. His son worked at a New York counting-house, but felt inferior because of his diminutive size. Bailey encouraged him by telling him that the most valuable articles come in small packages, and that "merit will always meet reward." Finally, although Bailey sought advancement as much as anyone, he lacked the hypersensitivity and touchiness that infected so many naval officers.[9]

Most high-ranking Union Navy officers had little formal education, so it was not surprising that they possessed only the most rudimentary knowledge of grand strategy. Many retained a superficial understanding of the blockade's purpose, but not of its larger connection to the war of attrition that the Union increasingly resorted to as the conflict progressed. Bailey, however, was an exception to this rule. Upon assuming his new command on 9 December 1862, he provided as good an explanation of the blockade's objective as any Union officer during the war: "The outward pressure of our Navy, in barring the enemy's ports, crippling their power, and exhausting the resources of the States in rebellion; in depriving them of a market for their peculiar productions, and of the facilities for importing many vital requisites for the use of their Army and peoples, is slowly, surely, and unostentatiously reducing the rebellion to such straits as must result in their unconditional submission."[10]

Bailey never had more than forty warships in his squadron, many of which were slow sailing vessels, but he was determined to enforce the blockade as rigorously as possible. Like his friend Farragut, before he assumed command of his squadron, Bailey took careful inventory of the vessels at his disposal and the available intelligence on Confederate harbors in his jurisdiction. While he was successful in seizing a good number of sailing Confederate blockade-runners, most of the few steamers the rebels employed in his theater eluded capture because of their fast speed. Even so, his warships ultimately snagged around 150 blockade-runners during his twenty-one-month tenure. Moreover, Bailey recognized that the Navy could contribute more to the destruction of the Confederacy's resources than merely maintaining the blockade. He cooperated with Brig. Gen. Daniel Woodbury, army commander of the District of Key West and Tortugas, in destroying rebel resources up and down the western side of the Florida Peninsula. Bailey's warships assisted bluecoats in seizing Confederate saltworks and cattle, rescuing runaway slaves, and recruiting a small unit of local unionists. In the big scheme of things, these activities contributed little to Union victory, but in undertaking them, Bailey waged war in an imaginative way unknown to many naval officers.[11]

Bailey's pride in his squadron, which he called his "cotton plantation," was offset by his frustration with the constraints he labored under. In May 1864 Fox rejected his plea for more warships. Fox believed that 90 percent of all blockade-running now went through Wilmington, so he wanted to focus the bulk of the Navy's resources there, not in the remote waters off the Florida coast. In addition to reinforcements, Bailey also sought promotion to rear admiral. Like many of his colleagues, he resented the advancement of well-connected officers such as Davis and Dahlgren. He lamented that he had neither the political strings to pull nor the opportunity to distinguish himself in battle that could propel him into one of the few available active-list rear admiral slots. When one of his friends, who apparently did not realize that Bailey was not yet of retirement age, suggested to Fox that the Navy Department make Bailey a rear admiral on the retired list, the assistant secretary replied that doing so would anger equally deserving officers such as Cornelius Stribling, Frederick Engle, and Charles Bell who also had their eyes on that coveted rank. Besides, continued Fox, the Navy Department had already rewarded Bailey by giving him a squadron and the opportunity to partake in prize money. Despite these aggravations, Bailey still had much to be thankful for. Welles and Fox were both pleased with his efforts, and in the summer of 1864 Fox told one of Bailey's friends that the Navy Department had no intention of relieving him unless he wanted to come home. In addition, by the end of May 1864 Bailey's prize money amounted to $8,255, enough for him and his wife to purchase a little retirement house in Oyster Bay, New York. Finally, Bailey could enjoy the parties in his Key West offices, a welcome if raucous visit from his adolescent daughters, the support of his subordinates, the warm climate, and the knowledge that he was in Welles' good graces.[12]

Yellow fever is a mosquito-borne infectious viral disease characterized by sudden high fever, prostration, ennui, and jaundice. In advanced cases patients vomit black grainy material and bleed through every orifice in the body. It had long been the scourge of soldiers and sailors stationed in southern climates. It had stricken Key West in 1862 and forced Lardner to surrender the East Gulf Squadron to Bailey, but was mercifully absent the following year. In March 1864 workmen on the island began to fall ill. Although some suspected "yellow jack," local doctors said it was something else. Whatever the disease was, it abated after a cool wind swept through the Florida Keys, and Bailey hoped that was the end of that. In June, however, yellow fever descended on the island in earnest. Bailey hurriedly dispatched all his vessels at Key West up the coast before their crews sickened, but he remained behind to keep up the flow of paperwork upon which the squadron depended. He assured concerned relatives that he was acclimated to the disease because he had contracted it back in 1822. He was wrong. In July he became ill, and for several days his survival was in doubt. He slowly started to recover, but all that time in bed sapped his spirit. Reports indicated that the island would not be healthy again until November at the earliest, five long and dangerous months away. Bailey had planned to participate in his friend Farragut's assault on Mobile Bay, but that was now out of the question, and with it, in all probability, his last chance to distinguish himself in battle. On 19 July Welles gave him permission to return north to recuperate his health, and Bailey decided to accept the offer. He turned the squadron over temporarily to Capt. Theodore Greene on 7 August and steamed to New York City on *San Jacinto*. He expected to return to the Gulf in October or November, so he was shocked when on 12 September Welles detached him from the East Gulf Squadron and made him commandant of Portsmouth Navy Yard in Kittery, Maine. Bailey hastened to the Navy Department to tell Welles that he preferred to resume his old command, but he pleaded his case poorly. Welles thought he was doing Bailey a favor by giving him such a cushy and plum assignment, and mistakenly and ungraciously attributed Bailey's appeal to his greedy desire for more prize money. He refused to change his orders, ending Bailey's active role in the war.[13]

Like Lardner before him, Bailey regretted his decision to leave the East Gulf Squadron. When he learned that the Navy was preparing for a major operation against Fort Fisher, outside Wilmington, he wrote David Porter, now commanding the North Atlantic Squadron, and asked to participate in the assault. Porter responded that he would be delighted to have him so he could earn one of the rear admiral slots, and added that he would do all he could to secure Bailey's services. Nothing ever came of it though. There is no record that Porter followed through and asked the Navy Department for Bailey, but if he did it is also possible that Welles refused the request. Bailey remained commandant of Portsmouth Navy Yard until

1867, after which he served on various boards until his retirement in 1873. He died four years later in Washington.[14]

Eleven days after he reassigned Bailey, Welles appointed Commo. Cornelius Stribling as the new commander of the East Gulf Squadron, with a temporary rank of rear admiral. It was an odd selection for numerous reasons. Stribling was a South Carolinian born back in 1796. He entered the Navy in 1812, and his career highlights included stints as superintendent of the Naval Academy, Efficiency Board member, and chief of the East India Squadron. Like William McKean, he was a classic old-navy officer. One observer described him as "a fine seaman of the old school, of rigid Presbyterian stock, stern, grim, and precise, with curt manners, sharp and incisive voice that seemed to know no softening, and whose methods of duty and conception of discipline smacked of the 'true blue' ideal of the Covenanters of the old in their enforcement of obedience and conservation of morals."[15] On the other hand, although he once signed a temperance pledge, a short time later one sailor saw him drunk with several other officers. When the war broke out, he was leading the East India Squadron from Hong Kong on board *Hartford*. As a South Carolinian, he knew his loyalties were suspect, especially after his son resigned his commission to serve in the rebel navy. He wrote to Welles to assure him of his devotion to the Union, but the navy secretary ordered him home anyway. Welles eventually got over his doubts about those Southern-born officers such as Farragut and Drayton who stayed in the service, and in November 1862 he appointed Stribling commandant of the Philadelphia Navy Yard, from which he ascended to head of the East Gulf Squadron. He ran the yard diligently enough, but that hardly qualified him for squadron command. He had seen no combat in the war, was at sixty-eight comparatively old, and suffered from rheumatism. In addition, he had lamented Du Pont's removal and treatment at the Navy Department's hands, and had criticized Dahlgren's promotion and performance outside of Charleston, so he was hardly a Welles loyalist. There were certainly other younger and more dynamic high-ranking officers—John Rodgers and Percival Drayton, for example—who deserved the post more.[16]

Considering Stribling's numerous shortcomings, it is hard to understand exactly why Welles chose him to lead the East Gulf Squadron. Welles made no mention of his decision in his diary, but there is evidence in his personal papers that political considerations may have motivated him. The previous year, a local Republican official had complained to the president that Stribling was hostile to his efforts to organize and rally

Philadelphia Navy Yard workers to vote the right way, and Lincoln had passed these concerns on to Welles. With the crucial presidential election fast approaching, Welles may have concluded that removing Stribling was the politically expedient thing to do. Bailey's discomfiture, moreover, enabled Welles to transfer him to a new post which, while strategically remote, was every bit as prestigious as running a naval yard. Finally, although there may have been other officers more worthy of the position than Stribling, he was the most readily available.[17]

Stribling arrived at Key West on 12 October and superseded Theodore Greene, who was just getting over his own bout of yellow fever. Welles' marching orders to Stribling stressed the need to enforce the blockade and avoid international incidents, but said nothing about cooperating with the Army or destroying the Confederate infrastructure. Like Bailey before him, Stribling soon concluded that he needed more warships to carry out his mission. Welles, however, explained that preparations for a planned assault on Wilmington were sucking up an inordinate number of vessels, but he promised to give Stribling's request due consideration. In the end, though, Stribling had to make do with the thirty to forty vessels currently available. Although Welles' directive said nothing about it, Stribling opted to retain Bailey's strategy of raiding rebel saltworks, but he tempered such activities with his concerns that his officers were taking too many risks in the hopes of getting their names in the newspapers or official reports. He also continued cooperation with the Army, especially in its efforts to overrun the Florida Panhandle toward the end of the war. Finally, he reorganized his squadron into two subdivisions to give local officers more power to respond quickly to problems. Stribling handled his responsibilities with aplomb and professionalism, despite a terrible case of neuralgia that afflicted him throughout the winter. On 30 May 1865, as hostilities ceased, he asked Welles to remove him from his command and assign him to direct the Naval Asylum, citing health reasons. Welles agreed to let him come home and told Stribling to begin the process of dissolving his squadron. On 9 July Stribling left Key West on *Powhatan*, and when he reached Boston he received a letter from Welles officially detaching him from his post and congratulating him on a job well done. Stribling served on the Lighthouse Board for five years until he retired, and died in Martinsburg, West Virginia, in 1880.[18]

Because the East Gulf Squadron contained no major ports with railroad lines to the Confederate interior, it was the least important of the four blockading squadrons. It is therefore not surprising that the Navy Department invested so few resources in and expected so little return from the squadron. Nothing attests to this attitude more than Welles' willingness to use the squadron as a dumping ground for the lackluster McKean in January 1862. Indeed, Welles barely mentioned the squadron in his diary, and he devoted no pages whatsoever in it to his selection of its commanders. Despite this inattention, the East Gulf Squadron was not irrelevant to the Union war effort. The Navy had opportunities there to enforce the blockade, contact local unionists, and destroy rebel infrastructure. Of the East Gulf Squadron leaders, Lardner, Stribling, and especially Bailey recognized and exploited these opportunities with their limited resources. By doing so, they contributed marginally to the Confederacy's defeat, especially through the destruction of local saltworks that the rebels depended upon to preserve their meat products. Investing more resources in the squadron, though, probably would not have brought the Union victory any sooner. In fact, one of Welles' greatest personnel mistakes was assigning the capable and insightful Bailey to such a strategically marginal theater. On the whole the East Gulf Squadron got about as much attention as it deserved in light of the Navy's limited available resources.

The West India Squadron

Like John Rodgers and David Porter, Cdr. George Preble came from a distinguished American naval family. His uncle, Commo. Edward Preble, was a hero of the war with Tripoli and an inspiration for an entire generation of American naval officers. George Preble entered the Navy in 1835 as a midshipman, and saw extensive service in the Seminole War, in the Mediterranean, with Matthew Perry in Japan, and along the Chinese coast. In early 1862 the Navy Department put him in charge of the gunboat USS *Katahdin* and sent him to the Gulf of Mexico with Farragut. He participated in the run past Fort Jackson and Fort St. Philip, and then in the frustratingly drawn-out campaign against Vicksburg. He succeeded Phillips Lee as skipper of *Oneida,* and in early September 1862 was temporarily in charge of the Mobile blockade. On 4 September, the new British-constructed Confederate raider *Oreto,* also called *Florida,* attempted to run the blockade into Mobile in broad daylight flying British colors. Preble had plenty of time to react, but he hesitated out of fear of precipitating an international incident. After *Florida* ignored three warning shots, however, Preble opened fire with all his guns. Although he hit the rebel warship, *Florida*'s superior speed enabled her to slip into Mobile more or less intact.

Needless to day, Farragut was mortified by *Florida*'s successful run. His chagrin, however, paled in comparison to Welles' fury. The latter half of 1862 was a difficult time for the navy secretary. Critics in and out of Congress attacked him for the blockade's leakiness, and there were even untrue reports that blockade-runners were bribing naval officers to permit their cargoes to enter and leave Confederate ports unmolested. A series of Union military reverses—the failed naval campaign against Vicksburg and army defeats in Virginia at the Seven Days' Battles and Second Bull Run—spawned allegations of disloyalty among the officer corps. Finally, Welles' three-year-old son, Herbert, succumbed to diphtheria that fall, leaving Welles and his wife disconsolate. All this accumulated stress no doubt contributed to Welles' anger with Preble. As far as he was concerned, Preble's lethargy led him to commit an error in judgment that damaged the Union war effort. In fact, Welles concluded that Preble's failure was symptomatic of an officer corps that lacked the proper ruthlessness and energy to enforce the rigorous blockade that victory required. To Welles, too many officers were more concerned with diplomatic niceties than in taking whatever actions were necessary to win the war. Welles felt that the State Department was already treating the British in particular with kid gloves, and Preble's excuse for his hesitation outside Mobile demonstrated that this counterproductive attitude had seeped into the Navy. In an effort to purge the Navy of this mental complacency, Welles determined to make an example of Preble. After consulting the president, on 20 September he ordered Preble dismissed from the service.

Preble did not meekly accept his fate, but instead waged a vigorous campaign for reinstatement. Fellow officers such as Farragut were shocked by Welles' draconian decision, and lobbied for a court-martial or a court of inquiry that would enable Preble to clear his name. One of Preble's friends, Senator William Fessenden of Maine, appealed on his behalf to both Lincoln and Welles. Although Lincoln sighed to Welles that he could not dismiss or punish anyone in the military without bringing out a bunch of sympathizers to agitate on his behalf, he promised to support whatever decision Welles made. After several months Welles figured that he had made his point, and he grudgingly agreed to let the president and the Senate restore Preble to duty. Resorting to poetic justice, Welles thereupon placed Preble in command of the old sailing sloop USS *St. Louis* and exiled him to the high seas to search for Confederate raiders such as *Florida*, which had recently slipped out of Mobile to begin her career seizing Union merchant ships.[19]

Florida's run into Mobile underscored Welles' frustration with rebel raiders. Confederate thinking called for using government-owned cruisers to attack Union merchant shipping. It was a familiar strategy that the United States had resorted to during the Revolutionary War and the War of 1812. On the surface, it was remarkably successful. During the Civil War, the score or so raiders the Confederacy commissioned seized around three hundred merchant ships totaling 120,000 tons, or 5 percent of the entire Union merchant marine. By the end of the conflict, half of the Union merchant ships had transferred their registry to foreign flags out of fear of these rebel cruisers, leading to a decline from which the American merchant marine never recovered. Substantial though these losses were, they did not have a major impact on the overall Union war effort; the Union was still able to import and export everything it wanted or needed. Confederate cruisers did, however, contribute somewhat to the weakening of the blockade. Rebel successes on the high seas embarrassed the Navy and hurt Welles politically. For example, after the Confederate raider *Nashville* slipped into Morehead City, North Carolina, in April 1862, outraged insurers and merchant ship owners circulated petitions in Boston, New York City, and Philadelphia calling for Welles' removal. Welles believed that enforcing the blockade was the Navy's number one priority, but this did not prevent him from diverting resources to pursue Confederate raiders. During the war, he ordered fifty-eight warships to search for rebel cruisers at one time or another, including twenty-nine deployed exclusively for this task. No doubt some of these vessels were unsuited for blockade duty anyway, but crewing them siphoned desperately needed manpower from the blockading squadrons. Even as *Florida* was preparing for her run into Mobile, the Navy Department ratcheted up its war against the cruisers. To capture *Florida* and another Confederate raider, *Alabama*, rumored to be in the West Indies, Welles and Fox decided to detach a half dozen of the speediest available Union warships and organize them into a "flying squadron" that would steam into the Caribbean and neutralize these rebel cruisers.[20]

To command this new West India Squadron, Welles selected Commo. Charles Wilkes, one of the Union Navy's most controversial and celebrated officers. On paper, there was much to admire about Wilkes. He was born in New York City in 1798, the grandnephew of radical English politician John Wilkes. His prosperous businessman father made sure he got an education, and he entered the Navy as a midshipman in 1818 after a short stint in the merchant marine. He saw duty afloat in the Pacific and the Mediterranean, but spent much of his career engaged in surveying and scientific work. Most prominently, from 1838 to 1842 he led the Navy's six-ship Exploring Expedition to Antarctica, the remote Pacific, and the Oregon and California shores. During the voyage, he surveyed sixteen hundred miles of Antarctic coast and 280 Pacific islands. He got the job because no one else wanted it, but it made him famous. Writing up and publishing his data consumed most of his time and energy until the Civil War. When the war began, the Navy Department first dispatched him to the Norfolk Navy Yard, where he witnessed its destruction, and then to the African coast to take over *San Jacinto* from Capt. Thomas Dornin, whose loyalty Welles suspected because of his Southern connections. In December 1861 *San Jacinto* seized as contraband two Confederate diplomats, James Mason and John Slidell, off the British mail steamer *Trent* in the West Indies. Provoking John Bull made Wilkes a hero in the Union, but the outraged British government saw the incident as a violation of its neutrality and threatened war unless the Union released the diplomats. In the end, the Lincoln administration opted to free Mason and Slidell rather than risk hostilities with Great Britain. In July 1862 Welles appointed Wilkes head of the James River Flotilla. After the Army of the Potomac evacuated the Peninsula, Welles dissolved the James River Flotilla and briefly assigned Wilkes to the Potomac River Flotilla, from which he ascended to command the West India Squadron.

Wilkes was a lean, erect, and clean-shaven man. Despite his accomplishments, he was among the most unpopular officers in the Navy. He was quarrelsome and contentious, opinionated and critical, and carping and complaining. Many officers questioned his basic integrity. During his Exploring Expedition, he antagonized so many people that when he returned he was court-martialed, convicted, and publicly reprimanded for illegally punishing his men. Although he was quick to find fault with others, he had little introspection or self-awareness. Something of a martinet, he expected immediate obedience from his subordinates, but he also interpreted his orders from his superiors as suggestions, so his relationship with Navy Department authorities was always tenuous. This hypocrisy, combined with his greediness, unhealthy ambition, and unthinking willfulness, gained him few friends. Phillips Lee, Louis Goldsborough, Percival Drayton, Charles Davis, and John and Raymond Rodgers all disliked him. Even Frank Du Pont, who was more sympathetic toward him than most, referred to Wilkes as an Ishmaelite. While he gained considerable public acclaim for precipitating the *Trent* affair, most of his fellow naval officers regarded his actions as a cheap ploy to gain glory. He was not, however, without ability, and had over the years mellowed somewhat in his personal relations, so by the Civil War his petulant reputation probably exceeded reality. When Wilkes assumed command of the James River Flotilla, for instance, John Rodgers noted that he tried to be kind, but it just was not in him. Rodgers objected not so much to Wilkes' personality, but rather to his inefficiency and laziness. Rodgers believed that Wilkes spent more time with his wife at Fortress Monroe than fighting the rebels on the James.[21]

Welles shared many of Rodgers' concerns about and criticisms of Wilkes. Indeed, Welles had had his misgivings about Wilkes ever since the *Trent* affair. He believed that Wilkes had erred in removing Mason and Slidell and releasing the vessel. Instead, he thought that Wilkes should have seized the ship and brought it into port for a prize court to determine the fate of it and its human cargo. Although Welles had congratulated Wilkes for his actions in a guarded letter, he thereafter saw him as a loose and troublesome cannon. Nor had Welles been happy with Wilkes' performance on the James River, mostly because he disliked Wilkes' tendency to exceed his authority in matters big and small. Wilkes, for example, had tried to dismiss naval personnel from the service without the Navy Department's permission. Welles constantly worried that Wilkes would commit some rash act that would damage the Union cause.[22]

Considering his doubts about Wilkes' judgment, Welles' willingness to assign him to a post that required autonomy, discretion, and tact was hard to fathom. Welles believed that Wilkes had ability, however, and he initially got along with him better than he expected. Moreover, now that Welles had dissolved the James River Flotilla following McClellan's evacuation of the Peninsula, Wilkes was available for a new mission. But there was more to Welles' thinking than that. Welles made his decision only after meeting with the president and Seward, and then against his better judgment. Seward was so adamant that Wilkes get an important command somewhere that Welles suspected incorrectly that there was some sort of private agreement between the two men. Actually, Seward probably wanted to demonstrate to the British that the Union believed it had acted properly during the *Trent* affair. The international community might interpret sidelining Wilkes as an implied admission of guilt that would hurt Seward's efforts to conduct an effective foreign policy. Lincoln for some reason had a soft spot for Wilkes, so he was amenable to Seward's request. In the face of this pressure, Welles gave way and made the appointment. At Wilkes' insistence Welles also promoted him to acting rear admiral, figuring that the special circumstances—specifically, dealing with foreign officials at various ports of call—merited it. Welles was not especially happy about any of this, but he felt it would be worth all the aggravation if Wilkes could capture *Alabama*.[23]

Unhappily, the communication problems and misunderstandings between Welles and Wilkes that ultimately characterized and marred their relationship began even before the West India Squadron set sail. Welles was under heavy pressure from commercial interests to capture or destroy *Alabama,* so he wanted Wilkes to get under way as soon as possible, preferably by 11 September. When that target date came and went with Wilkes still at Hampton Roads, Welles grew increasingly frustrated and displeased. Although he knew as well as anyone that there was much to dislike and distrust about Wilkes, he had believed that Wilkes would at least be prompt in his duties. In fact, Welles concluded that Wilkes' dilatoriness was due to his vain desire to show off his new rank and command to his fellow officers at Hampton Roads. On 11 September Welles issued Wilkes peremptory orders to get going, but this merely provoked a letter from Wilkes explaining the reasons for his delay. Wilkes had small use for authority to begin with, and he had lost confidence in Welles during the *Trent* affair because the navy secretary had not supported him as enthusiastically as he expected. As far as Wilkes was concerned, Welles' ignorance made it impossible for him to understand his difficulties. His warships were broken down and in need of repair, and many of his officers were slow in returning from their leaves. Wilkes did not want to depart until he was good and ready, and he resented the pressure Welles exerted on him. Finally, on 21 September, Wilkes and his flagship, the sloop USS *Wachusett*, embarked from Hampton Roads.[24]

Wilkes deployed his half-dozen warships at various West Indies choke points in an effort to catch *Alabama* and any other rebel raiders in the area. It soon became apparent to him, though, that he lacked sufficient strength to carry out his orders; a handful of vessels was simply not enough to patrol effectively the vast region for which he was responsible. To make matters worse, the few warships he possessed had serious mechanical problems that required constant attention and repair. On 11 November, for instance, Wilkes noted that *Wachusett* had broken down twelve times since she left Washington. Welles had promised Wilkes more vessels when possible and if available, but they were not forthcoming as the weeks turned into months and success remained elusive. Wilkes became convinced that in denying him assistance and hamstringing his plans, the Navy Department was condemning him to failure and ignominy. Wilkes was a big believer in the squeaky-wheel approach to bureaucratic allocation, however, so asking for additional warships became a regular and increasingly strident feature in his ample correspondence with Welles. Welles responded on 15 December that while he would like to send Wilkes reinforcements, enforcing the blockade remained the Navy's top priority, and Lee, Du Pont, Bailey, and Farragut needed every vessel they could get to do so. This logic did not stop Wilkes' pleas for help, which convinced a suspicious Fox that he was creating a paper trail against the Navy Department for future use to justify any setbacks he might incur as squadron commander. Indeed, Wilkes later noted to Welles, "The failure to capture these vessels in the West Indies is to be attributed solely to the want of . . . force. The blame, therefore, does not rest with me, but elsewhere."[25]

If Wilkes was unable to attain additional warships for his squadron the proper way, he was fully prepared to resort to more underhanded means to secure them. On the night of 15–16 January 1863, *Florida* slipped out of Mobile through the seven blockading Union vessels Farragut had deployed offshore. Although the gunboat USS *R. R. Cuyler* gave chase, *Florida* got away and began her career seizing Union merchant ships. On 20 January, at Cabo San Antonio off Cuba's westernmost tip, *Cuyler* fell in with Wilkes and *Wachusett*. Upon learning of *Florida*'s escape, Wilkes confiscated *Cuyler* and dispatched her to scour Cuba's southern coast in search of the rebel raider. Wilkes then proceeded to Havana, where he found *Oneida*. After *Florida* broke out of Mobile, *Oneida* had steamed to Key West to sound the alarm, and from there Bailey sent her to Havana to look for the Confederate cruiser. On 23 January Wilkes appropriated her as well for his squadron. Despite this additional strength, Wilkes was unable to find either *Florida* or *Alabama*, which was also in the area. He finally released *Cuyler* in March, but held onto *Oneida* even after Farragut ordered her return. Farragut did not object to Wilkes borrowing his warships for the hot pursuit of *Florida,* but he wanted them back after the trail turned cold. He argued that not only had Wilkes weakened the Union blockade of Mobile, but he had also set a bad precedent. If squadron commanders could get away with poaching warships, Farragut's distant squadron would never receive any reinforcements. In a letter to the navy secretary, Farragut wryly noted that he was tempted to apply these same tactics to Wilkes personally if he entered the West Gulf Squadron's jurisdiction.[26]

Unlike Farragut, Welles found little humor in Wilkes' impressment of *Cuyler* and *Oneida.* He was even less amused with Wilkes' handling of the side-wheeler USS *Vanderbilt* a month later. When the Navy Department got information that *Alabama* was heading toward the South American coast, Welles on 27 January 1863 dispatched *Vanderbilt* there under Acting Lt. Charles Baldwin to intercept the rebel raider. On 26 February, *Vanderbilt* encountered Wilkes and *Wachusett* at Saint Thomas in the Danish Virgin Islands. After looking *Vanderbilt* over and reading Baldwin's orders, Wilkes opted to transfer his flag to the vessel and incorporate her into his squadron. As Wilkes later explained to Welles, *Vanderbilt* had been slated to eventually join the West India Squadron anyway, and *Wachusett* was in such poor state of repair as to render her inoperative. Moreover, Wilkes resented a warship operating within his jurisdiction independent of his control. Wilkes cruised around the West Indies in *Vanderbilt,* and then along the northern coast of South America, but without finding *Alabama.* He finally steamed to Key West and relinquished control of the vessel. He eventually claimed that he followed Baldwin's orders by making sure *Alabama* was not in the West Indies before going to South American waters, but neither Baldwin nor Welles saw it that way. To Welles, by hijacking *Vanderbilt* for his own selfish purposes, Wilkes had interfered with and bungled a well-conceived plan to stop *Alabama.*[27]

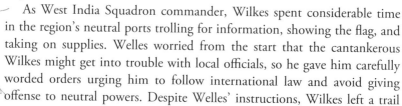

As West India Squadron commander, Wilkes spent considerable time in the region's neutral ports trolling for information, showing the flag, and taking on supplies. Welles worried from the start that the cantankerous Wilkes might get into trouble with local officials, so he gave him carefully worded orders urging him to follow international law and avoid giving offense to neutral powers. Despite Welles' instructions, Wilkes left a trail of international incidents as he crisscrossed the Caribbean in search of rebel raiders. The Danish, Mexican, Spanish, and especially British governments all lodged complaints to the State Department against Wilkes during his time in their ports or waters. The British minister to Washington, Lord Lyons, protested almost monthly to Seward about Wilkes' actions. According to Lyons, while in Bermuda Wilkes illegally blocked access to the harbor with his warships, stationed sentinels in British territory, violated rules governing the coaling of vessels, and treated the governor with disrespect. Lyons later registered additional charges against Wilkes. He stated that Wilkes refused to exchange proper courtesies with a British warship in the Bahamas, threatened to capture British mail packet ships despite the U.S. government's assurances that he would not do so, and wrongly seized a British steamship. Seward passed these allegations along to Welles, who dutifully asked Wilkes for his side of the story. Wilkes defended his activities with long and detailed explanations, and stated that he was doing his best to get along with local officials and avoid British ports. Upon weighing the evidence provided, Welles decided that most of the charges against Wilkes were baseless, although in all fairness the navy secretary was biased against the British. Welles felt that the Danes and the Mexicans might have minor grievances, but not the British. The reality was that Wilkes was outraged by the sympathy that local British officials showed toward rebels and blockade-runners, and was not shy in expressing it. The British disliked him anyway for his role in the *Trent* affair, read the worst into his behavior, and subjected him to numerous slights. Welles eventually concluded, "I had seen nothing in his conduct thus far, in his present command, towards the English deserving of censure, and that the irritation and prejudice against him were unworthy."[28]

Considering Welles' attitude on the subject, it is therefore surprising and ironic that Wilkes' downfall flowed directly from one of those international incidents Welles discounted. On the morning of 25 March 1863, *Wachusett*, now under the command of Lt. Cdr. Charles Fleming after Wilkes procured *Vanderbilt* for his flagship, captured the British steamer *Dolphin* in the waters between Saint Thomas and Puerto Rico. Because *Dolphin* had eluded pursuit and ignored three warning shots, Fleming was confident that she was a blockade-runner. The British, however, saw things differently. On 29 April Lord Lyons protested *Dolphin*'s seizure to Seward, and added details to the complaint a week later. According to Lyons, Fleming had not only wrongfully detained the British ship, but had also placed her chief engineer in irons and illegally used Saint Thomas as a base. In addition, Lyons provided Seward with a confidential dispatch that he had received from British foreign secretary John Russell expressing his concerns about the episode and its impact on already-tense Anglo-American relations. After absorbing Lyons' written and verbal statements, Seward promised to place the matter before the president.[29]

On the evening of 11 May, Welles received an unexpected knock at his door from Seward, who had come by to discuss Charles Wilkes. Seward explained to Welles that Wilkes had become a major source of friction in Anglo-American relations, and asked if Welles could do anything about him. Welles already knew something about high-level British anger toward Wilkes, having recently received word to that effect from his contacts across the Atlantic. Although he stated to Seward that in his opinion Wilkes was more sinned against than sinning in his dealings with the British, the navy secretary still had plenty of other grievances against his West India Squadron commander. Welles had by now grown tired of Wilkes' constant complaining and grumbling about his lack of warships. He was also angry about Wilkes' unauthorized impressment of *Cuyler, Oneida,* and especially *Vanderbilt.* Welles felt that *Vanderbilt* could have destroyed *Alabama* had Wilkes not interfered with Navy Department plans. Most damning of all, Wilkes had plainly failed in his primary mission of seizing rebel raiders in the region. Instead, it seemed to Welles that Wilkes was most interested in accruing prize money by capturing blockade-runners. Notwithstanding his unhappiness with Wilkes, Welles had refrained from relieving him in large part because he did not want to antagonize Seward, who had lobbied for Wilkes' appointment in the first place. Now, however, Seward was withdrawing his support for Wilkes for the sake of improved Anglo-American relations, giving Welles a green light to remove him from his post. Welles initially considered transferring Wilkes to lead the Pacific Squadron, a less-active but equally honorable position. The more Welles thought about it, though, the more he realized that would be a mistake. Commanding the remote Pacific Squadron required lots of tact and discretion, which Wilkes lacked. After more consultations with Seward, Welles opted to simply order Wilkes home without explanation. Despite the dislike many naval officers had for him, Wilkes was popular among the public for his role in the *Trent* affair, so Welles expected the press to censure him for his decision, but he was willing to sustain their slings and arrows to rid himself of his troublesome

West India Squadron chief. On 1 June Welles issued the necessary orders.[30]

The Civil War had changed the Navy in all sorts of dramatic ways, but some things remained the same. Despite the wartime need for secrecy, the Navy retained an informal information network among officers and their families about personnel matters. Officers talked to their wives, who talked to each other. The network operated from the top of the Navy Department down to the lowliest lieutenant and his spouse. Fox in particular was a gossipy and flirtatious man always willing to tell concerned wives who accosted him at parties or church the latest news on their husbands. Thus Welles could hardly have been surprised when he sat down for breakfast on the morning of 4 June to find a note from Mrs. Wilkes to his wife inquiring about rumors that the Navy Department had ordered her husband home. Mrs. Welles' response must have been unsatisfactory; later that day Mrs. Wilkes sent a note to Welles at his office asking about the report, and directed the messenger to wait for an answer. Welles was fond of Mrs. Wilkes, with whom he and his wife socialized, so he opted to confirm the rumor even though he knew that it might end their friendship. Wilkes later claimed that he was grateful to learn of his recall because he was fed up with the constraints under which he had operated. When Wilkes returned home in July, he traveled to Washington and talked with Welles. The meeting was cordial enough. Welles explained somewhat disingenuously that Wilkes' relief was due to the administration's desire to placate the British. He said nothing about his grievances toward Wilkes, which had played just as important a role in his removal. When Wilkes visited Seward, though, the secretary of state declared that he had approved of Wilkes' conduct toward the British, and that his relief from command had nothing to do with Anglo-American diplomacy. Somewhat suspicious, Wilkes went north to recover his health, and while there he became convinced that Welles and the Navy Department were trying to defraud him out of some of the prize money he thought he deserved.[31]

Wilkes' simmering resentment toward Welles finally boiled over in anger when he read the navy secretary's annual report at the end of the year. In it, Welles referred to Wilkes' unauthorized seizure of *Vanderbilt* as the reason for the Navy's failure to destroy *Alabama*. To Wilkes this was nothing short of slander. He believed that the Navy Department's refusal to send him adequate reinforcements was the real reason for *Alabama*'s escape. In fact, Wilkes counted his mission as successful because he had driven all rebel raiders out of the Caribbean. To protect his honor, on 11 December 1863 Wilkes wrote Welles a letter defending his actions, and asked Welles to send all departmental correspondence relating to the West India Squadron to Congress for its perusal. Welles labeled Wilkes' letter impudent, untruthful, and un-officer-like. On 15 December Welles responded with a hasty and full-scale broadside of his own. He listed his innumerable grievances against Wilkes, emphasizing in particular his impressment of *Vanderbilt*. He noted that he had tried to limit his review of Wilkes' performance in his annual report to as few words as possible to spare him any embarrassment, so Wilkes had received better treatment from the department than he deserved. Finally, he rejected Wilkes' request to forward all departmental correspondence to Congress as improper, and snidely added that Wilkes' inability to capture *Alabama* and his dereliction of duty were two different matters.[32]

The mutual recriminations might have ended there, except that someone violated naval regulations by leaking Wilkes' letter to the press, so it soon appeared in the *New York Times* and the *Philadelphia Inquirer*. Welles immediately concluded that Wilkes was the culprit, and sent him a letter asking if this was true. Wilkes denied all knowledge of the disclosure, however. In fact, in his memoirs he blamed Welles, though it is hard to see what the navy secretary had to gain by making Wilkes' letter public. Wilkes, on the other hand, had a substantial motive to bring his case to the public's attention, so Welles' suspicions were probably accurate. Welles believed that Wilkes was lying, and by failing to admit his guilt, he was casting doubt on the honor of everyone in the Navy Department building who had had access to his letter. In mid-January 1864 he convened a court of inquiry made up of the current Lighthouse Board— William Shubrick, Charles Davis, and John Rodgers—to investigate Wilkes' leaked letter. Although Wilkes declined to testify, the court established that the press got possession of the letter a day before the Navy Department did. With this evidence in hand, Welles ordered Wilkes court-martialed. He did so reluctantly, in part because he did not want to irretrievably destroy Wilkes' career, but also because he hoped to avoid more controversy in the Navy such as that generated by Du Pont's relief. At the same time, Welles felt that Wilkes' flagrant and unrepentant flaunting of Navy regulations, as well as his manifest dishonesty, required some sort of remedy. Wilkes' court-martial consisted of thirteen officers who met in March and April 1864. Wilkes again refused to testify, but he vigorously protested every facet of the trial in an effort to cloud its proceedings. In particular, he objected to the presence of Louis Goldsborough and Hiram Paulding on the court because, he argued, both men were prejudiced against him. On 30 April the court found Wilkes guilty of disobedience, insubordination, and conduct unbecoming an officer, and sentenced him to three years' suspension and an official reprimand. Considering his unpopularity in the officer corps, it was a surprisingly light sentence.[33]

Wilkes may have been unpopular within the officer corps—one exception was Du Pont, who now shared Wilkes' hatred of Welles and sent him a friendly letter congratulating him on his courtroom defense—but he had friends among the general public and in the Lincoln administration. These supporters included the pastor of St. Paul's Episcopal Church in Washington and the secretary of state. Even before Wilkes' court-martial began, Mrs. Wilkes asked Seward to intervene on her husband's behalf. Seward dutifully dispatched his son to talk with Welles, but Welles insisted that the court-martial proceed. After the court-martial reached its verdict, Wilkes' friends focused their efforts on persuading the president to mitigate his punishment. In December Lincoln told Welles that despite Wilkes' hostility toward both of them, he was inclined to remit the sentence. Lincoln had always had a soft spot for Wilkes, and no doubt he did not want to alienate Wilkes' supporters. Welles advised against it, but Lincoln did so anyway by limiting Wilkes' suspension to a year. Although this enabled Wilkes to claim victory in his memoirs, he played no further part in the war. He was, however, promoted to rear admiral on the retired list in 1866, and spent the rest of his career working on the results of his Exploring Expedition. He died in 1877.[34]

To supplant Wilkes as head of the West India Squadron, Welles initially considered Commo. Charles Bell, current commander of the Pacific Squadron. Bell had plenty of stature and experience. Of the loyal captains, he ranked twenty-fifth on the naval register when the war began, and had previously led the Mediterranean Squadron. After recalling Mervine, Welles had thought about appointing Bell to command the Gulf Squadron, but gave the job to McKean instead. Since he knew that McKean's health was uncertain, Welles kept Bell in reserve for several months just in case, but at Paulding's suggestion he dispatched him to the tiny Pacific Squadron in January 1862. Ultimately, though, Welles opted against ordering Bell to the West Indies, probably because it would take too long to get him there, and Welles needed someone to replace Wilkes as soon as possible. Instead, Welles tapped James Lardner. Lardner was one of the heroes of Du Pont's Port Royal expedition, and had subsequently led the East Gulf Squadron until a bout of yellow fever forced him to step down in November 1862. He had recovered even before he returned north, so he was available and ready for new duty when Welles summoned him from Philadelphia to give him his new assignment. Unfortunately, Lardner did not impress Welles during their first meeting. To the navy secretary, Lardner seemed discreet and prudent enough, but lacking in energy, efficiency, and forcefulness. Perhaps because these first impressions so belied Lardner's firebrand reputation, Welles refrained from immediately passing judgment on him.[35]

Lardner left Philadelphia on board the sloop USS *Ticonderoga* on 6 June 1863, and relieved Wilkes at Saint Thomas two weeks later. In an effort to help Lardner avoid some of Wilkes' mistakes, Welles gave the new commander long and carefully worded orders to protect Union commerce, destroy rebel raiders, seize suspected blockade-runners, and avoid problems with neutral countries. Lardner's tenure, however, proved quiet and anticlimactic. During his sixteen months as West India Squadron commander, Lardner never saw a Confederate raider or captured a blockade-runner. Except for a short visit to Martinique by *Florida* in May 1864, no rebel cruisers entered his jurisdiction. Therefore, Lardner's efforts to capture them by disguising his flagship, *Powhatan,* as a French warship accomplished nothing. As the weeks stretched into months, the grinding tedium and inaction took a toll on the crews. Sailors became bored, irritable, and discontented. When Lardner gave *Powhatan*'s crew liberty at Saint Thomas, for example, a riot broke out between Union and British sailors. Yellow fever also put in an appearance, but Lardner dealt with it on *Powhatan* by stopping her engines, hoisting sail, and cruising in the wind away from the shore until it abated. Despite all these difficulties and frustrations, life was not altogether bad. Lardner and his officers enjoyed visiting local dignitaries, entertaining guests, watching the sparkling scenery, and strolling along sandy white Caribbean beaches. Lardner communicated with Welles in short messages that lacked the grumbling and complaining that characterized Wilkes' reports. In light of Lardner's inactivity, Welles eventually reduced his squadron from eight vessels to three. Indeed, at times *Powhatan* was the only steamer available to Lardner. Meanwhile, other Union naval officers elsewhere hunted down rebel raiders one by one. On 19 June 1864, Capt. John Winslow in the sloop USS *Kearsarge* sank *Alabama* off the French coast. Less than four months later, on 7 October, Cdr. Napoleon Collins and *Wachusett* captured *Florida* in the harbor at Bahia, Brazil. On 12 September 1864, with the threat posed by rebel cruisers receding, Welles dissolved the West India Squadron and ordered Lardner home, and he and *Powhatan* arrived in Philadelphia three weeks later. Welles placed Lardner on the retired list shortly thereafter, and he did not play an active role in the rest of the war. He served on various boards until 1869, when Welles made him governor of the Philadelphia Naval Asylum. He retired fully three years later and died in Philadelphia in 1881.[36]

In retrospect, it is hard to justify the West India Squadron's existence. Welles and Fox established it without much forethought and in response to public pressure for naval action against rebel raiders. Welles wanted a "flying squadron" of swift vessels to scour the Caribbean for *Alabama* and *Florida,* but it did not work out that way. Although the conception was reasonable enough, the implementation was seriously flawed. Welles believed in the primacy of the blockade, so he only detailed a half-dozen warships in various states of disrepair to the squadron, hardly enough to flush out and capture nimble Confederate cruisers operating in the region. In addition, Welles erred in succumbing to Seward's pressure and appointing Wilkes to lead the squadron. Commanding the West India Squadron required a naval officer with sufficient tact and discretion to deal with complicated questions of international law and neutral rights. Not only did Wilkes lack these traits, but the British in particular were predisposed against him because of his role in the *Trent* affair. Wilkes recognized accurately and quickly enough that he did not possess the resources to complete his mission, and sought to absolve himself of responsibility for failure by shifting the blame onto the Navy Department. He alienated Welles by petulantly demanding reinforcements and arbitrarily impressing warships over which he had no authority. Once Seward had withdrawn his support for Wilkes for diplomatic reasons, Welles moved swiftly to remove him. Lardner accomplished just as little, but by then Welles did not expect much from him or his squadron, and Lardner was smart enough not to pester the Navy Department. In the end, the West India Squadron served as little more than a cautionary tale of the dangers inherent in mixing military and diplomatic matters.

CHAPTER SIX

Turning the Tide

David Porter and the Mississippi Squadron

F OR THE UNION the summer of 1862 was a season of military frustration. In the East Gen. Robert E. Lee and his Confederate Army of Northern Virginia defeated Maj. Gen. George McClellan's Army of the Potomac in the Seven Days' Battles in late July, terminating the Union offensive against Richmond. The rebels took advantage of McClellan's subsequent quiescence to focus on defeating Maj. Gen. John Pope's Army of Virginia at the Battle of Second Bull Run the following month. Although McClellan and his Army of the Potomac thwarted Lee's subsequent invasion of Maryland at the Battle of Antietam on 17 September, this strategic success did not bring the Union much closer to victory, at least not in a military sense. In the West the rebels also grasped the initiative that summer by invading Kentucky and assailing Union forces in northern Mississippi. Here too Union forces managed to repulse the Confederate offenses, first at the battles of Iuka and Corinth in northern Mississippi on 19 September and 3–4 October, respectively, and then at the Battle of Perryville in central Kentucky on 8 October. These engagements were mostly army affairs, but the Navy had its own disappointments that summer. Two efforts to seize the Confederate stronghold of Vicksburg, Mississippi, failed. There were plenty of reasons for the Navy's setbacks, but the lack of army support was certainly among the most important. In the aftermath of these difficulties, David Farragut fell back to New Orleans to avoid becoming stranded by the Mississippi's diminishing waters. Charles Davis and his Western Flotilla retreated too, upriver to Helena, Arkansas. Vicksburg remained in rebel hands, serving as a stubborn nail that held the two halves of the Confederacy together.

On 26 July 1862, David Porter and his mortar flotilla reached Hampton Roads after a two-week voyage from the fetid Mississippi River. Like most of his sailors, Porter was sick with one of the many tropical diseases that plagued Union forces on the Mississippi, but this did not prevent him from dashing off a quick letter to Fox denouncing Charles Davis for his contribution to the *Arkansas* fiasco. In response to Welles' summons, Porter hurried to Washington, where the navy secretary took one look at his haggard frame and gave him two weeks' leave of absence to recuperate at his home in Newport, Rhode Island. Although Porter was happy to see his family again, he chafed at his forced inactivity. He wrote to Fox that he had given up drinking, smoking, and overeating, and was soon healthy enough to unwind at the fashionable Newport Club. One evening there, while playing billiards, he became embroiled in a heated debate between local copperheads and Radical Republicans about the war. Always a big talker, whether or not he knew much about the subject at hand, Porter freely voiced his many opinions about and criticisms of the Union war effort. Next day a friend warned him that rumor had it that he had uttered many treasonous things the night before. Word of his performance apparently reached the Navy Department; shortly thereafter Welles instructed him to report to Washington at once.

When he reached the capital, Fox informed a shocked Porter that the Navy Department had decided to break up his mortar flotilla and send him to St. Louis to oversee ironclad construction. Porter protested this unglamorous assignment for all he was worth, claiming that Welles and Fox were exiling and sidelining him. He would, he said, resign before he accepted such a position. He stomped out of the Navy Department building and made a beeline across the street to the White House to appeal to the president. Lincoln was talking with Seward, but he invited Porter in to hear his grievance. Seward had been tight with Porter since the war's start, so he quickly came to his defense. Lincoln had small use for Porter's bombast, however, and he remembered Porter's part in the *Powhatan* fiasco during the Fort Sumter crisis. On the other hand, after months of dealing with timid and incompetent army officers, it was refreshing to see someone eager to come to grips with the enemy. Lincoln told Porter that he would see what he could do, and sent for Fox. Porter had enough sense to scuttle out of the White House before Fox arrived, and he returned to Newport to await the results of his blatant string pulling. While there, Porter brazenly rejected Fox's offer to command the Potomac River Flotilla, secure in the knowledge that the president wanted him to take a more active role in the war. On 22 September Welles ordered Porter back to Washington. When he arrived at the Navy Department building, he was ushered into Welles' office. The navy secretary smiled benignly, extended two fingers for Porter to shake, offered him a seat, and handed him sealed orders. With as much nonchalance as he could muster, Porter opened them and discovered that Welles was putting him in charge of the newly upgraded Mississippi Squadron, with the rank of acting rear admiral. Welles congratulated him and invited him to his house later to discuss the details. As he left the building, the bureau chiefs, Fox, and Faxon all offered their congratulations too, though Faxon added that Porter ought to direct his gratitude to the president.[1]

David Porter came from a renowned American naval family. He was born in 1813 in Chester, Pennsylvania, the third of ten children. His father and namesake fought gallantly in the Quasi War, the Tripolitan War, and especially the War of 1812. He resigned his commission in 1826 after the Navy court-martialed and suspended him from duty for alienating the Spanish while leading the West India Squadron against local pirates. He took David with him after he accepted an offer to command the Mexican navy. Young Porter was captured after a terrific battle between his Mexican warship and a Spanish frigate and taken prisoner to Havana. Upon his release, he returned to the United States and entered the Navy as a midshipman in 1829. He spent much of his early career in the Coast Survey mapping the Atlantic shoreline and compiling his data in Washington. Along the way he married Commo. Daniel Todd Patterson's daughter, whom he met when both were on board USS *United States* during a Mediterranean cruise. He also spent four months in 1846 in the Dominican Republic gathering information on local conditions for the State Department. During the Mexican War, he led a landing party against a fort at Tabasco, and later skippered USS *Spitfire*. After the conflict he secured a prolonged leave of absence to captain various merchant ships. He rejoined the Navy in 1855 and took USS *Supply* to the Mediterranean to secure camels for the Army. From 1857 to 1860 he was on routine duty at the Portsmouth Navy Yard, during which he seriously considered returning to the merchant marine. Before he could do so, however, the Southern states seceded. Although he was friends with Jefferson Davis and many other Southerners, he remained loyal to the Union and was involved in the machinations that contributed to the war's outbreak at Fort Sumter. He conceptualized, constructed, and commanded the mortar flotilla that played such a prominent role in Farragut's operations on the Mississippi River until the Navy Department ordered him home in the summer of 1862.

Porter was of medium build and height, with a craggy face hidden behind a large black shovel beard. Although he emerged from the Civil War as one of the Union Navy's greatest heroes, there was much to dislike about him. He was a verbose man willing to opine even on topics about which he was completely ignorant. To Porter, truth was malleable and relative. He frequently exaggerated his accomplishments and promised a good deal more than he could deliver, so many people questioned his basic integrity. Porter was aware of this perception and often prefaced his statements by stating, "Now, you think I'm lying but I am not."[2] Like many naval officers, he was more than willing to pull strings to get what he wanted. He often took advantage of his private correspondence with his good friend Fox to denigrate his brother officers' abilities, and usually accompanied his diatribes with thinly veiled hints that he could do a better job if given the opportunity. For example, he freely criticized William Mervine, William McKean, Charles Davis, Henry Bell, and David Farragut when they became obstacles to his ambitions. Porter augmented these lamentable characteristics with ample doses of selfishness, vanity, and egotism. As a result, he had innumerable enemies in and out of the officer corps. In fact, he and his older brother and fellow officer William were on such bad terms that at the beginning of the war they each denounced the other to Welles as a secessionist sympathizer. On the other hand, Porter also possessed some admirable traits that made him an effective squadron commander. He had plenty of physical courage and coolness under fire. There was nothing unusual about that among Union naval officers, but he combined his bravery with a resourcefulness, imagination, persistence, and energy that put him in a league of his own. Porter never suffered from the timidity and caution that infected so many Union officers, and was always willing to engage the enemy with little fear of the consequences. Unlike many of his naval colleagues, Porter could get along with army officers, especially those who shared his can-do attitude, which served him well in his many interservice ventures. Finally, while he squirmed under the authority of others, he shared Du Pont's and Farragut's knack for inspiring loyalty and encouraging teamwork among his subordinates with his hospitality, optimism, and good humor. Squadron command occasionally brought out his better nature, enabling him to exhibit a charity and generosity he rarely displayed earlier in the conflict in subordinate roles.[3]

In selecting Porter to command the Mississippi Squadron, Welles made his boldest personnel decision of the war. Welles knew that Porter had plenty of enemies and doubters. For instance, neither Lincoln nor Secretary of War Edwin Stanton thought much of Porter's bluster, although the president was willing to defer to Welles' judgment on the matter. Moreover, Porter was only a commander, so Welles elevated him over dozens of more senior officers. Indeed, Porter's appointment represented a complete abandonment of the seniority principle. Welles, however, had his reasons. For one thing, Lincoln wanted Porter to play a major role in the Navy's war, and he was commander in chief. Welles by now had dealt with Porter on numerous occasions, and was under no illusions about the man. He was a pretty good judge of character, and his evaluation of Porter in his diary was right on target. Even so, he was convinced that Porter's attributes outweighed his faults, and he hoped that Porter's determination, ingenuity, ambition, and combativeness would serve him and the Union cause well on the Mississippi. Welles recognized that the Mississippi River was a unique theater of war for many reasons, and required a unique squadron commander such as Porter, one with experience in that region and capable of thinking creatively. Besides, many senior officers had no desire to serve in these unfamiliar surroundings, so Porter's promotion was less likely to cause an uproar than if Welles had assigned him to lead one of the prestigious blockading squadrons. Welles concluded that choosing Porter was an "experiment" that, if successful, would inspire motivated junior officers to emulate Porter's daring in the hopes of similar rewards. Before Porter headed out west, Welles met with him and tried to encourage his virtues and discourage his vices. Whether Porter would accept this advice remained to be seen, but Welles was certain that if he failed, the public would blame the Navy Department.[4]

Welles' prediction that Porter's appointment would elicit extensive comment proved accurate. Charles Davis, for instance, wrote, "This is one of those sudden events that takes place in the naval world—like earthquakes in the physical, not only creating surprise, but bringing in such waves, and stirring up howling eddies, in which great ships, as much as little boats, get tossed about in the most disturbing and perplexing manner."[5] John Dahlgren was one of those disturbed great vessels. Dahlgren was then angling for his own squadron command in an effort to achieve the glory he so desperately wanted, and he complained to Welles of the partiality shown to Porter. If, Dahlgren stated, Welles was willing to advance Porter over the heads of so many senior officers, then the navy secretary should permit him to lead an assault on Charleston instead of Du Pont. Welles dismissed Dahlgren's argument as invalid by noting that few officers wanted to serve on the Mississippi.[6]

Lt. Sam Phelps was also unhappy with Porter's promotion. Phelps commanded the ironclad USS *Eastport,* and had served as both Foote's and Davis' right-hand man during their tenures with the Western Flotilla. When he realized that Davis' days as flotilla commander were numbered, Phelps decided to campaign for the position. Phelps had been fighting on the Mississippi since the conflict's earliest days, so he figured that he had as much or more experience in riverine warfare than any other officer, despite his junior rank. Foote and Davis both attested to his abilities, and Phelps asked influential men such as Iowa senator James Grimes to lobby the Navy Department on his behalf. Phelps did such a good job selling himself that Welles eventually complained of the "extraordinary influences [that] have been brought to bear upon the Department with a view of influencing its action in relation to the command of the Western Flotilla."[7] Welles and Fox were accustomed to officers and their friends pulling strings, but not to such blatant and heavy-handed string pulling by someone so low on the totem pole. Fox wrote to Davis, who was still out west, to ask him to tell Phelps that he was too junior for the post to which he aspired. Welles, on the other hand, took a more direct approach. In a friendly but blunt letter to Phelps, he acknowledged Phelps' merit, but chastised him for attempting to influence the department through unofficial channels. He also reiterated Fox's assertion that Phelps was too junior for the post. Davis and Foote each wrote to Phelps to assure him of Welles' sympathy and support, and suggested that he might someday succeed Porter if he performed well under his command. Although Phelps worked capably with Porter, he eventually resigned from the Navy in October 1864 to become an agent for the Pacific Mail Steamship Company after Welles replaced Porter with Phillips Lee.[8]

Porter got to work as soon as he reached Cairo on 15 October, often putting in eighteen-hour days. In fact, Porter had much to do. The newly christened Mississippi Squadron was the most complicated organization in the Union Navy, containing a weird conglomeration of 125 vessels—river-going ironclads, tugs, supply steamers, mortar scows, and tenders—crewed by ten thousand sailors and thirteen hundred officers deployed along numerous rivers that had to be supplied far from the Navy's prewar yards. The previous July, Congress had given the Navy full authority over all vessels operating on inland waters, but Stanton and Henry Halleck, the army general in chief, had tried to retain army control over Charles Ellet's ram fleet, now led by his brother, Alfred Ellet. Porter saw the value in Ellet's rams immediately and asked Welles to secure their transfer to the Navy. Welles did so, but only after he appealed to the president to get the recalcitrant Army to obey the law. Porter assigned naval personnel to the rams and turned Ellet's men into a marine brigade that would operate along rivers against rebel guerrillas. In addition, Porter also tackled the health problems that plagued Union Navy crews on the Mississippi and its tributaries. He improved hygiene, rented a Mound City, Illinois, hotel for use as a hospital, and ordered fresh meat and vegetables served three times a week. As for his warships, he rearmed many of them with 6-inch Parrott rifled cannon. Fortunately for Porter, he had Capt. Alexander Pennock's assistance. Pennock commanded the new Mound City Navy Yard, and proved so valuable that Porter gave him authority to communicate directly with the Navy Department in his absence. Porter's whirlwind changes cost a good deal of money, but Welles and Fox were willing to pay that price if it bought the Navy control of the Mississippi, which as far as they were concerned was the Union's number one strategic objective.[9]

While Porter was reorganizing his squadron, the Union Army was preparing for a full-scale offensive against Vicksburg. Maj. Gen. Ulysses Grant, commander of the Army of the Tennessee, planned a two-pronged assault on the rebel city. He wanted to invade northern Mississippi along the Mississippi Central Railroad with part of his Army in an effort to divert rebel forces in that direction while Maj. Gen. William Sherman, his chief lieutenant, took his 32,000-man corps down the Mississippi River to directly assail the hopefully scantly defended Confederate citadel. Despite Porter's best efforts, he was hard-pressed to scrape up enough healthy sailors to man the warships necessary to accompany downriver the fifty-odd transports containing Sherman's men. Escort duty was more dangerous than expected. On 12 December, for example, two enemy mines sank the ironclad *Cairo* while she was clearing out Confederate opposition on the Yazoo River north of Vicksburg. *Cairo*'s young skipper, Lt. Cdr. Thomas Selfridge, fully expected career-ending repercussions when he met Porter on his flagship, *Benton*, at the Yazoo's mouth. Instead, he got an illuminating glimpse at his new squadron commander's way of thinking. Upon seeing Porter, Selfridge approached, informed him of *Cairo*'s fate, and stiffly asked for a court-martial. "Court!" Porter roared. "I have no time to order courts; I can't blame an officer who seeks to put his ship close to the enemy. Is there any other vessel you would like to have?"[10] On 26 December Sherman disembarked his soldiers and deployed them for an attack on Chickasaw Bluffs, which according to Grant's scheme should have been denuded of rebels. Unfortunately for Sherman and his soldiers, Confederate cavalry raids on Union supply lines in northern Mississippi and western Tennessee had forced Grant to withdraw, permitting the rebels to send reinforcements to Vicksburg to counter Sherman. Sherman, though, did not yet know of Grant's setback when he ordered his men forward into action on 29 December. Sherman's charge up Chickasaw Bluffs failed miserably, resulting in nearly 1,800 Union casualties, as opposed to a mere 200 Confederate killed and wounded.

After the battle, a demoralized Sherman huddled with Porter in the cabin of Porter's floating headquarters, the side-wheeler USS *Black Hawk*. The two men lamented the Union defeat over hot whiskey punch. They conversed in whispers to avoid disturbing Cdr. William Gwin, who lay dying in Porter's bed after a rebel round shot had mortally wounded him several days earlier on *Benton*'s deck. Porter and Sherman had only recently met, but they hit it off immediately. Both were talkative, impatient, fidgety men full of ideas and energy. In a determined yet macabre effort to lift Sherman's sagging spirits, Porter rationalized the setback: "[It's] simply an episode of the war. You'll lose 17,000 before the war is over and think nothing of it. We'll have Vicksburg yet, before we die."[11]

Porter also brought to Sherman the unwelcome information that Maj. Gen. John McClernand had arrived. McClernand was an Illinois political general who had seen action at Fort Donelson and Shiloh. Several months earlier, he had traveled to Washington and persuaded his friend the president to authorize him to recruit an Army of Midwesterners that he would lead against Vicksburg. He had done a good job raising the troops, but was angry that Grant and Sherman had appropriated them for their failed operation, and he had come downriver to supersede Sherman and reclaim what he believed was rightfully his. He also brought the unhappy news that Grant's offensive in northern Mississippi had gone into reverse. Porter had been impressed with McClernand when they talked in Washington the previous October, in part because he considered West Pointers impractical, pedantic, and uncooperative. His short time with the dynamic Sherman, however, had dramatically altered his perspective on McClernand. Despite an atmosphere tinged with suspicion and resentment, the three men decided to assault Fort Hindman at Arkansas Post, fifty miles up the Arkansas River. Seizing this position would help seal off the Transmississippi states from the rest of the Confederacy, and provide an upbeat coda for a hitherto unsuccessful campaign. The expedition departed on 5 January 1863, with Porter's thirteen warships escorting McClernand's 30,000 soldiers aboard their transports. The troops disembarked four days later, and after a combined attack that saw Porter's ironclads engage the enemy at point-blank range, the outnumbered and outgunned rebel garrison of around 5,500 men surrendered on 11 January.[12]

At sunset on 28 January 1863, Grant arrived at Young's Point, opposite Vicksburg, and assumed control over the 47,000 Union soldiers in the region or on their way there, thus relegating McClernand to a corps commander. Grant knew as well as anyone that military operations are invariably subject to geographic constraints, and Vicksburg had plenty of those. Chickasaw Bluffs and the swampy and impassible Yazoo River delta protected the city from the north. Vicksburg itself was perched high atop a bluff, so batteries emplaced there guarded against any Union attempt to move directly downstream along the Mississippi River. Although Farragut had demonstrated the previous summer that Union warships could successfully steam past those batteries, the transports and supply vessels the Union Army required to get to and operate on the Mississippi's eastern bank below the citadel were far more vulnerable to artillery fire. Grant's problem, in short, was to find some way to overcome or circumvent Vicksburg's natural and man-made obstacles so he could place his troops on the dry ground to the south. Once there he could finally engage and hopefully defeat the enemy's forces out in the open. Porter's self-assigned role was to assist these efforts with his squadron. Throughout that long winter, Porter searched the maze of rivers, creeks, and bayous in the region for a water route that would enable him to bypass Vicksburg and get the vessels south of the city necessary to transport Grant's troops across the Mississippi. There was nothing easy about it. Heavy winter rains flooded the countryside and filled the waterways with every kind of debris imaginable. Pneumonia, malaria, and other diseases struck down soldiers and sailors alike. Warships steaming along spooky, narrow, and twisty creeks fell prey to mechanical breakdowns, mudflats, stumps, rebel snipers, cannon, mines, and even willow switches that fouled paddle wheels and trapped vessels the way Gulliver was trapped by the Lilliputians. Sailors sometimes had to sweep decks clear of termites, rats, lizards, snakes, weasels, possums, raccoons, and even cougars knocked out of overhanging trees. The Confederates dammed creeks and burned cotton bales along the banks to keep them out of enemy hands, forcing Union warships to run fiery gauntlets. Even naval officers accustomed to the idiosyncrasies of riverine warfare were unnerved. Worst of all, however, was that none of these endeavors succeeded in finding the route Grant and Porter were looking for. There was plenty of water in the area that winter for the warships, but never enough in the right places at the right time.

While Porter was scouring the region's creeks and bayous for a water route around Vicksburg's formidable batteries, he also undertook efforts to disrupt Confederate river communications south of the citadel. On 2 February 1863, he sent the ram *Queen of the West* under Charles Rivers Ellet to run downstream and destroy the rebel steamer *City of Vicksburg,* then unloading at Vicksburg's docks. Ellet was unable to sink *City of Vicksburg,* but he succeeded in getting past Vicksburg's batteries and capturing several rebel vessels before returning upriver three days later. Encouraged by Ellet's accomplishments, over the course of the following week Porter dispatched the tug USS *De Soto,* Ellet in *Queen of the West,* and the new ironclad USS *Indianola,* under Lt. Cdr. George Brown, downriver past Vicksburg to do as much damage as possible to rebel commerce. After some initial successes, however, Union sailors soon discovered that they had stumbled into a rebel beehive. On 13 February *Queen of the West* ran aground near Fort DeRussy on the Red River. Confederate fire forced Ellet to abandon his vessel and transfer his flag to *De Soto.* Unhappily, *De Soto* soon lost her rudder, so Ellet burned her and moved his operations to one of his prizes, *Era No. 5.* Near Natchez, Mississippi, Ellet encountered *Indianola,* which scared off rebel pursuers so that *Era No. 5* could safely ascend the river and rejoin Porter's squadron. *Indianola* remained behind to blockade the entrance to the Red River, but on 24

February four rebel warships, including the repaired *Queen of the West,* rammed and sank *Indianola* after a confused free-for-all fight. Once the Confederates raised and patched *Indianola,* they could, with their two new powerful warships, challenge Porter for control of the Mississippi River. Worse yet, Porter's squadron was dispersed throughout the region looking for ways to circumvent Vicksburg, so it would have a hard time resisting a rebel attack.

Welles once noted in his diary that Porter was professionally unlucky, but now Porter's luck—though he would have called it providence— changed for the better. He had noticed that some of Vicksburg's cannon burst after repeated firing, so in an attempt to get the rebel batteries to open up without endangering his own sailors, Porter ordered his men to construct a phony ironclad dubbed *Black Terror.* Sailors took an old coal barge, installed false bulwarks, attached empty paddle-wheel boxes, placed smudge pots filled with burning oakum and pitch under a faux smokestack of barrels, and covered the vessel with a coat of tar, all for less than $9. After tacking a sign on the wheelhouse reading "Deluded People Cave In," the sailors pushed *Black Terror* downstream on the night of 24–25 February. She floated silently past Vicksburg's roaring cannon without injury, and, after a push from Union soldiers on the west bank, headed downstream. Unwilling to engage such an apparently powerful ironclad, the rebel flotilla scurried downstream, and *Indianola*'s salvage crew blew her up to prevent her from returning to her original owners before they discovered the embarrassing truth. Because of the ruse, Union naval supremacy on the Mississippi remained intact.[13]

For Porter, it was both a rewarding and a difficult winter. To secure his present command, he had undercut and denounced superior officers, made lots of extravagant promises, and disrupted Navy Department routines. He had been dishonorable, unscrupulous, disloyal, duplicitous, and self-serving. Indeed, he had devoted as much energy to gaining advancement as he had to waging the war. Now that he had finally achieved a prestigious position, he needed to demonstrate that he was worthy of the responsibility with which Welles had entrusted him. Congress had voted him its thanks for his role in seizing Arkansas Post, and Fox promised him promotion to permanent rear admiral when and if Vicksburg fell. Spurred on by the prospect for further glory, Porter remained busy throughout the winter, even though he grumbled that he was too often tied to his desk and unable to exercise. Although widely disliked by senior officers, he proved popular among the squadron's subordinates, many of whom were, like him, aggressive younger men willing to take the war to the enemy no matter what. Porter had always had a knack for cultivating loyalty among his subordinates, but he also got along surprisingly well with Grant and Sherman, who appreciated his positive attitude. Both generals frequently visited Porter's well-stocked and comparatively luxurious headquarters vessel for dinner, cigars, and talk. This greatly facilitated the interservice cooperation necessary for success along the Mississippi, and was in stark contrast with, for instance, Du Pont's and Dahlgren's troubles with local army commanders outside Charleston. The three men saw eye to eye on most issues, and were especially united in their hatred of McClernand, who was still angling for command of Grant's army and would ultimately be relieved for insubordination.

Unfortunately, this conviviality did not erase the numerous difficulties under which Porter labored that season. He had many of the problems common to the blockading squadrons, plus a good many more unique to the Mississippi. He complained of poor-quality warships, stifling red tape, insufficient funds, slow turnaround in repairs, and especially the lack of sailors. In fact, Porter was so short of manpower that Fox told him to recruit local blacks to make up the shortfall. He was also pained and humiliated by the fate of *Queen of the West* and *Indianola*. As usual, Porter sought to escape accountability for the loss of these warships by shifting the blame elsewhere, in this case onto Ellet and Brown. Worst of all, despite all his efforts, Porter simply could not find a way around the cannon frowning down from Vicksburg's high bluffs. He was an ebullient man under most circumstances, but on 26 March he wrote a pessimistic letter to Welles practically admitting failure. The only way to get at Vicksburg, he explained, was for Grant to retreat, reassemble his army in western Tennessee, and attack the place through northern Mississippi along the Mississippi Central Railroad, away from the swamps and bayous that had caused so much trouble.[14]

Back in Washington, far from the waterlogged floodplains along the Mississippi, Welles and Fox were coping with their most stressful months of the war. Both men were frustrated that their star squadron commanders were not meeting their stipulated objectives. Although Du Pont's planned assault on Charleston had top priority, Welles and Fox considered Porter's operations against Vicksburg every bit as important. Success on the Mississippi, however, was slower in coming than they had anticipated.

Welles had viewed Porter's victory at Arkansas Post not as the minor strategic diversion it in fact was, but as a harbinger of good things to come in that theater. As the weeks stretched into months and Porter made little headway against Vicksburg, Welles increasingly questioned his decision to assign such a junior officer to such a significant post. His doubts blossomed upon learning of the loss of *Queen of the West* and *Indianola,* which Welles labeled a "disgrace and disaster" that required immediate rectification. Nor was Welles amused with Porter's seemingly flippant and nonchalant attitude. Throughout the winter, Porter sent humorous anecdotes and caricatures to his brother-in-law, a Navy Department clerk. These stories made their way around the Navy Department, and the somber Welles was not the only person who saw them as buffoonish. Worse yet, the navy secretary was hardly the only policy maker with misgivings about Porter. The president was also closely watching events along the Mississippi, and he was unimpressed with Porter's efforts. Lincoln did not think much of Porter anyway, as a person, and responded unenthusiastically whenever Welles defended him. Secretary of War Stanton, for his part, considered Porter a blustery gasbag who took credit for the actions of others. Fortunately for Porter, he retained Fox's backing and support. Fox reassured Welles that Porter would fulfill his expectations and open the Mississippi. Even so, as spring approached, Welles was convinced that it was now time for Porter to step up and demonstrate once and for all whether Welles' confidence in him was correctly placed. Porter did not quite realize it, but his time was rapidly running out.[15]

Unvexing the Mississippi River

Welles may have had increasing doubts about Porter, but he maintained unlimited confidence in David Farragut. Farragut, though, had had a winter every bit as frustrating as Porter's. For one thing, his West Gulf Squadron had serious logistical problems. Farragut was responsible for the Gulf Coast from Texas' border with Mexico to Florida's Panhandle, as well as the lower Mississippi up to the Red River. Unfortunately, he lacked sufficient warships to effectively blockade such a long coastline. Moreover, those vessels he did possess were plagued with engine, boiler, and structural problems that sometimes rendered them unserviceable. He also needed more sailors to crew his ships, and he resented the Navy Department poaching some of his most promising young officers for new duties elsewhere. Worst of all, Farragut's squadron suffered several military setbacks that winter, the most serious being the successful Confederate assault on Galveston on New Year's Day. Reestablishing the blockade there required Farragut to divert much-needed warships from the lower Mississippi. Farragut was an astute man, and he recognized that robbing Peter to pay Paul was the sign of an officer who had lost the strategic initiative. Regaining it, though, proved more easily said than done. Aggressive as ever, Farragut wanted to assail Mobile, but he did not have the means to do so. Although Fox promised him some of Du Pont's ironclads after the Charleston operation was over just in case the rumored armored rebel vessel at Mobile was real, it would be months before all the pieces were in place for an assault. Besides, Welles and Fox insisted that his primary mission was opening up the Mississippi River. Finally, Farragut had learned at Vicksburg and elsewhere that assuming the offensive usually required army support that was not immediately forthcoming because Maj. Gen. Nathaniel Banks, Benjamin Butler's recent replacement as commander of the Department of the Gulf, seemed bereft of military skills or ideas. Despite newspaper reports to the contrary, Farragut was eager to cooperate with Banks, but he was also willing to go it alone if the opportunity presented itself.[16]

Despite these burdens, Farragut was not about to let them crush his spirit. He was an optimistic man who liked to enjoy himself, and Union-occupied New Orleans offered plenty of amusement and diversion. He went ashore almost every evening to visit army officers and old acquaintances. He even bought an opera box for himself and his staff as part of a charity drive, though he wryly noted that the secessionist ladies sponsoring the campaign would never return his kindness. Even so, the stress he was under occasionally bubbled to the surface. One day he and a group of officers heard some British sailors singing the rebel anthem "Bonnie Blue Flag" on their ship down at the docks. Farragut turned to James Palmer and exclaimed, "If it *isn't* stopped, we shall have to drop down and blow him out of the water."[17]

Reports that *Queen of the West* and the ironclad *Indianola* were in rebel hands galvanized Farragut into action. If the Confederates succeeded in raising and patching *Indianola,* their growing flotilla would threaten his vessels on the Mississippi as well as Porter's. Farragut was not one to wait around for danger to come to him, so he decided to go upriver and destroy the gathering rebel naval force. He hurriedly scraped together and assembled a dozen warships off Profit Island near Baton Rouge on 11 March. There was, however, one major obstacle between him and the rebel warships he was after: Port Hudson, located on a hundred-foot-high bluff twenty miles upstream from Baton Rouge at a hairpin turn in the river. The rebels had established the post after the Union Navy's failed assaults on Vicksburg, and had by now emplaced approximately 15,000 troops and twenty-one cannon there to block any Union ascent up the Mississippi. Farragut planned to deploy his mortar schooners, a few gunboats, and the underpowered ironclad *Essex* downstream to provide fire support while he took his remaining seven vessels past Port Hudson. Although he would have ninety-five cannon at his disposal on those seven warships, the Mississippi's fast current would slow the vessels to a crawl, giving the rebels plenty of time to cripple them. To help compensate for this, Farragut lashed three gunboats to the port sides of the three large sloops of war. This would not only help protect the vulnerable gunboats from enemy fire, but would also provide the coupled vessels with more speed against the current. By now Farragut's sailors were familiar and proficient in the art of running past rebel batteries, so he let his officers make their own preparations. They removed the rigging from their warships, hung chain garlands over the sides for additional protection, hoisted small howitzers to mast tops to blast any rebel boarders, and spread sawdust on the decks to prevent crews from slipping on any blood that might be spilled. Throughout 14 March, the crews completed their last-minute preparations for their run. An anxious Farragut and his staff stood on the deck of his flagship, *Hartford,* watching the activity and listening to the engines thump. Farragut's young son had recently joined him, and when Farragut noticed him nearby, he knelt down to give him some impromptu and last-minute advice on properly applying tourniquets.[18]

Happily for young Farragut, he did not get the opportunity to ply his new trade when his father ordered the Union warships upriver that windless and moonless evening. The surgeons in the other warships, on the other hand, more than earned their keep that night. As *Hartford* got under way, the rebels lit large bonfires that illuminated the river for their gunners. Soon the Mississippi reverberated with the sound of discharging cannon, wooden hulls cracking and shattering, the screaming wounded, and officers shouting orders and encouragement to their men. *Hartford* and her attached gunboat, USS *Albatross,* had the lead, and they made it past the enemy batteries relatively unscathed. The other warships, however, had no such luck. The next vessel in line, the sloop USS *Richmond,* took a shot in her engine room that demolished the valves controlling boiler pressure. Unable to make steam, she drifted downriver with her gunboat still lashed alongside. The sloop USS *Monongahela,* coming up behind *Richmond,* went aground on the Mississippi's west bank. Rebel fire destroyed her bridge and dumped Capt. James McKinstry onto the deck. Although the executive officer got the ship afloat, the crank pin overheated, rendering the engine useless. Like *Richmond, Monongahela* staggered downriver. The last vessel, the side-wheeler USS *Mississippi,* suffered most of all. *Mississippi* had a proud tradition, having conveyed Commo. Matthew Perry to Japan a decade earlier. Her paddle wheels made it impossible to strap a gunboat to her, so she was on her own when she ran into a mudflat. For a half hour under increasingly heavy Confederate fire, Capt. Melancton Smith tried without success to refloat her. By the time he ordered her evacuated, more than sixty of her crew had been killed or wounded. Fortunately for American imperialistic ambitions, Lt. George Dewey, the future victor of the Battle of Manila Bay during the Spanish-American War, survived. *Mississippi* later broke free of the mudflat and glided downstream, but exploded on the tip of Profit Island the next morning.

When dawn broke the following morning, a shocked Farragut discovered that *Hartford* and *Albatross* were the only vessels that had successfully passed Port Hudson. Scanning the empty river around him for the other five warships, he gasped, "My God! What has happened to them?"[19] Indeed, Farragut initially believed that the engagement had been nothing short of a disaster. Moreover, in the calm aftermath of battle, it suddenly occurred to him that Welles might not understand or share his sense of urgency in ascending the river. Farragut worried that the navy secretary might even condemn him for rashly and unnecessarily exposing his men, 112 of whom were now dead or injured. Finally, he was now isolated from the rest of his squadron and unable to exert any authority over it. He did not enjoy operating on the cramped river anyway, and yearned for the Gulf's wide and predictable expanses. Despite his apprehensions, he was not about to take counsel of his fears. After learning that the rebels had destroyed *Indianola* and fled up the Red River in *Queen of the West,* he pushed on past the enemy batteries at Grand Gulf and on 25 March established contact with Porter. Although the main reason for his mission had been nullified, Farragut perceived other advantages to his situation. He now controlled the strip of water between Vicksburg and Port Hudson, and could by blockading the Red River prevent the Confederates from sending reinforcements and supplies to either place from the Transmississippi. He did not like being there, but he decided to stay until Porter could relieve him. In the meantime he was comforted by the knowledge that Lincoln, Welles, and Fox, having recognized the benefits of placing Union warships between the two rebel strongholds even before he did, heartily approved of his daring mission.[20]

Farragut was not the only Union officer along the Mississippi River willing to resort to desperate measures that spring. Ulysses Grant was too. After failing seven times to come to grips with the Confederates in Vicksburg, Grant determined on a radically simple but risky scheme. He proposed to march his army along the Mississippi's west bank to New Carthage, from where transports would ferry his soldiers across the big river to the dry ground south of Vicksburg. Once there, he could hope-fully engage the rebel army, defeat it, and capture the city. Despite its straightforwardness, the plan was fraught with peril. Grant's men would require naval protection while the vulnerable transports conveyed them across the river, but the only way for Porter to provide it was to take his warships downstream past Vicksburg's batteries. Porter had no doubt he could do so, but he warned Grant that his vessels' underpowered engines would preclude them from returning upstream until Vicksburg was in Union hands. Even so, Porter was willing to take the chance. He had just about run out of ideas of his own, and Welles was pressuring him to do something dramatic. Although his warships were still recovering from a recent expedition to Steele Bayou in another unsuccessful search for a water route around Vicksburg, Porter put his sailors to work making the necessary repairs and preparations. On the moonless evening of 16 April, his eight ironclads, accompanied by three supply-laden army transports, slipped downstream from the Yazoo River. Moments after Porter wryly noted, "The rebels seem to keep a very poor watch," the Confederate bat-teries opened fire. The rebels hurled 525 shots at the slow-moving Union flotilla, but, except for one of the transports, all the vessels made it safely past Vicksburg.[21]

Grant hoped to land his troops at Grand Gulf, but rebel batteries there pummeled Porter's warships during their 29 April assault. Four of his ironclads sustained heavy damage, and the squadron suffered seventy-eight casualties, six times as many as during the run past Vicksburg. Stymied there, Grant instead crossed the river the next day at undefended

Bruinsburg and began his march inland that culminated in his placing Vicksburg under siege three weeks later. The Navy played an important, albeit secondary, role in these operations. After Grant's army disappeared into the Mississippi interior, Porter took some of his warships downstream to relieve Farragut and raid up the Red River to Alexandria, Louisiana. Porter returned in time to add his squadron's firepower to the army cannon that pounded Vicksburg around the clock during the six-week siege. On 4 July 1863, the city's 30,000 defenders surrendered. Five days later, Nathaniel Banks' army marched into Port Hudson after a siege of its own, thus terminating Confederate control of the Mississippi.

On 7 July Welles returned to the Navy Department building after a cabinet meeting. A delegation consisting of Vice President Hannibal Hamlin and several senators was waiting for him there to discuss coastal defense and the protection of New England fishermen from rebel raiders. Before Welles could address them, he received a telegraph from Porter announcing Vicksburg's surrender. Welles immediately excused himself and rushed across the street to the White House. There he found Lincoln discussing Grant's latest deployments with Secretary of the Treasury Salmon Chase and a number of others. Welles could not contain his joy, and he quickly broke the news to the president. Lincoln decided to go to the War Department's telegraph office immediately to send word of the victory to Maj. Gen. George Meade, current commander of the Army of the Potomac. Before doing so, Lincoln threw an arm around Welles' shoulders and exclaimed, "What can we do for the Secretary of the Navy for this glorious intelligence? He is always giving us good news. I cannot, in words, tell you my joy over this result. It is great, Mr. Welles, it is great!" As the two men walked together across the White House lawn, Lincoln said, "This will relieve Banks. It will inspire me."[22]

Indeed, Lincoln had good reason to be inspired. Vicksburg's conquest was the Union's most significant military triumph of the war so far. Not only did it remove an entire rebel field army from the strategic chessboard, but it also severed the Confederacy in two. The three Transmississippi states were now isolated from the rest of the Confederacy and could contribute little to its survival. The victory opened the Mississippi River to Union shipping, exposed the Mississippi and Alabama interiors to attack, and helped convince Lincoln that Grant was the general who would win the war. Vicksburg's fall was one of several important Union successes that summer. The same day Vicksburg surrendered, Robert E. Lee's Army of

Northern Virginia began its retreat back to Virginia after losing the Battle of Gettysburg in Pennsylvania, and Union soldiers repulsed a rebel assault on Helena, Arkansas. In Ohio, Confederate general John Hunt Morgan was captured with the remnants of his force on 26 July, ending his three-week-long raid. Finally, Maj. Gen. William Rosecrans' Army of the Cumberland maneuvered the rebels right out of Tennessee in an almost bloodless campaign that culminated in the occupation of Chattanooga on 9 September.

The Army could justifiably claim the bulk of the Vicksburg laurels, but Welles still had plenty of reasons to be pleased with the campaign's outcome. Most obviously, it vindicated his controversial selection of Porter as Mississippi Squadron commander. Except for Arkansas Post, Porter's record prior to his run past Vicksburg's batteries was one frenetic failure after another. By April Welles had just about lost confidence in him, and was increasingly critical of him in his diary. Victory at Vicksburg, however, put an entirely different face on things. Throughout the campaign, Porter demonstrated considerable tenacity, ingenuity, aggressiveness, and initiative. Moreover, his ability to cooperate effectively with Grant and Sherman stood in marked contrast with Du Pont's and Dahlgren's problems with army generals outside Charleston. Welles was a firm believer in rewarding success, so on 13 July he asked Lincoln to elevate Porter to permanent rear admiral. The president cheerfully agreed, even though Welles doubted that Lincoln's jaded view of Porter had changed much. Welles informed Porter of the good news the same day, and added his congratulations for a job well done. Fox followed up with a note of his own in which he referred to Porter's promotion as a "king's reward." Now that he was receiving the adulation and glory he had long sought, Porter could afford to be gracious. In a letter to Fox, he wrote, "I feel also that I am indebted to you, and the kind partiality of Mr. Welles for being where I am now; not forgetting the kind Providence, without whose aid all efforts are unavailing."[23] He also willingly gave the credit to his subordinates, especially Alexander Pennock, who had minded shop back at Mound City while Porter waged war. All this praise had not affected Porter's sense of humor. He confessed to Sam Phelps that his promotion made him feel like an old fogy, and he speculated that the other rear admirals, especially Goldsborough, would be horrified that he was now one of them. The big difference, though, was that Porter's star, unlike Goldsborough's or Davis' or Du Pont's, was still rising.[24]

Henry Bell's Interregnum

After he successfully ran past Vicksburg's batteries, Porter steamed downstream to end Farragut's lonely vigil at the Red River's mouth. This did not, however, terminate Farragut's responsibilities on the Mississippi. Nathaniel Banks' army began its siege of Port Hudson on 23 May, and Farragut felt obligated to support Union operations there with his warships. Although Farragut understood Port Hudson's strategic value and was determined to see the siege through to its end, he still found his mission unsettling. He had never really reconciled himself to riverine warfare, and he felt that he was neglecting his many other commitments in the Gulf. Galveston remained in Confederate hands, and the blockade along the Mexican border at the Rio Grande was particularly leaky. Moreover, on 6 April Welles warned him that the rebels might sortie out of Mobile with an ironclad to attack his blockaders. For now, though, all he could do was trust his subordinates in the Gulf and hope that Port Hudson fell soon. As the siege progressed, Farragut became increasingly tired and worn out. He had commanded the West Gulf Squadron for eighteen grueling months, and the continuous stress was taking its toll. His wife urged him to return home, and chided him for placing his ambition ahead of all else. Farragut denied that he was motivated by anything other than his duty—in fact, he dryly noted that in his line of work he was more likely to lose rather than gain additional honors—but he was coming around to her way of thinking. Indeed, as early as 27 March he had told Porter that he planned to ask the Navy Department to relieve him after he seized Mobile. Taking the rebel port would require oceangoing ironclads to combat the armored vessel the Confederates were supposedly constructing there, but on 3 June Welles informed him that Dahlgren's campaign against Charleston precluded their use elsewhere anytime soon. Farragut excelled at tackling specific problems such as assailing rebel ports, but he was less happy and effective overseeing the day-to-day routine of blockade duty. With no prospect of immediate action after Port Hudson fell, Farragut asked Welles for a furlough, which the navy secretary granted on 15 June. After Port Hudson finally surrendered on 9 July, Farragut put his affairs in order and left New Orleans on *Hartford* on 1 August, arriving in New York City nine days later. He expected to return by autumn to storm Mobile and Galveston and put an end to Confederate opposition along the Gulf Coast once and for all.[25]

Farragut originally intended to appoint Commo. Henry Morris to lead the West Gulf Squadron in his absence. Morris was the squadron's most senior officer after Farragut, and had overseen naval affairs in New Orleans while Farragut besieged Port Hudson. Unfortunately, Morris fell gravely ill and died on his way north with Farragut on *Hartford*. With Morris unavailable, Farragut instead turned to Commo. Henry Bell. Bell was currently skipper of *Brooklyn*, and had served faithfully as Farragut's best friend, chief lieutenant, sounding board, and voice of reason. Bell had been in the Gulf as long as Farragut, so Farragut had hoped Bell would accompany him north for a few months' rest. Indeed, Bell's wife was very disappointed when her husband did not walk down the gangplank at the Brooklyn Navy Yard with Farragut after *Hartford* docked. Once he realized the seriousness of Morris' sickness, Farragut determined that Bell was the logical choice to command the squadron while he was gone. When he arrived at New York City, Farragut immediately informed Welles of his decision. Welles, however, had never been as impressed with Bell as Farragut. It was not so much that Bell was a Southerner by birth, but rather that Welles did not believe that he had been sufficiently aggressive and enthusiastic. Welles had complained to Farragut of Bell's timidity off Galveston after the rebels reoccupied the town, and of rumors that Bell had fraternized with enemy officers in his cabin. Farragut had defended his friend to Welles, explaining that Bell was a loyal, prompt, dependable, and ardent officer who had fought well in the New Orleans campaign and during the unsuccessful assaults against Vicksburg. Despite his doubts about Bell, Welles confirmed Farragut's selection. After all, he expected Farragut to return to the Gulf in a few months, and the Navy had no major campaigns planned there until he got back, so there was little reason to be especially picky about who ran the squadron in Farragut's absence.[26]

On 31 July Bell received Farragut's orders to sail to New Orleans from his Galveston station to assume temporary command of the West Gulf Squadron. He neither expected nor wanted the assignment. Bell realized that, with Farragut gone, he would merely be a caretaker with lots of responsibilities and few opportunities for distinction, and he worried about again contracting one of the Mississippi River's many diseases. He confided in his diary that the post was "a barren honor," but dutifully added, "God's will be done."[27] Indeed, honor was part and parcel of Bell's makeup. Born in North Carolina in 1808, Bell joined the Navy as a midshipman in 1823. He fought Caribbean pirates later in the decade, and participated in storming the Chinese barrier forts outside Canton with Foote in 1856, but on the whole his prewar career was uneventful. When the Southern states seceded, everyone expected Bell to resign his commission and join the Confederacy. After all, he and his wife were both Southerners, and Bell was an outspoken Democrat who hated abolitionists. For Bell, however, it was a matter of duty, so he opted to remain loyal. In fact he asked the Navy Department to change his place of residence in its records from North Carolina to New York. As he stated emphatically to one person, "I have made up my mind. *I shall stand by the flag.*" By way of elaboration, he explained, "You are a civilian; but that flag which you regard only in the light of political principles, is to me the symbol of a sworn and solemn duty. It represents the Government that is to me both master and benefactor. I have my own political opinions and as much freedom in them as any other citizen; but when the Government once gives the *word of command,* all questions of politics stop there for me, and I must do my duty. . . . I cannot, shall not make my duty as an officer a question of moral casuistry; if so, there would be no Government."[28]

Since then, Bell had had an active war. He had gone to the Gulf as Farragut's chief of staff, and participated in the New Orleans campaign, the two fruitless assaults on Vicksburg later that summer, and the blockades of Galveston and Mobile. He had demonstrated his courage by working under fire to sever the chain boom across the Mississippi River before the run past Fort Jackson and Fort St. Philip, and later in leading the Marines through an angry mob to occupy New Orleans' city hall. His biggest contribution to Union success, though, was tempering Farragut's impulsive nature and convincing him to see reason. For instance, Bell helped dissuade Farragut from assailing Vicksburg during his first expedition up the Mississippi in May 1862. Like Farragut, he disliked operating on the river, and had hoped for a transfer to another theater. Farragut instead placed him in command of *Brooklyn.* Bell was a ruddy, clean-shaven man with thatched hair. Although a severe disciplinarian, he was respected by his men. He possessed a dry sense of humor, was easily exasperated, and had little patience for blustering officers such as David Porter. Bell had proven himself a good second in command, but it remained to be seen how he would perform in the top spot.[29]

Bell's suspicions that his new post would be an unrewarding ordeal personally and professionally proved accurate. He lived ashore at a friend's house in New Orleans, but kept his office on board the sloop USS

Pensacola. Bell was amazed by the enormous amount of paperwork maintaining the West Gulf Squadron required. Unlike Farragut, who had such a blasé attitude about such things that Welles complained of his indifference to rules and procedures, Bell was a diligent officer committed to running a tight bureaucratic ship. He did not initially have a chief of staff to help him, so the work almost overwhelmed him. One officer discovered Bell scribbling at his desk behind a huge stack of papers that almost completely concealed him. Happily for his health, in September he acquired a chief of staff who relieved him of much of this burden. This enabled him to concentrate on the numerous problems he inherited from Farragut. Although Porter had now assumed control of the entire Mississippi, Bell's geographic obligations remained immense. He had to keep a close eye on Mobile, the remote Texas coast, and the especially porous Rio Grande. Unfortunately, Bell lacked the necessary resources to maintain the tight blockade that the Navy Department desired. He did not have enough warships to begin with, and at any given time one fourth of those on hand were under repair. A shortage of boiler plate iron, plate iron, bar iron, boiler rivets, and boilermaker and blacksmith tools retarded turnaround time, so other barely seaworthy vessels had to remain on station despite their severe mechanical problems. Bell was also short of men, and at one point he wrote to Welles that he needed five hundred sailors just to fully crew his serviceable vessels. Coping with this manpower shortage forced Bell to keep sailors on board their warships even after their enlistment terms had expired. To exacerbate the problem, sickness broke out at the Pensacola Navy Yard, in New Orleans, and on some of the warships, further reducing the number of men available for duty. All in all, it was small wonder that Bell longed for Farragut's return.[30]

In addition to these interminable logistical problems, Bell also had to cope with a humiliating defeat on 8 September. For political and diplomatic reasons stemming from the French invasion of Mexico, Lincoln and the War Department wanted to establish a Union foothold in Texas, so Nathaniel Banks planned to land four thousand troops under Maj. Gen. William Franklin at Sabine Pass for an overland march to Galveston. Bell dispatched his four available gunboats under Lt. Frederick Crocker to provide the necessary naval support. Intelligence reports indicated that Sabine Pass was scantly defended, so Bell did not expect significant opposition. In fact, the rebels had only a half-dozen light guns manned by fifty artillerymen in a ramshackle earthwork, but they made up in spirit what they lacked in numbers and firepower. In a short half hour, the Confederates captured two of the gunboats after one ran aground and rebel fire punctured the boiler of the other. The third gunboat beat a hasty retreat, and the last one never even made it into the fight. Without effective naval assistance, Franklin decided to abandon the operation and return to New Orleans.[31]

Bell exaggerated when he labeled the defeat a disaster, but it certainly did nothing to improve his standing with Welles. Farragut was visiting the Navy Department building when Welles received Bell's dispatch announcing the proposed expedition. After reading it, Farragut predicted to Welles that the operation would fail. He explained that the Army always expected the Navy to smother enemy batteries so soldiers could march unopposed into rebel fortifications and claim all the credit. The gunboats, Farragut continued, lacked sufficient firepower to do the job. Thus forewarned, Welles was not altogether surprised with the repulse, but he was still very disappointed. After receiving the details, Welles concluded that the operation was poorly conceived, planned, and executed. Bell's attempt to deflect blame onto the Army and Crocker failed to prevent Welles from issuing him a sharp reprimand. Although Welles acknowledged Bell's zealousness and willingness to cooperate with the Army, he criticized him for committing too few vessels under such a junior officer into a narrow and shallow channel. The fiasco no doubt helped persuade Welles that Bell was unfit for permanent wartime squadron command.[32]

On 22 January 1864, Bell happily surrendered command of the West Gulf Squadron to Farragut and returned north for a reunion with his family after nearly two years of hard and continuous service. Welles appointed him commandant of the Brooklyn Navy Yard, but his health had broken down, and for a while some doubted his survival. Fortunately, Bell recovered, and shortly after the war ended Welles assigned him to lead the East India Squadron. In 1867, while awaiting his successor, Bell decided to visit the American minister in Osaka, Japan. He drowned when his barge capsized in a river on his way to shore, and he was buried in Hiogo, Japan.

"Damn the Torpedoes!"

While Bell watched the paperwork pile up around him, Farragut was basking in the adoration of a grateful Union public. A crowd gathered dockside when *Hartford* reached the Brooklyn Navy Yard, and curious onlookers counted the 240 shot and shell marks that peppered the vessel. After discharging the crew and informing a disappointed Mrs. Bell that her husband was still in the Gulf, Farragut and his wife traveled to their home at Hastings-on-Hudson. Mrs. Farragut's contribution to the happy reunion was news that their son, who had returned north after the Port Hudson run, had been appointed to West Point and passed the admission exam. Farragut also received a kind note from Welles congratulating him on a job well done, and inviting him to Washington for a visit after the heat broke and he had rested with his family. Farragut intended to relax and unwind, but the locals did not give him much of a chance. He was the Union Navy's biggest hero to date, and people bombarded him with invitations to dinners and speeches. Prominent New York citizens signed a petition thanking him for his services, and the New York Chamber of Commerce passed a resolution honoring him. Farragut was a convivial, friendly, and generous man naturally drawn to such gatherings, but even he admitted after a while that all the attention was wearying. Despite his success and adulation, he maintained his humility. For example, he donated his first $500 of prize money toward the construction of a local Episcopal church as an expression of gratitude to the Lord for his safe delivery from danger. After the weather cooled, he journeyed to Washington twice to consult with Navy Department officials and catch up with old colleagues. Welles had him and some friends over for dinner, and noted in his diary that the more he saw of Farragut, the more impressed he was with him. He liked Farragut's sincerity, ardor, and gentleness, and felt vindicated in his decision to appoint him to squadron command.[33]

On the afternoon of 5 January 1864, Farragut stood on *Hartford*'s bridge during a blinding snowstorm and watched a pilot take the warship out through the upper and lower reaches of New York Bay to Sandy Hook for her trip down the coast. He had recently injured his foot, so he hobbled about as he watched his officers and crew go about their duties. Although he had enjoyed his furlough, he was eager to return to the Gulf to finally assail Mobile. Farragut had wanted to storm the Alabama city after he occupied New Orleans, but Welles and Fox ordered him to help open up the Mississippi River instead. Doing that consumed a year, but after Port Hudson fell Farragut again focused his attention on Mobile. Unfortunately, he made the mistake of assuring the Navy Department that few blockade-runners made it through the Union Navy cordon outside the city. This inadvertently helped dissuade Welles and Fox of Mobile's strategic importance, and they gave top priority to Dahlgren's campaign against Charleston. Rather than cool his heels in the Gulf, Farragut came home for his well-deserved leave. Throughout the winter, however, the Navy Department heard persistent rumors that the rebels were constructing a monstrous ironclad up the Alabama River at Selma, named *Tennessee,* that when completed could destroy Union Navy blockaders offshore. This motivated an alarmed Welles to authorize Farragut to assault Mobile as soon as possible. Manning *Hartford* took longer than anticipated, and in fact Farragut finally had to press sailors from *Niagara* to attain a full complement, so he did not depart until after the New Year. Whatever he had lost in time, though, he made up with the acquisition of a new chief of staff, Capt. Percival Drayton. After Du Pont's failed attack on Charleston, Welles had transferred Drayton to the Brooklyn Navy Yard as chief ordnance officer, where Farragut found him. Like most people, Farragut respected Drayton's thoughtfulness and efficiency, so he asked him to become his chief of staff. Drayton liked shore duty, but he believed that a naval officer's place in wartime was afloat, so he agreed. Like Bell before him, Drayton helped compensate for Farragut's bureaucratic inadequacies, and would play a key role in the upcoming campaign.[34]

After he arrived in New Orleans and relieved the beleaguered Bell, Farragut threw himself into his duties, both naval and social. While New Orleans remained a largely secessionist town, Union occupation had not dampened its festive atmosphere for long. Masked balls were the rage that season, and Farragut took advantage of the city's diversions despite his rheumatism. Drayton was accustomed to Du Pont's monastic shipboard existence while commanding the South Atlantic Squadron at Port Royal, so he found Farragut's revelry disconcerting. In fact, Farragut himself admitted that New Orleans' distractions could ruin a good officer, so he made sure to get out of the city often and focus on his naval responsibilities. He inspected all points of his far-flung squadron, from the Texas coast to the Pensacola Navy Yard, and did his best to rectify its numerous logistical deficiencies by complaining to the Navy Department about shortages in supplies, coal, and men. As usual, Farragut found maintaining the blockade tedious, but he tried to look at it philosophically, explaining to his family: "I confess blockading is a most disagreeable business; but, if we had nothing but agreeable things to do in war, everybody would be in the Navy, and no one would be worthy of reward or promotion. We must take the world as it comes. This is a state of civil war, and God has dealt with us most generously thus far. My duty is arduous mentally only."[35] But even mental strain could be taxing; Drayton, for example, was like Bell before him amazed by the huge amount of paperwork the squadron generated for him to manage. Although Farragut tried to avoid these bureaucratic tasks as much as possible, he was deluged with letters from autograph seekers as a kind of penance for his accomplishments.[36]

Despite Farragut's innumerable social and naval obligations, he understood that shutting down Mobile was his top priority. In fact, now that he had a specific mission on which to focus his attention, he was in his element. On 7 February he placed James Palmer in charge of the West Gulf Squadron's day-to-day operations so he could concentrate on the main objective. Palmer's social network in New Orleans rivaled even Farragut's, which may have explained his willingness to remain behind in the city to attend to the squadron's routine business. Even before he officially relieved Bell of his command, Farragut had stopped outside Mobile Bay, transferred to the gunboat USS *Octorara*, crossed the harbor bar, and reconnoitered its defenses. His observations convinced him that he could successfully storm the bay with the assistance of five thousand soldiers and at least one ironclad. Unfortunately, acquiring these necessary resources took much longer than he anticipated. In March Nathaniel Banks undertook a disastrous campaign up the Red River in an effort to conquer western Louisiana and eastern Texas that ended in defeat at the Battle of Sabine Crossroads on 8 April. By the time Banks finally slunk back into New Orleans in May, his army had suffered so many casualties that it was incapable of launching a full-scale offensive against Mobile anytime soon. Happily, on 8 July, Banks' successor, Maj. Gen. Edward Canby, agreed to provide two thousand bluecoats under Maj. Gen. Gordon Granger to help Farragut by assailing the forts guarding the harbor, though not the city itself. Granger and fifteen hundred soldiers arrived on 1 August and landed immediately on Dauphin Island to besiege Fort Gaines.

Obtaining the ironclads was just as frustrating. Farragut did not think much of these clumsy and claustrophobic warships, explaining: "If a shell strikes the side of the *Hartford*, it goes clean through. Unless somebody happens to be directly in the path, there is no damage excepting a couple of easily plugged holes. But when a shell makes its way into one of those damned tea-kettles, it can't get out again."[37] Having witnessed *Arkansas'* rampage two summers ago on the Mississippi, though, he recognized that he needed ironclads to subdue *Tennessee*. To increase the odds in his favor, in mid-January Farragut asked for four ironclads, two from Porter's Mississippi Squadron and two from the Atlantic Coast. Porter had a couple of new riverine ironclads, USS *Chickasaw* and USS *Winnebago,* near completion in St. Louis, and he promised them to Farragut in March or April. Farragut knew his adoptive brother well enough, however, to estimate that he would not get them until May or June. Retrofitting them for operating in the Gulf in fact delayed their appearance until 30 July. For his part, Welles did not order two new oceangoing monitor ironclads, USS *Manhattan* and USS *Tecumseh,* to the Gulf until he learned in June that *Tennessee* had reached Mobile Bay. Although *Manhattan* was on hand on 20 July, *Tecumseh* did not join the rest of the fleet until the evening of 4 August, the day before the battle. Until he had all these ducks in a row, Farragut could do little but plan, prepare, and fret.[38]

Farragut's frustration that summer was matched by that of the Lincoln administration and the Union public. In March Lincoln appointed newly promoted Lt. Gen. Ulysses Grant commander of the entire Union Army. To bring the rebels to their knees, Grant developed a grand strategic plan that called for simultaneous advances by all the Union armies into the Confederate hinterland. In the West William Sherman's army group would attack Atlanta, Georgia, and the Army of Tennessee, which defended the city. In the East Grant intended to accompany George Meade's Army of the Potomac in its campaign against Richmond and Robert E. Lee's Army of Northern Virginia. Grant's sweeping offensive began in early May, but it quickly degenerated into a brutal war of attrition. Although the Army of the Potomac slowly pushed Lee's troops southward, it suffered 55,000 casualties in a series of continuous engagements that included the battles of the Wilderness, Spotsylvania, and Cold Harbor. In mid-June Grant attempted to sever Richmond's rail lines to the rest of the Confederacy by seizing Petersburg, Virginia. Unfortunately, the Army of the Potomac botched the operation, which led to a nine-month-long siege of the city. In Georgia Sherman's army group sustained fewer losses, but it too had little luck in occupying Atlanta and destroying its rebel protectors. This emerging and bloody stalemate came at a particularly inopportune time for the president. Lincoln was running for reelection in the fall, and his political future was tied to the progress of the Union military. By late summer victory seemed as far away as ever, which gave plenty of fodder to Democrats who argued that the war was a failure. Unless the Army and the Navy could win some battles soon, chances were good that George McClellan, the Democratic presidential candidate, would prevail in the upcoming election.

Mobile was located at the northern end of a shallow thirty-mile-long bay. Farragut's objective was not the city itself, whose reduction was an army responsibility if and when that service got around to fulfilling it, but rather the bay. Once the bay was in Union hands, Mobile would be no more useful to blockade-runners than any other landlocked Alabama city. Seizing it, however, would not be easy; the Confederates had constructed an intricate and multilayered defensive network designed to thwart just the kind of naval assault Farragut envisioned. Adm. Frank Buchanan led the rebel's naval flotilla, consisting of *Tennessee* and three small wooden gunboats, and he was determined to fight as hard here as he had at Hampton Roads while commanding *Merrimack*. The Confederates had also fortified a series of islands at the head of the bay: Fort Powell on tiny Tower Island, Fort Gaines on Dauphin Island, and Fort Morgan on Mobile Point. Although Fort Powell was just a small earthwork, Fort Gaines and Fort Morgan were solid brick structures from the prewar era, mounting a combined total of sixty-three cannon. Finally, the rebels laced the channels connecting the islands with pilings and, halfway between Dauphin Island and Mobile Point, a triple line of mines.

The only unobstructed way through these obstacles was via a small two-hundred-yard gap under Fort Morgan's guns that blockade-runners used to slip in and out of the bay. Farragut planned to take his warships through the gap and destroy Buchanan's flotilla, thus isolating the forts and closing down the harbor. Although he had conducted his previous runs at night, he decided to make this one by day so he could better see and control the action. It was a simple enough scheme, but that did not make it easy. Even if Farragut's vessels survived the gauntlet run past Fort Morgan, *Tennessee* might blow them out of the water shortly thereafter. To deal with these problems, Farragut decided to attack in two columns. The four ironclads constituting the starboard column would steam closest to Fort Morgan to draw its fire away from the more vulnerable wooden vessels advancing to the west. As at Port Hudson, Farragut ordered a gunboat lashed onto the port side of each of the seven larger wooden warships to provide additional propulsion. Once the warships entered the bay, Farragut expected the ironclads to destroy *Tennessee,* but he attached boiler plate to the bows of some of his wooden vessels in case they had the

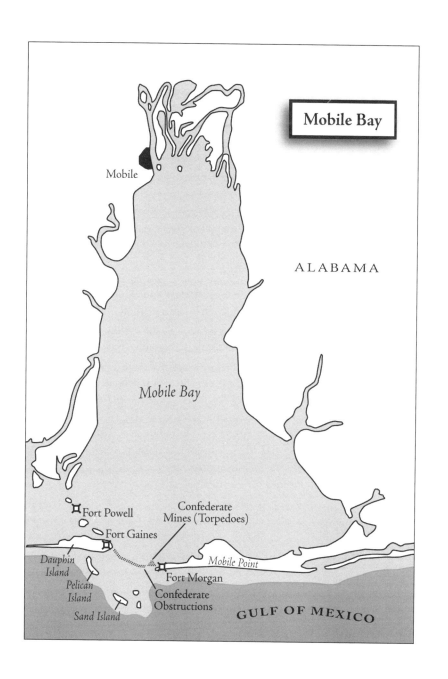

Mobile Bay

Mobile

ALABAMA

Mobile Bay

Fort Powell

Confederate
Mines (Torpedoes)

Fort Gaines

Dauphin
Island

Mobile Point

Pelican
Island

Fort Morgan

Sand Island

Confederate
Obstructions

GULF OF MEXICO

opportunity to ram the rebel behemoth. Against his better judgment, Farragut let his officers dissuade him from spearheading the charge in *Hartford*. *Brooklyn* had a minesweeping device called a cowcatcher attached to her bow, and the officers felt she was better suited to lead the fleet into the bay. As far back as mid-July, Farragut had instructed his captains to prepare for the assault by removing most of their spars and rigging, placing chains and sandbags around vital machinery, stringing chain garlands over the sides to protect the hulls, and making any other necessary adjustments for the mission. The night before the operation, Farragut took his skippers on board a tender for a last look at the rebel defenses. Satisfied that everyone knew what was expected of him and that he had done all he could, he retired to his cabin and wrote to his wife, "I am going into Mobile Bay in the morning, if God is my leader, as I hope He is, and in Him I place my trust."[39]

On the morning of 5 August, Farragut and Drayton breakfasted together while their officers and men maneuvered their vessels into position and made last-minute preparations for the upcoming battle. At 5:30 AM Farragut took a sip from his teacup and said, "Well, Drayton, we might as well get under way." The warships steamed toward the bay while Farragut clambered to the port main rigging for a good look at his fleet. At 6:20 AM *Tecumseh* fired a ranging short at Fort Morgan, but it took nearly an hour for the action to become general. As the cannon exchanges grew increasingly heated and smoke settled over the Union warships, Farragut climbed higher and higher, ratline by ratline, for a better view, until he settled himself just below the maintop. Seeing this, an alarmed Drayton finally sent a sailor aloft to tie Farragut to the rigging. So far, the battle was going according to plan, even though *Tennessee* and her consorts had taken up a position behind the minefield to pump shells into the approaching Union vessels. Shortly after 7:30 AM, however, *Brooklyn* suddenly backed engines and stopped because lookouts had spotted mines straight ahead. The other vessels coming up behind her did likewise, compressing the column and bringing it to a standstill. Farragut did not understand *Brooklyn*'s actions, and shouted through his brass trumpet for her to go forward. To the right, in the ironclad column, *Tecumseh* had poured on steam in an effort to come to grips with *Tennessee*. At about the same time *Brooklyn* stalled, *Tecumseh* hit a mine. The resulting explosion sent a huge fountain of water into the air, and *Tecumseh* rolled over and plunged to the bottom in less than a minute. Of her 144-man crew, only a score or so managed to scurry out of the turret in time. All the while, Confederate batteries ashore and afloat continued to pulverize the now stationary wooden warships. One sailor on *Hartford* later remembered: "The sight on the deck was sickening beyond the power of words to portray. Shot after shot came through the side, mowing down the men, deluging the decks with blood, and scattering the mangled fragments of humanity so thickly that it was difficult to stand on the deck, so slippery was it."[40] Despite *Tecumseh*'s cautionary experience only moments earlier, Farragut did not hesitate, but instead decided to take the lead with *Hartford* and plow directly through the minefield into Mobile Bay. Using contemporary parlance for mines, he allegedly called to those below, "Damn the torpedoes! Full speed ahead!" *Hartford* and her attached gunboat, USS *Metacomet*, steamed past *Brooklyn*'s port side and plunged into the minefield, followed by the remaining dozen warships in the column. Here, as elsewhere in his career, Farragut's impetuosity paid off. Although some heard the popping of percussion caps under *Hartford*'s hull, water corrosion had rendered the underwater devices useless, so the vessels emerged from the minefield unscathed. Once in the bay, Farragut ordered the gunboats unlashed, and they ran off or ran down their Confederate counterparts. *Tennessee* attempted to engage and ram Farragut's wooden warships, but the three surviving Union ironclads intervened, forcing Buchanan to draw off toward Fort Morgan.[41]

Farragut understood that *Tennessee* posed a threat to the new Union foothold in Mobile Bay. He intended to attack her, but he first wanted to give his tired crews a chance to catch their collective breath, clear the decks of casualties and debris, and eat breakfast. Before they could do so, however, lookouts spotted the rebel ironclad chugging across the bay to reengage the Union fleet. Farragut was surprised that Buchanan would do something so foolish, but Buchanan had his reasons. *Tennessee* was running short of coal, and she drew too much water to recross the sandbar and take refuge up the Alabama River. Rather than give up without a fight, Buchanan planned to inflict as much damage as possible on the Union warships, despite the odds against him. As the rebel ironclad approached, Union vessels raised their anchors and got up steam. In the ensuing melee, Union warships rammed and pounded *Tennessee* until the battered ironclad finally struck her colors. Union losses for the day totaled 172 killed—93 of whom went down with *Tecumseh*—and 170 wounded. The Confederates, for their part, suffered only 12 dead and 19 injured.

Once safely ensconced in Mobile Bay, Farragut turned his attention to the now isolated rebel forts. The Confederates evacuated Fort Powell that night, and Fort Gaines surrendered three days later. The rebels in Fort Morgan, on the other hand, withstood a combined army and naval bombardment for two weeks until they finally ran up the white flag on 23 August.

Welles was elated when on the evening of 8 August he received word of Farragut's triumph in Mobile Bay. From a purely military perspective, it was the Union's most important victory in months. Mobile was still in Confederate hands, but without the bay it was of no use to blockade-runners. Except for the remote and isolated Transmississippi, only Wilmington and Charleston now connected the Confederacy to the outside world, and Welles was working hard to shut them down too. Moreover, the battle raised Union morale during a bloody and gloomy period in the war, giving Lincoln's reelection campaign a much-needed boost. Although later generations remembered Farragut's call to "Damn the torpedoes," it was the news that he had been strapped to the rigging during the battle that caught the fancy of most contemporaries. Welles was hardly surprised when Halleck and other army officers in the War Department expressed little excitement about Farragut's success, but he expected more than he got from the president. Lincoln merely commented that Farragut's accomplishment might relieve pressure on Sherman's ongoing operations against Atlanta. The president and the Army may have lacked sufficient appreciation for Farragut's efforts, but his naval colleagues knew better. Charles Steedman, for example, called it a "glorious achievement." Congress agreed, and in December voted for the establishment of the new rank of vice admiral with Farragut in mind, which Lincoln immediately bestowed on him.[42]

No doubt the tears Farragut shed upon seeing *Hartford*'s dead laid out on deck were a genuine expression of grief, but he also likely wept from the release of months of tension and stress. Victory at Mobile Bay, though, did not improve his declining health. Farragut worked best when he concentrated on one particular problem at a time, be it surmounting New Orleans' defenses or storming Mobile Bay. Now that Mobile Bay was in Union hands and there were no immediate plans to attack the city itself, Farragut had nothing specific on which to direct his energies. To be sure, he still faced innumerable logistical problems that his recent hard-fought battle did little to alleviate. Many of his vessels, especially *Brooklyn*, required refitting, and he still had to keep an eye on the extensive Texas coastline. Without a definite mission to tackle, however, Farragut's mental and physical condition deteriorated. In the weeks after the battle, he complained of exhaustion, and regretted that he was unable to leave his ship more and stretch his legs. Worse yet, he was afflicted by a terrible case of the boils. He was in so much pain that he fainted while talking to another officer. On 27 August he wrote to Welles to ask for a furlough so he could recover from his recent exertions. After some confusion over Farragut's role in the planned assault on Wilmington, on 9 November Welles gave him permission to come home. Farragut's constitution had improved by then, but, without any prospect of immediate action, he opted to accept Welles' offer. He left on *Hartford* from Pensacola Bay on 30 November and reached New York thirteen days later.[43]

Despite its Democratic antiwar leanings, New York City had always been kind to Farragut. Not only did leading citizens greet his return with a formal reception, but they also raised $50,000 in bonds for him to purchase a home. In response to their generosity, Farragut explained with his usual humility: "I don't think, however, that I particularly deserve anything from your hands. I can merely say that I have done my duty to the best of my abilities. I have been devoted to the service of my country since I was eight years of age, and my father was devoted to it before me. I have not specially deserved these demonstrations of your regard. I owe everything, perhaps, to chance, and to the praiseworthy exertions of my brother officers serving with me. That I have been fortunate is most true, and I am thankful, deeply thankful for it, for my country's sake."[44]

A month after his arrival, Farragut and his wife visited Washington to receive his appointment to vice admiral, where he was further lionized and feted. Welles was as usual kind and solicitous. He invited Farragut and his wife to his home for dinner, and took him over to the White House to see the president. The Senate officially recognized him, and he and Mrs. Farragut accompanied Lincoln and his wife to the opera, even though Farragut did not especially like that kind of entertainment. As things turned out, Farragut played no further part in the fighting. He and his wife remained in Washington for several months enjoying the capital's social activities that surrounded the president's second inauguration. In fact, a concerned Drayton wrote to Mrs. Farragut to warn her against letting her husband's life degenerate in lassitude. As the conflict drew to a close, they traveled to Norfolk so Mrs. Farragut could see her relatives, but not surprisingly they received a chilly welcome. After the war Congress voted Farragut its thanks for his actions at Mobile Bay, and in July 1866 President Andrew Johnson appointed him a full admiral. He commanded the European Squadron in 1867–68, and the following year journeyed to California. On his way back east, he suffered a heart attack in Chicago from which he never fully recovered. He died of a stroke at the Portsmouth Navy Yard in August 1870 while visiting its commandant, his friend Alexander Pennock. President Ulysses Grant, his cabinet, and thousands of others walked in his New York City funeral procession in a blinding rainstorm, a moving testament to his basic decency and his reputation as the country's greatest naval officer.[45]

Before he left Pensacola Bay, Farragut had assigned command of the West Gulf Squadron to Commo. James Palmer. For Farragut, it was an easy and obvious choice. Not only was Palmer Farragut's most senior officer, but he had also run the squadron's day-to-day functions while Farragut concentrated on storming Mobile Bay. Palmer had arrived in the Gulf shortly after the New Orleans campaign as commander of *Iroquois,* and had served faithfully and continuously since then. He had participated in the two failed attacks on Vicksburg and in operations against Port Hudson. He first came to Farragut's attention during the run past Vicksburg's batteries by closely supporting *Hartford* while under heavy fire. The two men shared aggressive instincts and convivial natures, and they had developed a warm friendship. Farragut thought these reasons were more than sufficient to justify Palmer's selection, but he also wanted to give him the opportunity to prove himself in an independent command. Despite the valor and competence that Palmer had consistently exhibited in the Gulf, he had also suffered from uncommon bad luck earlier in the war that had tainted his record and almost ruined his career.[46]

Palmer was born in 1810 in New Jersey, and entered the Navy as a midshipman fifteen years later. He participated in the war against Sumatran pirates in the East Indies, and later commanded a schooner on blockade duty during the Mexican War. His health declined after the conflict, and in 1855 the Efficiency Board recommended placing him on the reserved list. Palmer appealed the verdict and was reinstated, and the Civil War's start found him skippering *Iroquois* in the Mediterranean. After returning to New York City, the Navy Department sent him and *Iroquois* to the West Indies in search of the rebel raider *Sumter.* Palmer found the Confederate vessel at Saint-Pierre, Martinique, on 14 November 1861, and stationed *Iroquois* outside the harbor to intercept her when she left. Palmer had so much water to cover, though, that he doubted his chances from the beginning, so he was disappointed but not surprised when *Sumter* slipped away on the evening of 23 November. Welles, on the other hand, was neither understanding nor forgiving of Palmer's failure. He removed Palmer from his command and ordered a court of inquiry into his conduct. Fortunately for Palmer, the officer corps rallied to his defense, and the court cleared him of any wrongdoing. Welles thereupon returned him to *Iroquois* and sent him to join Farragut in the Gulf, where he served capably for two and a half years until Farragut put him in charge of the squadron. Although Palmer had had opportunities to return north, he turned them down. As a bachelor, he had no family waiting for him, and had embraced the squadron as his family. Palmer was a clean-shaven, spare, and dignified man with a sense of rectitude that bordered on pomposity. Farragut's son remembered that Palmer entered battle immaculately dressed in full uniform, right down to his white gloves. A stickler for protocol, he was outraged by John Dahlgren's promotion to rear admiral when there were so many more deserving officers who had seen much more combat. Underneath his starchy manner, however, Palmer was kind and generous, and enjoyed a wide circle of friends of whom Farragut was simply the most famous. [47]

Like Henry Bell before him, Palmer inherited many of Farragut's intractable problems. As usual, the West Gulf Squadron was short of sailors and Marines, and Palmer complained that he had to borrow soldiers from the Army to guard naval storehouses and hospitals. Warships patrolling the Texas coast were in desperate need of repair, and there was confusion over Pensacola's status as an open port. Palmer moved his headquarters to New Orleans and worked closely and effectively with Edward Canby on plans for an interservice campaign against Mobile. Just before the offensive began, though, Palmer was deeply surprised and disappointed to learn on 10 February 1865 that Welles had dispatched Acting Rear Adm. Henry Thatcher to supersede him as squadron commander. Although Welles authorized Palmer to return home upon Thatcher's arrival if he so chose, he opted to remain and pledged to work cheerfully with his new boss. In a war full of prima donnas who nursed bruised egos and hurt feelings, Palmer demonstrated a rare selflessness. He served loyally as Thatcher's right-hand man during the siege of Mobile in March–April 1865 by leading the ironclads. After Mobile fell on 12 April,

Thatcher wrote to Welles that Palmer "rendered me most efficient and untiring service throughout the attack upon the defenses of the city, which has resulted so favorably to our arms, and I am indebted to him for the admirable manner in which the vessels to be employed for this service were prepared under his supervision previous to my arrival on the station, and I part with him with reluctance and regret."[48] Palmer returned to New York in May, but in December Welles assigned him to command the resurrected West India Squadron. He was at Saint Thomas in *Susquehanna* when an earthquake and tidal wave devastated the island. He contracted yellow fever while helping the survivors, and died a few days later, on 7 December 1867.[49]

Palmer's replacement, Acting Rear Adm. Henry Thatcher, was born in Maine in 1806. He was the grandson of Henry Knox, who commanded George Washington's artillery during the Revolutionary War and later became the country's first secretary of war. Thatcher was educated in Boston, and briefly attended West Point before accepting a midshipman's warrant in 1823. He fought with Porter's father against the West Indies pirates, and subsequently served in the Pacific, the Mediterranean, and off the Brazilian coast. Like so many other Union naval officers, Thatcher eschewed the fashionable beards of the era, preferring a clean-shaven look, which, combined with a large nose and a sad look about his eyes, made him appear to have the world's weight on his shoulders. When the Civil War began, he was executive officer at the Boston Navy Yard, but in February 1862 Welles assigned him to skipper the sloop USS *Constellation* and sent him to the Mediterranean in search of rebel raiders. *Constellation*'s mission lasted eighteen months, and after he returned in September 1863, Welles gave him *Colorado* and dispatched him to the West Gulf Squadron, where he ran the Mobile blockade during Bell's tenure. From there the Navy Department eventually transferred him to the North Atlantic Squadron. During Porter's assaults on Fort Fisher, Thatcher ably led a division of warships in the thick of the fighting. He was at the Brooklyn Navy Yard with *Colorado* when he received Welles' orders to take over the West Gulf Squadron, and arrived at Mobile Bay with that warship on 20 February 1865.

Welles' selection of Thatcher was puzzling. Thatcher had served as long as anyone, but mostly on the war's periphery, away from the great events that had shaped the course of the conflict. There were other available high-ranking officers who had seen more action and possessed more administrative experience, such as John Rodgers, Percival Drayton, and of course James Palmer. Drayton speculated that Welles gave Thatcher the assignment to avoid alienating either him or Rodgers, though there is no evidence that Rodgers campaigned for the post. Palmer had been in the Gulf for years, was familiar with its geography, had the necessary administrative background, and was thoroughly battle-tested. Moreover, no doubt Farragut vouched for him when Welles consulted him on the subject. On the other hand, it is possible that Welles still remembered Palmer's failure to capture *Sumter,* or that he believed that Palmer needed a rest after his long ordeal in the Gulf. As for Thatcher, he had a number of attributes that recommended him. He was a diligent, resolute, and efficient man with strong convictions, considerable integrity, good judgment, and plenty of kindness—the very embodiment of the Christian gentleman to which so many nineteenth-century naval officers aspired. Although he had not fought under Farragut, Farragut had heard good things about his performance running the Mobile blockade that he conveyed to Welles. Thatcher also enjoyed plenty of seniority as a commodore who had ranked sixteenth on the list of loyal commanders at the war's start. Most important of all, however, Porter gave Thatcher a glowing review for his role leading a division of warships during the two assaults on Fort Fisher in December 1864 and January 1865. Porter wrote to Welles:

> First and foremost on the list of commodores is Commodore H. K. Thatcher, full of honest zeal and patriotism; his vessel was always ready for action, and when he did go into it, his ship was handled with admirable skill. No vessel in the squadron was so much cut up as the *Colorado*. . . . I believe Commodore Thatcher would have fought his ship until she went to the bottom, and went into the fight with a full determination to conquer or die. There is no reward too great for this gallant officer; he has shown the kind of ability naval leaders should possess—a love for fighting and an invincible courage.[50]

Nothing impressed Welles more than an officer willing to fight to the bitter end. Whatever Welles' doubts about Porter's character, he respected his combativeness, and Thatcher's ability to measure up to Porter's exacting standards in that regard certainly impacted the navy secretary's thinking. Finally, Thatcher was comparatively fresh, whereas officers such as Rodgers, Drayton, and Palmer had been on station in war zones for years.

Drayton himself admitted afterward that while he would have been honored to command the West Gulf Squadron, he was equally grateful for the opportunity to remain at home. There may have been a logic to Welles' thinking, but his decision to appoint a man with minimal administrative experience to a position that required directing, supplying, crewing, deploying, and managing more than seventy-five warships in the Navy's most remote squadron was questionable.[51]

Thatcher's primary mission was to support the Army's long-awaited assault on Mobile. When he arrived at Mobile Bay, he was pleased to learn that Palmer had already coordinated and organized the Navy's contribution to the campaign with Canby. Canby's offensive got under way on 17 March 1865, a month behind schedule, and he and his 45,000 troops reached Mobile's defenses ten days later. Thatcher's vessels played a key role by transporting soldiers and bombarding enemy positions along the shore. The Navy's biggest challenge, as things turned out, was not rebel cannon, but rather rebel mines. Although Thatcher had ordered the waters around Mobile swept for the underwater devices, his crews did not get them all. Mines sank three ironclads and several smaller ships before Mobile finally fell on 12 April. Welles was pleased, but hardly surprised, with the victory. Indeed, by that time the war was almost over. Robert E. Lee's Army of Northern Virginia had capitulated at Appomattox Court House on 9 April, and other Confederate armies were in the process of following suit. Thatcher's mission therefore changed to one of occupying surrendered Confederate posts along the Texas coast and demobilizing portions of his squadron. On 9 June Welles directed Thatcher to assume command of the newly consolidated Gulf Squadron. After the Navy Department dissolved the Gulf Squadron in May 1866, Welles assigned Thatcher to take over the North Pacific Squadron, which he led until August 1868. He finished his active career as port admiral at Portsmouth, New Hampshire, from 1869 to 1870, and died in Boston a decade later.[52]

Phillips Lee's Denouement

Throughout 1863 and 1864, Acting Rear Adm. Samuel Phillips Lee continued his unheralded and thankless chore overseeing the Wilmington blockade as North Atlantic Squadron commander. While officers such as Farragut and Porter basked in the glow of public and departmental approval for their dramatic naval exploits, Lee was usually ensconced in his

flagship, *Minnesota,* devising more effective ways to keep blockade-runners
from delivering their desperately needed cargoes to the Confederates. It
was undoubtedly vital work; by 1864 Wilmington was by far the most
important of the three remaining rebel-held ports east of the Mississippi,
and the Confederacy could scarcely survive without it. Lee understood this
as well as anyone, but he got little satisfaction from his responsibility. He
was an ambitious man who craved professional recognition and advancement, and he realized that the day-to-day drudgery of blockade duty, no
matter how well performed, was unlikely to secure him one of the few
available rear admiral slots he sought. He wanted desperately to lead an
attack on Wilmington's imposing defenses not only to help win the war,
but also prove to his superiors that he deserved a promotion to permanent
rear admiral. Unfortunately, no such assault was possible without army
support that was not forthcoming. During his tenure as general in chief,
Henry Halleck had resisted navy requests for army assistance in a joint
assault on the port because he believed that the Army had too many other
irons in the fire already. Although Ulysses Grant supplanted Halleck in
March 1864, he initially shared Halleck's strategic myopia about
Wilmington and refused to divert army resources in that direction.[53]

If Lee could not rely on an offensive against Wilmington to propel him
into the pantheon of Union Navy heroes, he would instead try to make
his mark by prosecuting the blockade of the port as vigorously and effectively as possible. Closely blockading Wilmington might be less glamorous
than Farragut's run into Mobile Bay, but that did not make it any easier.
Wilmington's geography was tailor-made for blockade-running, and the
rebels were supplementing Fort Caswell with the even more formidable
Fort Fisher, both of which contained numerous big guns to keep Union
warships at a respectable distance. Despite these obstacles, Lee directed his
meticulous mind toward sealing the port off from the rest of the world.
He repeatedly asked the Navy Department for additional resources. He
had over seventy-five vessels in his squadron, but usually a third of them
were coaling or under repair, and he had to dispatch others to the North
Carolina sounds and Newport News, Virginia. By November 1863, however, he could station sixteen steamers off Wilmington at any given time,
and he deployed them in increasingly sophisticated and intricate ways.
Although Welles and Fox periodically expressed concerns about the blockade's continued leakiness, Lee achieved some success. In the five months
prior to Lee's appointment, Goldsborough's warships captured only

twenty-seven blockade-runners. From November 1862 to May 1863, on the other hand, Lee's crews seized or destroyed forty-one vessels. This number jumped to sixty-four in the first nine months of 1864. Lee was justifiably proud of these accomplishments, though he made clear to Welles and Fox that the blockade could never be totally foolproof. Nevertheless, he claimed what credit he could in the hope that it might sufficiently impress the Lincoln administration and Congress to promote him. On 22 December 1863, for example, he wrote to Welles to summarize his wartime career, and then added inarticulately:

> Having done all that was within my means and opportunity I find myself the only acting rear-admiral on the active list of the navy not promoted but subject to, if not in danger of losing even the acting rank I now hold, which, whilst being eligible for promotion under the limitations of, would be a reproach to me. It has not been my fault that I have not had the cooperation of a grand army to capture Richmond or Wilmington. . . . I throw myself upon your indulgence to excuse any sensibility I may have shown respecting confirmation to the grade I hold under an acting appointment with existing vacancies.[54]

While Lee pleaded his case from Hampton Roads, his wife and father-in-law did the same in the capital. Elizabeth Blair Lee had no doubt that her husband deserved a vote of thanks and elevation to rear admiral, and she did not understand Welles' and Fox's reluctance to recommend it to Congress and the president. After all, both men had on numerous occasions praised Lee and his performance as North Atlantic Squadron commander. What more, she wondered, did he have to do to gain their endorsement? In an effort to discover the answer, she asked former Ohio governor William Dennison to find out exactly what the Navy Department's criteria for rear admirals were. Dennison queried Fox, who replied that they were the same as they had always been: distinguished and outstanding service. To Elizabeth, though, this did not ring true. Welles had the previous year acquiesced to John Dahlgren's promotion to rear admiral, and his wartime record was nowhere near as stellar as her husband's. This was accurate enough, but in that exceptional case Welles had bowed against his better judgment to pressure from Lincoln. In a rare display of naïveté, Elizabeth did not recognize that others could play the lobbying game as well as—or, in Dahlgren's case, better than—herself. Instead, she increasingly questioned Welles' and Fox's integrity and considered their actions hypocritical. Despite her growing qualms about them, she continued to treat them kindly because they controlled her husband's fate, but it went against her grain. Welles and Fox, on the other hand, saw no ambiguity, furtiveness, or hypocrisy in their dealings with Lee. They felt that they had always been open and honest with Lee and his kinfolk about their conditions for promotion. In March 1864, for example, Welles summarized in his diary a conversation with Francis Blair about Lee in which he stated the matter clearly: "Old Mr. Blair called on me on Sunday evening to look to the interests of Acting Rear Admiral Lee, his son-in-law, who is uneasy lest he shall not obtain promotion. I told Mr. B[lair] that L[ee] could not have the vote of thanks with the President's recommendation without some marked event to justify it. That the higher appointments must be kept open to induce and stimulate our heroes. That Lee was doing his duty well, and, should there be no others to have earned the great distinction when the war is over, he would be among those who would compete for the prize."[55] This position was plain enough to any objective and dispassionate observer, but no one ever accused the Blairs of objectivity and dispassion.[56]

Unfortunately, events in the summer of 1864 stretched the strained relationship between the Blairs and the Navy Department to the breaking point. In early July Gen. Jubal Early and his ten-thousand-man Confederate army crossed the Potomac River from the Shenandoah Valley to raid Washington. Grant had transferred many of the available Union forces in the region down to the Petersburg trenches, so the capital was almost unguarded. Lee had already dispatched four warships, including an ironclad, to help protect the city, but when on 13 July he learned that the rebels had severed its communications with Annapolis, he decided to go to Washington himself on the fast steamer USS *Malvern*. Welles disapproved of Lee's mission and told him to stay put, but Lee did not get the message in time. He therefore received a frosty reception from Welles when he appeared at the Navy Department building, and Welles ordered him back to Hampton Roads at once. Lee hastily recalled *Malvern*'s officers, who had made a beeline to Willard's Hotel to indulge in the amenities there, and steamed back to Hampton Roads. On his way down the Potomac, he wrote a note to Welles explaining his actions, though he never made clear exactly what he believed his presence would accomplish. As it was, sufficient Union reinforcements arrived in Washington in time to push Early's troops back to the Shenandoah Valley. Lee's relief at this news, however, was tempered with the anger he felt when he received Welles' blistering reply on 19 July. Welles called Lee's behavior "unsatisfactory," and stated that he could have contributed little to Washington's defense. While Welles admitted that the rebel threat to the capital had been real, he thought that Lee had played into the panic the Confederates undoubtedly hoped to create by listening to unfounded and exaggerated rumors. This, Welles continued, reflected poorly on the Navy, especially because Grant refused to be stampeded into returning to Washington. From now on, Lee must obey the Navy Department, not idle gossip. He concluded, "To stand firm in such an exigency is important, and I regret that the rebels, or the rumors, should have moved you at such a time or led you to leave your post." [57]

Predictably, Welles' impertinent letter infuriated Lee and his family. Their sources indicated that Fox had actually written the note, and Welles simply signed his name to it. This hardly surprised Elizabeth, who had always suspected that for all his outward friendliness, Fox was really a snake in their midst out to destroy Lee's career. Her father had a more generous view of the assistant secretary, but even he was angered by Fox's alleged authorship of the letter. Most galling of all, Fox continued to treat the Blairs with his customary joviality. After some hurried consultations and correspondence, Lee and the Blairs opted to swallow the insult and avoid an overt confrontation with the Navy Department. If this was completely out of character for the family, of whom it was said, "When the Blairs go in for a fight, they go in for a funeral," they had their reasons. For one thing, Fox was family and was close to Elizabeth's brother Montgomery, the postmaster general. There was already plenty of long-standing antagonism between Lee and Montgomery, so a war with Fox could quickly get out of hand, and Francis Blair had no desire to see further strife in his collective household. In addition, it was an election year, and the Blairs understood that Republicans of all stripes needed to remain focused and united to help secure Lincoln's victory. If the Democrats won, Lee's future would become the least of the Blairs' problems. Finally, there was a good chance Welles would defend Fox if the Blairs targeted him, and making an enemy of the navy secretary would certainly kill any chance of Lee ever becoming a permanent rear admiral. After mulling things over, Old Man Blair decided that the best strategy was to maintain a veneer of civility toward Fox, but to circumvent him by working directly with Welles and Lincoln to secure Lee's promotion.[58]

The Navy Department wanted to capture Wilmington every bit as much as Lee did. In fact, by the summer of 1864 Welles believed that the North Carolina city was the Navy's last really important strategic target left. He was right about that; Wilmington's value to the Confederacy grew in direct proportion to the effectiveness of the Navy's blockade elsewhere. From 26 October 1864 to 13 January 1865, for instance, thirty-one blockade-runners brought through Wilmington 4,000 tons of meat, 750 tons of lead, 950 tons of saltpeter, 546,000 pairs of shoes, 700,000 blankets, half a million pounds of coffee, 69,000 Lee-Enfield rifles, 43 artillery pieces, and sufficient cotton to pay for it all. On 30 August a frustrated Welles noted in his diary, "Something must be done to close the entrance to Cape Fear River and the port of Wilmington."[59] The problem, as usual, was to get the Army to provide the troops necessary to shut the city down. Lincoln and Secretary of War Edwin Stanton favored such a campaign, but they did not wish to overrule Grant, who remained focused on destroying the rebel armies defending Atlanta and Richmond. In early September, however, Fox journeyed down to Grant's headquarters at City Point, Virginia, to explain to Grant Wilmington's significance and persuade him to assign troops for an attack on Fort Fisher. Although Grant was initially reluctant, Sherman's victory at Atlanta helped convince him that the time was now ripe. On 10 September Grant assured Welles that he was ready to commence operations against Wilmington as soon as he could spare the soldiers and the Navy was ready.[60]

Grant's support, though, came with one condition. Fox told Welles that Grant did not want Lee to command the naval forces against Fort Fisher. It is hard to say what motivated Grant's thinking. He had met with Lee the previous April on board *Minnesota* during his visit to Hampton Roads, so the two men were not strangers. Several weeks later, Lee had pledged to Grant his full cooperation with the Army in any joint campaign against Wilmington. On the other hand, Lee may have alienated Grant by dismissing as impractical Grant's idea to lace with mines the Cape Fear River entrance. It is also possible that Fox poisoned Grant against Lee at their City Point meeting. Whatever the reason for Grant's stipulation, he believed that a new North Atlantic Squadron commander was in order. Under normal circumstances, Welles would have disapproved of an army general attempting to dictate naval personnel decisions, but Grant's reservations about Lee mirrored Welles' own. Although Welles believed that Lee was a diligent, careful, and intelligent administrator, he did not think that Lee possessed the relentless drive and combativeness necessary to lead an assault against a well-defended Confederate position such as Fort Fisher. Lee might suffice for routine blockade duty, but nothing beyond that. Welles was also put off with Lee's incessant direct and indirect lobbying for promotion, which contrasted unfavorably with men such as Farragut who earned their rank through performance. In so evaluating Lee, Welles overlooked the fact that Lee, through no fault of his own, had never really had the opportunity to prove himself in independent command in battle. Nevertheless, Welles' instincts told him that Lee was not the man for the job. Welles knew that his decision would outrage and upset the Blairs, but as usual he placed the country's welfare ahead of friendship and individual wants. On 17 September Welles issued the orders removing Lee from the North Atlantic Squadron.[61]

If not Lee, who, then, should lead the assault on Fort Fisher? Both Grant and Fox recommended David Farragut, who was still down in the Gulf celebrating his victory at Mobile Bay. Welles was not so sure, however. He certainly had no doubts about Farragut's ability to undertake such a mission. In fact, Farragut was now Welles' ideal naval officer, the standard by which he measured all others. Welles lauded Farragut's earnestness, devotion, selflessness, decisiveness, and tenacity. With two major and several minor victories to his credit, Farragut was the Navy's most successful officer, and Welles believed he could overcome Fort Fisher too. The problem for Welles was that if he recalled Farragut from the Gulf before the Army got around to seizing Mobile, the Union public might interpret it as a setback. Welles understood that Mobile's strategic importance vanished the moment Farragut's vessels steamed past Fort Morgan into the bay, but this might not be so obvious to others. With the presidential election rapidly approaching, Welles was loath to do anything to convey the message that the war was a failure. Besides, transferring Farragut to the North Atlantic Squadron could tip off the rebels that an attack on Fort Fisher was imminent. These were legitimate concerns, but, after mulling over the pros and cons, Welles decided to go with the proven winner and let any misperceptions take care of themselves. On 5 September Welles directed Farragut to report to Port Royal by the end of the month with all the warships he could spare from the Gulf to assume command of the North Atlantic Squadron in preparation for an offensive against Fort Fisher. To deceive the rebels as to the true intention of the Union naval buildup, Welles placed Dahlgren under Farragut's orders and told him to pretend another campaign against Charleston was in the works. Unfortunately for Welles' well-laid plans, Farragut balked at the appointment. For one thing, it was extremely short notice for such a big job in a region where he was unfamiliar with the officers and the local geography. In addition, his health was poor after months of arduous service in the Gulf. Farragut's firm devotion to duty made his inability to answer Welles' call painful, but he explained to the navy secretary that it would be unfair to the service and the country for him to agree to an assignment which he knew he was not physically capable of fulfilling. Although Welles was not ordinarily very tolerant of such excuses, he respected Farragut and took his word at face value.[62]

Farragut's rebuff greatly complicated Welles' planning. Naval preparations for assailing Fort Fisher were by now under way, and the Navy Department had already sent Lee his relief orders. Welles huddled with Fox to review their options. Since this promised to be the Navy's biggest operation so far, with dozens of high-ranking officers involved, Welles felt that he needed a permanent rear admiral in charge. Therefore, Welles disqualified capable commodores such as Percival Drayton and John Rodgers for the job. He took another look at Lee before again rejecting him for the same reasons as before. John Dahlgren was a possibility. He had already asked to be relieved as South Atlantic Squadron commander, and Welles had toyed with transferring him to the West Gulf as Farragut's replacement. Welles could instead send him to the North Atlantic. Dahlgren was intelligent and well connected, but the navy secretary concluded that he was too vain, cautious, and unpopular to lead the attack on Fort Fisher. Despite the hard feelings between them, Welles even considered the unemployed Frank Du Pont. Du Pont, though, lacked the necessary drive and aggressiveness. These criteria removed Louis Goldsborough and Charles Davis too. Thus, by process of elimination, Welles settled on David Porter. Fox as usual strongly supported him, though Welles suspected that his partiality toward his friend might in this case outweigh his objectivity. Welles did not share Fox's enthusiasm, but he realized that Porter was energetic, experienced, and well disposed toward Grant, possessed a winning record, and would probably welcome the challenge. Besides, there did not seem to be any good alternative, so on 22 September Welles sent the word.[63]

Settling on a new North Atlantic Squadron commander did not solve all of Welles' personnel problems. Lee remained at loose ends on *Malvern* at Beaufort, North Carolina, awaiting his successor. Although Welles had concluded that Lee was ill-suited for an assault on Fort Fisher, he still believed that Lee possessed administrative talents that the Navy could put to good use. Moreover, sidelining Lee would anger the politically powerful Blairs. Welles initially decided that Lee and Farragut would simply swap jobs so Lee would suffer no loss of status. Welles did so reluctantly, however, because he felt that Lee had already made more than his fair share of prize money, and operations along the Texas coast promised additional pecuniary gain. When Farragut's reassignment fell through and Porter was appointed in his place, Welles on 17 October chose to send Lee to the Mississippi Squadron in Porter's place as an acting rear admiral. Just in case Lee was tired after two continuous years of sea duty, Welles also gave him the option of taking over the Philadelphia Navy Yard in place of Cornelius Stribling, whom Welles had just dispatched to the East Gulf Squadron.[64]

There was plenty of chitchat in naval circles that the Navy Department was displeased with Lee, but his removal from command caught most people by surprise. As late as 6 September, Lee had written to Welles to ask for permission to consult directly with Grant about a joint campaign against Fort Fisher he proposed to undertake. Although Elizabeth Blair Lee heard rumors three days later that Welles planned to relieve Lee, her father told her not to worry. As Elizabeth explained to her husband: "You may be sure of one thing, that as far as we can see into the state of things Father can keep things all right, and he has *not* the least uneasy feeling about your position, thinks the Sec[retar]y heartily you're [*sic*] his and your friend, and gets irritated if I talk anxiously about these things to him. . . . [He] says all the time, 'Let Phil look to his blockade and I shall look out for him.' He said your case was for thanks [from Congress] as good as any man's who had received them, and he talks confidently and affectionately about your position. Only be steady at this point in the game and all will be well."[65]

 But everything was not well, and in fact Welles lowered his ax the same day Elizabeth wrote her encouraging letter. When Lee's father-in-law learned the bad news on 27 September, he hurried over to the Navy Department building for an explanation from Welles. The navy secretary was hardly surprised by the visit, having factored it into his calculations when he made his decision to replace Lee. Blair was deeply distressed, and noted that his family had suffered greatly in recent months for Lincoln and the Union cause. Indeed, that very day the president accepted Montgomery Blair's resignation as postmaster general as a peace offering to Radical Republicans who might otherwise support John Frémont's renegade presidential candidacy. Frank Blair also had attracted criticism from Radical Republicans for castigating Secretary of the Treasury Salmon Chase on the floor of Congress. To top off all these woes, Jubal Early's troops had ransacked Old Man Blair's Silver Spring estate and destroyed Montgomery's nearby country home during their recent raid on Washington. Now his son-in-law was an additional casualty of the political war within the Civil War. Welles had undoubtedly mentally rehearsed his response, and he did his best to reassure his old friend. Although he admitted that he did not believe Lee was the right man to lead the assault on Fort Fisher, he added that Lee had done a good job up until now and had made a lot of prize money in the process. Moreover, taking over another squadron was hardly a demotion. Seeing his opening, Blair again asked Welles to push for Lee's elevation to permanent rear admiral, but Welles as usual insisted that such promotions must be earned, not finagled.[66]

Down at Beaufort, Lee tried not to show his keen disappointment at his reassignment. He blamed his transfer on Fox, whom he saw as an evil Svengali manipulating the benign but gullible Welles. He explained to one officer that he would willingly give way to Farragut if he had sufficient clout to finally secure the resources necessary to successfully attack Fort Fisher. He was less serene, however, when he arrived at Hampton Roads on 11 October to find a large fleet gathering there and Porter waiting for him with orders to take over either the Mississippi Squadron or the Philadelphia Navy Yard. Lee respected Porter's ability, but he was displeased to learn about a naval buildup going on unbeknownst to him within his jurisdiction. Although Elizabeth hoped he would take the job at the nearby Philadelphia Navy Yard, he chose to go to the Mississippi instead. After a short leave, he reported for duty at Mound City on 1 November.[67]

Lee spent the remaining seven months of the war out west. Although the conflict's focus had shifted away from the Mississippi River by then, his work was not unimportant. His new squadron contained more than eighty vessels, and its geographic responsibilities included not just the

Mississippi River, but also its numerous tributaries, such as the Ohio, Cumberland, Tennessee, Arkansas, Red, and White rivers. Lee's warships kept open Union supply lines and provided support for Union troops, especially during Maj. Gen. George Thomas' successful defense of Nashville in December 1864. Lee established a good relationship with the Army, and administered his widely scattered squadron with his usual thoroughness. Although he tried to concentrate on his new duties, the manner in which the Navy Department had treated him still rankled, and for this he blamed Fox. He was also upset that some of his juniors—Porter, Rowan, and John Rodgers, for example—had been promoted over his head to commodore or rear admiral, even though he had seen as much action as they had. Back in Washington, his wife and father-in-law continued their efforts to secure his promotion to rear admiral. During a 21 February 1865 meeting with Welles, Blair as usual pressed for the navy secretary's endorsement for Lee's elevation. In his diary Welles claimed that he refused, arguing that doing so would anger the officer corps. Blair, on the other hand, alleged that Welles promised to recommend that the president appoint Lee to rear admiral. None of this would matter unless Congress voted Lee its thanks, and Blair was working hard there too. The Senate fell in line, but unfortunately the House refused to go along. It is unclear why, but it is possible that someone divulged the contents of a vituperative letter from Lee to Wisconsin senator James Doolittle that castigated Fox for all of Lee's wartime woes. The letter's harsh tone may have persuaded some congressmen to oppose the bill. As a result, Lee received neither Congress' thanks nor a promotion, and ended the war as a captain. He did a good job demobilizing the Mississippi Squadron, and hauled down his flag on 14 August 1865.[68]

Lee's subsequent career was as controversial as his wartime service. After a stint on promotion boards, in April 1866 Welles appointed him commandant of the Mare Island Navy Yard in California. Lee did not want to go so far from home, so he appealed all the way to President Andrew Johnson. Distraught at the thought of being separated from his daughter and grandson, Francis Blair added his objections. Even Montgomery, who had not spoken with Lee in nine years, weighed in on the issue on Lee's behalf. By then Welles had had enough of Lee's string pulling, but all the lobbying wore him down and he finally rescinded the order. He did, however, send Lee one of his patented scathing letters of condemnation. In the end Welles assigned Lee to the new Board of

Examiners to review the records of volunteer officers seeking permanent commissions. Later Lee served on the Yards and Docks Board, as the Navy's chief signal officer, and finally as commander of the North Atlantic Fleet. Along the way, he was at last promoted to rear admiral. Upon retiring in 1873, Lee settled down to run the Blair family estate at Silver Spring. He died in 1897, followed nine years later by his wife. Their son Blair entered Maryland Democratic politics and in 1913 became the country's first popularly elected senator.[69]

David Porter and Fort Fisher

As 1864 drew to a close, it was increasingly obvious that the Confederacy's days were numbered. In Virginia the hard-luck Army of the Potomac had not conquered Richmond, but it had driven Lee's army southward in the Overland campaign, inflicted heavy losses upon it, pinned it within its defensive network around Petersburg, and deprived it of its offensive power. To the northwest Maj. Gen. Philip Sheridan's troops had cleared the Shenandoah Valley of rebels and stripped it of its resources. Union success was not confined solely to Virginia. In the West, William Sherman's troops occupied Atlanta in early September, marched through Georgia against minimal opposition, and seized Savannah on 21 December. In doing so, the Union again divided the Confederacy in two. Less than two weeks before Savannah fell, Maj. Gen. George Thomas' army shattered the besieging rebels outside of Nashville and drove them southward in disarray. Politically, the Union situation was equally promising. Lincoln was reelected in November, demonstrating that a majority of the Northern population wanted the war to continue until the Confederacy and slavery were annihilated. Union resources seemed unending, while the Confederacy was rapidly running out of men, matériel, territory, and time.

Although Vicksburg's and Port Hudson's surrenders in July 1863 placed the entire Mississippi in Union hands, this did not mean that David Porter and his squadron were idle in the ensuing months. The rebels remained active on the big river's tributaries, and combating them required Porter's full attention. Most dramatically, in the spring of 1864 Porter participated in Nathaniel Banks' disastrous Red River expedition. In an effort to secure western Louisiana, Banks pushed his 30,000 bluecoats up the Red River toward Shreveport, Louisiana, accompanied by

Porter's warships. Porter doubted the campaign's strategic value from the start, but his skepticism turned into alarm as the weeks progressed and his vessels became increasingly exposed. After the battles of Sabine Crossroads and Pleasant Hill on 8–9 April, Banks retreated back downstream with his army. Even before this setback, Porter was already having trouble navigating on the twisting river and coping with frequent Confederate ambushes from the banks. For example, a mine damaged the ironclad USS *Eastport,* forcing Porter to order her destruction, and shortly afterward Porter himself had to steer the ironclad USS *Cricket* through a hail of Confederate gunfire from the shoreline after her pilot was mortally wounded. Worst of all, though, was that the Red's falling waters threatened to strand the entire Union flotilla behind rebel lines. Fortunately, one of Banks' engineering officers solved the problem by building a dam upstream, whose destruction released sufficient water to get the warships safely downriver. Although Porter had been dubious about the plan—"If damming would get the fleet off, we would have been afloat long ago"[70]—he was grateful for the results. The operation cost Porter several of his vessels and a good bit of his health, but at least he escaped the subsequent scapegoating that reduced Banks to a military figurehead. In fact, after corroborating Porter's version of events, Welles congratulated him for performing well under difficult circumstances.[71]

When Porter visited the Navy Department after a well-deserved leave of absence in late August, he told Welles that he wanted to remain with the Mississippi Squadron. Welles knew Porter well enough, however, to guess correctly that he would jump at the opportunity to lead what was shaping up to be the Navy's biggest operation of the war. After getting his affairs in order on the Mississippi, Porter left Cairo on 1 October, reached Washington four days later, and assumed command of the North Atlantic Squadron on 12 October. Porter got busy with his usual gusto. Although some officers complained that he did not conduct the Wilmington blockade with Lee's efficiency, Porter was more interested in shutting down the port once and for all than in the halfway measures Lee had utilized up to now. In fact, he claimed he was willing to lose half his ships to do so. The enormous naval buildup brought in officers who were senior to Porter in the prewar days, but he did a good job uniting them for the common cause, in part by dividing his force into five divisions under the squadron's most senior officers: Henry Thatcher, Joseph Lanman, Sylvanus Godon, William Radford, and James Schenck.[72]

Welles and Fox originally hoped to attack Fort Fisher in October, but, as the weeks passed, the soldiers Grant had pledged for the campaign failed to materialize. In fact Grant seemed to lose whatever interest he had once had for the operation. The naval buildup, on the other hand, continued apace, and by mid-October there were 150 vessels of various sorts at Hampton Roads and Beaufort waiting for orders and contributing little to the war effort. Porter grew increasingly frustrated with Grant's dilatoriness, and confided to Fox that he hoped he never had to deal with the Army again. On 28 October Welles wrote directly to the president to complain about the Army's tardiness. The Navy, he noted, had stripped its other blockading squadrons to provide the forces assailing Fort Fisher required, but if the Army did not intend to fulfill its part of the deal, he would disperse the vessels back to their previous stations. When this did not have its desired effect, Fox persuaded Grant to go to Hampton Roads for a conference with Porter in late November. There Grant finally agreed to provide a sixty-five-hundred-man division under Maj. Gen. Godfrey Weitzel for the assault. Unfortunately, there were continuing problems. Benjamin Butler, to whose Army of the James Weitzel's division belonged, suggested to his old friend Fox that the Army and the Navy employ a gunpowder-filled ship against Fort Fisher. Butler believed that if the ship was detonated near the fort, the resulting explosion might stun the garrison long enough for Union troops to land ashore intact. Although Welles and army engineers doubted it would work as Butler intended, Fox and Porter both thought it was worth a try. After a meeting with Butler, Porter's chief of staff and friend, Lt. Cdr. Kidder Randolph Breese, told Porter, "Admiral, you certainly don't believe in that idea of a powder-boat. It has about as much chance of blowing up the fort as I have of flying!" Porter disagreed, and when he left the cabin he heard Breese exclaim, "All bosh!"[73] Despite the scoffers, preparations for the powder boat went ahead, but doing so delayed the operation even longer. Even more ominous, in the long run, was that at the last moment Butler decided to accompany and lead the expedition personally. Not only did Butler have a long record of military incompetence, but there was also bad blood between him and Porter that hindered interservice cooperation, even though both men initially tried to set their personal differences aside.[74]

Not surprisingly, coordination between Porter's warships and Butler's transports quickly broke down. The two fleets left Beaufort separately—Butler on 12 December and Porter the next day—and lost touch. By the time they discovered each other off Fort Fisher on 18 December, the transports were low on coal and the soldiers stuffed in them were almost out of food, so Butler had to return to Beaufort to resupply his forces. A five-day gale further delayed things, but Porter was ready to commence operations on the night of 23–24 December by detonating the powder boat in which he and Butler placed so much faith. The powder boat was actually an old gunboat named *Louisiana,* stuffed with 215 tons of gunpowder and under the command of Cdr. Alexander Rhind, former skipper of the ironclad *Keokuk* during Du Pont's unsuccessful attack on Charleston a year and a half earlier. Rhind towed *Louisiana* toward shore, but in the moonless darkness he did not recognize that he was eight hundred yards off the fort, not three hundred as planned. After anchoring the vessel, he set the fuse and hastily departed. At 1:40 AM *Louisiana* exploded with a bright light and a loud rumble, but it failed to damage Fort Fisher. Those rebels who heard the blast assumed that one of the Union warships had ruptured a boiler. Porter later reported that a Confederate deserter told him, "It was dreadful; it woke up everybody in Fort Fisher!"[75] Porter was crestfallen. Nevertheless, the next morning around 11:30 AM, he opened up on Fort Fisher with his sixty warships, divided into five divisions. For five hours their 580 cannon pummeled the rebel fortress. The next day, Christmas, Porter's warships resumed fire to provide cover for the twenty-five hundred bluecoats Butler and Weitzel landed five miles up the beach from Fort Fisher. In all, Union warships lobbed more than 20,000 shells during the bombardment, but they dismounted only 3 enemy cannon and inflicted just sixty-four casualties. The Navy's losses, on the other hand, totaled eighty-three killed and wounded, most from accidental discharges of new Parrott cannon. When Weitzel and Butler reconnoitered Fort Fisher, they concluded that Porter's bombardment had not hurt its defenses enough to justify an assault with the troops at hand. Moreover, deteriorating weather threatened to isolate their men on the beach and render them vulnerable to rebel reinforcements rushed in to defend the Confederacy's most important strategic point. Instead of digging in as Grant had instructed, Butler evacuated his soldiers—except for seven hundred stranded by the high surf whom Porter rescued next day—and steamed to Hampton Roads.

The finger-pointing and scapegoating began before Butler's transports even reached Hampton Roads. Not unexpectedly, Butler and Porter blamed each other for the fiasco, and their ensuing public squabble led

Elizabeth Blair Lee to snidely comment on the "hot fight between two gas-bags—who may puncture each other."[76] Indeed, Congress eventually investigated the debacle. Butler argued that Porter's ineffective preliminary bombardment made a successful assault impossible, and that he had acted prudently and responsibly in withdrawing his men. Porter, on the other hand, maintained that his warships had done their job, but Butler lost his nerve and failed to take advantage of the opening the Navy developed. As Porter explained to Welles two days after the aborted attack:

> My dispatch of yesterday will give you an account of our operations, but will scarcely give you an idea of my disappointment at the conduct of the army authorities in not attempting to take possession of the forts, which had been so completely silenced by our guns; they were so blown up, burst up, and torn up that the people inside had no intention of fighting any longer. Had the army made a show of surrounding it, it would have been ours, but nothing of the kind was done. The men landed, reconnoitered, and hearing that the enemy were massing troops somewhere, the order was given to reembark.[77]

To buttress his position, Porter asked his chief subordinates to write reports refuting Butler's accusations. In this Porter had no difficulty; most naval officers were outraged and disappointed by the Army's retreat, and they readily rallied around their squadron commander. In fact, there was much to be said for Butler's charge, but he was still a good bit short of the full truth. Unfortunately for Butler, the veracity of his claim did little to spare him from the condemnation heaped upon him. He not only possessed a bad military record that undermined his credibility, but also had innumerable enemies in and out of the military who wanted him out of the Army. Finally, now that the presidential election was safely over, the Lincoln administration no longer needed Butler to attract the support of War Democrats. As one of Butler's many detractors, Welles placed little stock in his performance or opinions. When he received the first dispatches narrating the operation's failure, he quickly concluded that Porter and the Navy had acted properly and placed the blame on Butler's incompetence. He and Fox walked over to the White House to talk with the president about their options and concerns. Lincoln carefully read the dispatches, and then told them to take up the issue with Grant. Welles did so in a telegram that stated that the Navy was ready to assail Fort Fisher again if the Army was willing to provide the soldiers. Grant responded positively, and on 30 December he telegraphed Porter that he was prepared to resume the campaign with more soldiers and a different commander. Eight days later, he persuaded Lincoln to relieve Butler from his post, ending his controversial military career once and for all. As for Porter's fate, as soon as Welles received confirmation from trustworthy John Rodgers that Porter's version of events was accurate, the navy secretary wrote to him in support of his actions up to now.[78]

Grant assigned Brig. Gen. Alfred Terry to command the Army's contribution to the second assault on Fort Fisher. Terry was not a career soldier, but he had during the course of the war accumulated considerable experience in coastal operations. To improve Terry's odds of success, Grant reinforced his provisional corps to 8,000 troops and directed him to cooperate fully with Porter. Fortunately, Terry was a pleasant and unpretentious man who had no difficulty getting along with the gregarious Porter. Terry's nineteen transports reached Beaufort on 8 January 1865, and he immediately boarded the North Atlantic Squadron flagship, *Malvern,* to open his sealed orders and plan his attack with Porter. Another winter gale delayed their departure until 12 January, but the next morning Porter's sixty-plus vessels sporting 627 guns opened fire on Fort Fisher and its 1,550 defenders. For fifty-five hours Union warships pummeled the rebel fortress, with Porter often calmly and intently observing the action while smoking one cigar after another. Although he had vehemently refuted Butler's charge that his preliminary bombardment the previous month had been ineffective, Porter also rectified the errors he denied making. He instructed his subordinates to avoid targeting Fort Fisher's flagstaff, which had led to considerable overshooting during the first operation, and instead deliberately aim for the fort itself. As a result, while his warships fired 589 fewer shells than during their previous effort—19,682 this time, versus 20,271 in December—they were a good deal more accurate. Indeed, by 15 January only one cannon in the fort remained operational.

While Porter's warships worked over Fort Fisher, Terry's soldiers disembarked unopposed on 13 January five miles up the beach, dug in, and prepared for an assault on the fifteenth. To contribute to the Götterdämmerung they planned, Porter placed his chief of staff, Kidder Randolph Breese, in charge of a newly raised naval brigade of 1,600 sailors and 400 Marines that were to storm the fort's sea face while Terry's bluecoats attacked from its landward side. On the morning of 15 January, the naval brigade landed, and at 3:00 PM the men charged. Rebel fire cut down the inexperienced and underequipped sailors and Marines before they got within three hundred yards of the fort, inflicting nearly 400 casualties. Their sacrifices, however, diverted rebel attention from the soldiers, who reached the parapets undetected. During the ensuing vicious hand-to-hand fighting from traverse to traverse and bombproof to bombproof, Porter's warships provided the advancing bluecoats with crucial and, considering the technology available, surprisingly accurate fire support. By nightfall the last Confederates surrendered, finally placing Fort Fisher in Union hands. Union losses came to 955 soldiers and 386 sailors and Marines. Some 1,900 rebels—the 1,550 original members of the garrison, plus several hundred reinforcements who showed up during the fight— were either killed, wounded, or captured. Despite their exhaustion, the soldiers, sailors, and Marines still mustered the energy to celebrate their victory. Porter, though, was not among them; the stress of the past few days so weakened him that he had to be assisted back to his cabin. It was perhaps just as well that he chose not to go ashore to join the festivities. As dawn approached, a group of drunken New York soldiers entered the main magazine with torches, and the resulting explosion added another hundred men to the casualty list.[79]

Despite this unfortunate postscript, Fort Fisher's fall was still an important Union victory. Most obviously it put an end to large-scale, systematic blockade-running east of the Mississippi River. Blockade-runners could still slip into remote Texas ports, but they would contribute little to the Confederacy's defense. Without Fort Fisher, the Confederacy was now isolated from the outside world and unable to import the supplies, weapons, and equipment it needed to wage war. The Union victory made it increasingly obvious that the Confederacy's collapse was not far off. Union soldiers marched into Wilmington on 22 February, and from there pushed inland to link up with William Sherman's army burning its way through the Carolinas. The two forces met at Goldsboro, North Carolina, a month later. At the same time, another Union column was cutting its way through Alabama toward Montgomery. In early April the Army of the Potomac finally broke through rebel lines in front of Petersburg, seized Richmond, and ran Lee's army to earth at Appomattox Court House on 9 April, beginning a chain of events that led in a little more than a month to the surrender of all the Confederacy's armies and an end to the war.

Porter spent the rest of the war cooperating with the Army and enjoying the praise heaped upon him. Not only did Welles offer his warmest congratulations, but on 24 January Congress voted him its fourth thanks of the war. Generous as usual in victory, Porter gave credit to his subordinates for their roles in the final assault on Fort Fisher, especially Commo. Henry Thatcher, Commo. William Radford, and Commo. Sylvanus Godon. At the end of April, Welles put Radford in command of the North Atlantic Squadron, and later of its South Atlantic counterpart as well, so Radford received the demobilization responsibilities. As for Porter, despite his successful record—or, perhaps more accurately, because of his cocksure attitude—he remained controversial. Alluding to the *Monitor*-class ironclads' vulnerability on the high seas, one political enemy wrote, "Admiral Porter says some of the monitors can cross the ocean with perfect safety. Let him take one of them and accompany the fleet. I wish to God Porter would, provided the trip could have one result."[80] After the war Welles appointed Porter superintendent of the Naval Academy, where he instituted many necessary reforms. When Ulysses Grant became president in March 1869, he asked Porter to become adviser to the new navy secretary, Adolph Edward Borie. As Borie knew little of naval affairs, and as Porter thought he knew most everything, Porter virtually ran the department. Unfortunately his sometimes unwarranted self-assuredness spawned many enemies, especially among staff officers. Although he succeeded Farragut as vice admiral in 1866 and full admiral four years later, Porter's influence waned, in part because he had no authority over the bureau chiefs. He served as head of the Board of Inspection from 1877 until his death in 1891, but he devoted most of his time to socializing, writing fiction and nonfiction, and embellishing his Civil War adventures.[81]

Union Naval Victory

For the Union Navy, the second half of the war was much more difficult than the first. Early in the conflict, the Navy won a string of triumphs by successfully assailing numerous weakly defended Confederate positions with overwhelming force. Later in the war, as the number of Southern ports in rebel hands contracted and their defenses thickened, overcoming them required a proportionately greater naval investment. Doing so not only took more time, but it also necessitated army cooperation that was usually given grudgingly, if at all. Fortunately for the Navy, by this stage of the conflict it had some veteran officers capable of planning and under-taking such intricate interservice operations. These included not only Farragut and Porter, but also skilled subordinates such as Percival Drayton, James Palmer, and Henry Thatcher. As the Navy's campaigns became bigger and more important, Welles became even more ruthless in his personnel selections. Lee's experience demonstrated that mediocrity and plodding dependability no longer sufficed for the really important operations. Instead, Welles sought veterans who he believed possessed the necessary persistence, aggressiveness, courage, and intelligence for victory. Happily for the Union, the men he chose—Farragut and Porter, for exam-ple—were up to the jobs.

Conclusion

Selecting and Removing Squadron Commanders

GIDEON WELLES was burdened with countless vital responsibilities during the Civil War. His many jobs included overseeing the construction and purchasing of vessels, distributing patronage within the department, arming and equipping and provisioning warships, formulating strategy, determining new technology for development, and advising the president. Of all his tasks, though, selecting squadron commanders was undoubtedly among his most significant. After all, the squadron commanders implemented the strategic plans the Navy Department decided upon to wage the conflict, and these plans were no better or worse than the men chosen to carry them out. Certainly the talent pool from which Welles picked his squadron commanders was important, but the navy secretary's role as the man who selected these men was even more so.

As commander in chief, President Lincoln had the final say on all assignments to squadron command, but he almost always left the decision in Welles' hands. Lincoln's laissez-faire attitude was not due to his ignorance of or apathy toward the Navy. David Porter, for example, once noted that the president seemed to know something about every high-ranking naval officer, and usually had an illuminating anecdote about each one.[1] He also kept a close eye on the Navy's numerous operations, and had a keen interest in all things nautical. Instead, there were other factors that explained the president's behavior. For one thing, Lincoln ordinarily gave his cabinet officers considerable leeway in running their departments, and Seward's high-handed attempt to impose Samuel Barron on the Navy Department taught him that Welles would not brook excessive outside interference. Besides, as the war went on, Lincoln grew to trust Welles' personnel choices, and at one point remarked that they were much better than the Army's.[2] This does not mean that Lincoln never meddled in naval affairs; he sometimes did, though usually for personal, not political, reasons. He repeatedly pressured Welles to appoint his friend John Dahlgren a rear admiral because of the affection he felt toward him. Similarly, he asked the navy secretary to find Porter and Charles Wilkes appropriate positions not because of political demands, but rather because he thought that both men would be more useful to the war effort in more responsible posts.

Moreover, Lincoln had little to gain politically from getting deeply involved in the Navy's personnel matters. During the conflict he frequently commissioned prominent politicians as army generals in order to cement their constituencies to the war effort. These political generals, such as Benjamin Butler and Nathaniel Banks, were often incompetent on the battlefield, but they also attracted to Union ranks many men who might not have otherwise enlisted. There were, on the other hand, no political captains or admirals. While many politicians clamored for army positions because they naively felt that they were competent enough to lead soldiers into battle, not even the most egotistical among them believed he could skipper a warship. As a result, there was minimal political pressure for Lincoln to butt into Welles' personnel decisions because the Navy was simply not as lucrative a source of patronage as the Army.

Finally, Lincoln by and large left Welles alone because the Navy was not as large or as essential to Union victory as the Army. After all, the entire Navy contained fewer men than a single major field army, and all the naval triumphs in the world would not establish the Union's authority over the seceded Southern states; that required boots on the ground that only the Army could provide. Even though the Army and the Navy were bureaucratically equal in the federal government, Lincoln understood as well as anyone that this was not the case in terms of the two services' respective budgets, manpower, political influence, and, most important, contribution to the Union war effort.

From the very beginning of his tenure as navy secretary, Welles recognized the importance of assigning good squadron commanders. He was usually a keen judge of character, but unfortunately he initially possessed little knowledge of the naval officer corps. Like the fine newspaper editor and publisher he once was, however, Welles understood the value of information in making informed decisions. To compensate for his early ignorance of the human material at his disposal, he actively sought out people who were well acquainted with the Navy's officers. Gustavus Fox was chief among them. Although the Blairs pretty much imposed Fox on the Navy Department, he and Welles developed a good working relationship and learned to trust one another. The assistant secretary became a useful man to Welles in many ways, especially in his understanding of naval personnel. As a former naval officer, Fox knew all the Navy's principal players. With Welles' consent, he also deliberately cultivated an informal correspondence with many of them during the war. Welles and Fox realized that officers were more likely to write honestly to one of their own off the record than to the navy secretary in official messages that might become public. By keeping in touch with officers such as Frank Du Pont, David Farragut, Andrew Foote, and David Porter, Fox learned much about which officers were and were not performing well, and conveyed that information to Welles for his use. From such letters Fox and Welles gained insight into the performances of officers such as William Mervine, William McKean, Charles Davis, Theodorus Bailey, Henry Bell, and John Dahlgren.

To be sure, Fox's network was not always reliable. Some officers gave their subordinates unwarranted praise, or tried to exploit their relationship with Fox by exaggerating their own accomplishments and denigrating their colleagues. Du Pont, for example, never told Fox his concerns about Charles Davis' lethargy, but instead extolled and commended Davis' effectiveness in an ultimately successful effort to kick him upstairs and out of his squadron. On the other hand, Porter's insatiable ambition compelled him to unfairly malign Henry Bell and Farragut in his letters to the assistant secretary. Moreover, Fox was occasionally overly partial to friends such as Du Pont and Porter in his recommendations to Welles. Luckily, Welles built redundancy into his system by nurturing additional intelligence sources. He communicated with Hiram Paulding, Foote, and Silas Stringham on personnel matters, particularly early in the conflict. He also invited naval officers in Washington to his office or home for interviews, during which they discussed personnel. Such a meeting with Percival Drayton, for instance, helped convince Welles to remove Du Pont from his squadron. Welles placed all the data he gathered about naval officers into his mental hopper and sifted through it to determine his squadron commanders and make other crucial personnel determinations.

As a product of the New England middle class, Welles firmly believed in meritocracy. Ideally, he would have appointed squadron commanders solely on the basis of his evaluation of their character and record. He usually searched for officers whom he considered energetic, resourceful, uncomplaining, and ruthlessly aggressive—men such as Farragut, Foote, and Porter. As the conflict progressed, Welles could also scrutinize his officers' wartime records. There were a number of men whom Welles initially overlooked as potential squadron commanders, but later reconsidered after they demonstrated in battle the qualities he sought. These individuals included Theodorus Bailey, Henry Bell, Charles Davis, James Lardner, Phillips Lee, and Henry Thatcher. All these officers proved themselves in action in subordinate roles, which brought them to Welles' attention and convinced him that they deserved squadron command. Welles also understood that the different squadrons required different kinds of commanders, so he sought to marry the best available person to the proper position. The Mississippi Squadron of 1862–63, for instance, needed an unorthodox and pugnacious leader, such as Porter, who was capable of working closely with the Army. On the other hand, before the assault on Fort Fisher, the prestigious North Atlantic Squadron called for a commander, such as Lee, who had organizational ability, meticulousness, and an understanding of international law. To Welles, though, the important thing was that his squadron commanders possess an unquenchable desire to win victories, and to find them he was often willing to turn a blind eye to their less-admirable traits. For example, he tolerated Farragut's bureaucratic inefficiency, Foote's poor health, and Porter's boastfulness and duplicity because they usually achieved their objectives in spite of their foibles. In fact, Welles' poorer choices for squadron command often occurred when he compromised his principles and relied on criteria other than merit and record.

Although Welles would have preferred to use merit as his sole criterion for selecting squadron commanders, there were other factors that he had to keep in mind. Under normal circumstances, a navy secretary could take the loyalty of his officers for granted. The Civil War, however, was no ordinary time. Early in his tenure as navy secretary, Welles was astonished that so many Southern-born officers questioned or disregarded their oaths, which led many to resign their commissions to join the Confederacy. These individuals included well-respected and high-ranking officers such as Frank Buchanan, George Magruder, and Matthew Maury. Not surprisingly, Welles' experiences with these men led him to suspect all Southern-born officers of, if not downright treason, then certainly a lack of enthusiasm for the Union cause. His doubts were exacerbated by his lack of knowledge about the officer corps and who in it he could and could not trust. As a result, early in the war Welles took residency into account when selecting squadron commanders. He initially refused to give most Southern-born officers important posts, which prevented capable men such as Bell, Farragut, and Cornelius Stribling from playing active roles in the war's early months not because of anything they did, but simply because of their region of birth. Welles did not begin to tap this pool of talent until after the battle lines were clearly drawn and he got a feel for the officer corps. Until then, the Union Navy lost the services of some good officers, but Welles felt it was a necessary measure in uncertain times.

In addition to birthplace, Welles also contended with seniority. Squadron commanders had to possess sufficient rank to exert their authority over their colleagues, and the prewar Navy had a long history of seniority-based promotion. Theoretically, the next available slot went to whoever had been in the service the longest. Not surprisingly, mediocre officers high on the naval register liked this system because it guaranteed them rank and prestigious posts regardless of their competence. There was also a logic behind it. Seniority made things predictable and orderly, so every officer knew where he stood, and where he was likely to stand in the future. Welles and Fox, on the other hand, agreed with many younger officers that the emphasis on seniority rewarded longevity and conformity at the expense of innovation, creativity, initiative, and independent thinking. After all, if promotions were guaranteed by time in service, then there was little incentive for officers to prove their value to the Navy Department.

There had been efforts to chip away at seniority before the Civil War by removing some of the Navy's deadwood, and these accelerated when hostilities commenced and winning the conflict became more important to officers and congressmen than preserving the Navy's traditions and status quo. During the war Congress passed several laws that granted the navy secretary the authority to appoint high-ranking officers to whatever positions he saw fit, regardless of their seniority. Congress also created new ranks and left filling them largely to Welles' discretion. This legislation gave Welles the power to ignore seniority altogether if he wished, but he did not because he recognized that taking seniority into account was necessary to preserve harmony in the tradition-bound officer corps. Even the most reform-minded officers usually looked askance at the elevation of junior officers to prominent positions such as squadron command, and were likely to blame the Navy Department for their hurt feelings. Therefore, Welles usually factored seniority into his mental calculations to varying degrees when determining squadron commanders. This was especially true early in the war, before Congress passed the appropriate legislation, before officers had developed much of a combat record, and before Welles knew them very well. For example, Welles chose Louis Goldsborough, William McKean, William Mervine, and Silas Stringham primarily because of their positions on the naval register, and he ultimately regretted each selection. Even later in the war, when seniority had temporarily lost a good bit of its legal and institutional clout, Welles continued to consider it when determining squadron commanders. For instance, it played a role in Farragut's and Bailey's appointments. One of Welles' challenges in personnel matters was to give seniority its due, but not at the expense of more important concerns.

Seniority and loyalty were not the only factors that interfered with Welles' desire to use merit in his selection of squadron commanders. Personal and political connections were another. The prewar Navy was a tight-knit community in which officers and their influential friends lobbied their superiors for preferential treatment—usually securing plum or prestigious assignments. The war provided additional opportunities for string pulling by degrading the seniority principle, creating additional ranks, and establishing dozens of new command positions. Officers and their supporters pressured Welles to give them special consideration, though of course no one ever put it in those terms. While Welles claimed that such favoritism had little effect on his choice of squadron commanders, the record indicates otherwise. After all, the Navy was his constituency whose members he wanted to appease. If officers did not believe that Welles appreciated their talents, it might lower morale and undermine the war effort. As a result, several commanders owed their appointments and promotions at least in part to outside lobbying on their behalf. John Dahlgren, for example, never would have become a rear admiral and South Atlantic Squadron commander if the president had not successfully pressured Welles for his advancement. Although Welles later denied it, it is difficult to believe that Francis Blair's pleas did not play some role in the navy secretary's decision to assign Phillips Lee to the North Atlantic

Squadron. Other commanders had well-placed allies who looked after their interests in Washington while they fought the rebels. These included not only Dahlgren and Lee, but also Bailey, Du Pont, Porter, and Wilkes. There were officers, such as Farragut and John Rodgers, who eschewed the constant and intricate wire-pulling that was part and parcel of Navy Department routine—in fact, Rodgers once chastised his wife for speaking with Fox because he wanted to avoid even the appearance of influence peddling[3]—and still managed to attain squadron command, but they were the exceptions.

However, relying too much on connections to secure professional advancement had its downsides. Welles accepted it as a fact of life, but he found its most blatant manifestations distasteful and off-putting, which eventually soured him on Dahlgren and Lee in particular. Moreover, other officers resented colleagues who attained their posts through backstairs maneuvering instead of on a warship's quarterdeck. As both Dahlgren and Lee ultimately discovered, their connections could be a double-edged sword that hurt their careers in the long run. Either way, connections were an integral part of the Navy's wartime personnel matters.

Finally, Welles also took availability into account when determining squadron commanders. Because there was no real pension plan, naval officers remained in the service as long as possible in order to continue to draw their full pay. Consequently, at the beginning of the war the naval register contained lots of captains theoretically eligible for squadron command, but incapable of answering the call because of age and poor health. For example, Welles thought highly of Francis Gregory, who was sixth on the list of loyal captains at the war's onset, and twice considered him for squadron command, but rejected him each time because he was too old and sick for duty afloat. On the other hand, some officers became squadron commanders in part because they were healthy and on hand when a vacancy occurred. Most obviously, Welles permitted Bell and Palmer to continue as temporary leaders of the West Gulf Squadron because they were already on the scene when Farragut left for home. One reason he assigned Du Pont, Farragut, Goldsborough, Lardner, and Wilkes to their squadrons was because they were underutilized in their current positions, hence readily available for squadron command.

During the Civil War nineteen officers served as long-term temporary or permanent commanders of the six important naval squadrons. Of the nineteen, four of them led more than one squadron: Lardner, Lee,

McKean, and Porter. Farragut was alone in that he in effect oversaw his squadron twice, with Bell taking over during his extended furlough. Six of them—Bailey, Dahlgren, Du Pont, Farragut, Lardner, and Porter—ran their squadrons for more than a year, and only Lee had the distinction of reaching the two-year mark as chief of the North Atlantic Squadron. Lee and Porter each commanded two squadrons—the North Atlantic and the Mississippi squadrons—for about thirty months total, longer than anyone else. At the other end of the scale, six officers had tenures of less than six months as squadron commanders: Bell, Davis, Mervine, Palmer, Stringham, and Thatcher. Averaged out, the tenure of these Civil War squadron commanders was little more than ten months.

One reason for this high turnover rate was Welles' low threshold for failure. He expected results from his squadron commanders, and constantly evaluated their progress. Generally speaking, he was more forgiving of officers who attempted to overcome obstacles and failed than those who made excuses and never tried at all. For instance, he condemned Mervine's inaction in the Gulf, but accepted—grudgingly, to be sure—Farragut's defeats in front of Vicksburg because the latter's sins were ones of commission. When dissatisfied with squadron commanders, Welles resorted to a number of stratagems to replace them. During the war he relieved three squadron commanders outright for their lack of success: Du Pont, Mervine, and Wilkes. In addition, he transferred Lee to the Mississippi Squadron because he doubted his ability to lead the assault on Fort Fisher. Welles also successfully pressured Goldsborough and Stringham into quitting, thus placing on them the onus for their termination. Finally, Welles supplanted Bell, Davis, and Palmer in part because he considered their tenures temporary from the start, but also because he had not been terribly impressed with the performances of the former two anyway. On the other hand, although Welles was not usually tolerant of unsuccessful squadron commanders, on occasion he refrained from taking action against them. For example, he was disappointed with McKean and Dahlgren, but he left them at their backwater posts because he concluded that getting rid of them was not worth the effort.

There was, however, one other major reason why so many squadron commanders lost their posts. Although Foote was the only one wounded in action during the war, others succumbed to illness caused by microorganisms, bad food, overwork, lack of exercise, and stress. The fact that so many squadron commanders were over sixty years old no doubt contributed to this high sickness rate. Indeed, disease was at least partially responsible for the removal of nearly a third of the squadron commanders, including Bailey, Davis, Farragut, Foote, Lardner, and McKean.

Evaluation of Squadron Commanders

In September 1863 Welles recorded in his diary a conversation he had with the president, who was as usual at that stage in the war frustrated with the performance of so many of his generals: "Alluding to the failures of the generals, particularly those who commanded the armies of the Potomac, he [Lincoln] thought the selections, if unfortunate, were not imputable entirely to him. The generals-in-chief and secretary of war should, he said, know the men better than he. The Navy Department had given him no trouble in this respect." In particular, Lincoln praised Welles' choice of Farragut as the best appointment made by either the Army or the Navy so far. Welles was of course pleased with the compliment, but admitted to Lincoln that it was not entirely accurate, noting, "We had our troubles, but they were less conspicuous [than the Army's]."[4] In fact, although Welles could take credit for discovering David Farragut, he was also responsible for several mediocre and poor squadron commanders. Yet despite this less than perfect record, he escaped much of the criticism leveled against Lincoln, his war secretaries, and his generals for their personnel picks. As Welles realized, a major reason for this was the Navy's low public visibility, which, ironically enough, stemmed to a certain extent from its overwhelming superiority over its Confederate counterpart. While the Navy played a vital role in the Union's triumph, it was never as large or as important to the war effort as the Army. The Army could realistically lose the conflict, but, except for the *Merrimack*'s foray at Hampton Roads, the rebels never seriously challenged Union naval supremacy. As a result, poor squadron commanders such as William Mervine might not accomplish much to bring about Union victory, but no one could ever blame them for placing the Union in peril. On the other hand, good squadron commanders such as Farragut and Foote could and did advance the Union cause and generate positive publicity for themselves and the Navy. From this perspective, it was small wonder that Welles' reputation for assigning the right officers to the right posts was so good.

During the war Welles appointed two outstanding squadron commanders: David Farragut and David Porter. These two officers led Union naval forces to several of the conflict's most important victories. Farragut was most responsible for seizing strategically vital New Orleans and Mobile, and Porter played the key naval role in the Union conquests of Vicksburg and Fort Fisher. Neither officer had an unblemished record, but this should not divert attention from their significant accomplishments. Both men shared certain characteristics that contributed to their successes: audacity, an eagerness to engage the enemy without fear of the consequences, resourcefulness, meticulous planning, the knack of inspiring loyalty among their subordinates and molding them into an effective team, and an understanding of the value of interservice cooperation. Of course, both had character flaws. Porter undermined and denigrated his superiors, pulled every available string in his cold-blooded climb to the top, and bragged about achievements real and imagined. Farragut was a much more admirable and honorable man, but he had little patience for and understanding of the bureaucracy and paperwork that were necessary facts of life for squadron commanders. This damaged his effectiveness in the day-to-day enforcement of the blockade in the Gulf. Although in retrospect Porter's and Farragut's selection seems commonsensical, their attributes were hardly obvious at the time. After all, before he became a squadron commander, Farragut was an undistinguished captain with a Southern background. Porter, for his part, was a low-ranking troublemaker whose own brother had accused him of Southern sympathies, and who had repeatedly criticized his superiors and promised things he did not deliver. Welles, however, saw something beyond their run-of-the-mill records and personalities, and recognized their potential. In appointing them to squadron command, he helped the Union win the war.

Welles also selected some squadron commanders who provided solid service, but for various reasons never had the opportunities to become as outstanding leaders as Farragut and Porter. Andrew Foote, for example, fought aggressively at Fort Henry and Fort Donelson, but the wounds he received in the latter engagement gradually sapped his strength and spirit, and he eventually had to surrender his post before he could prove himself a great squadron commander. Theodorus Bailey and Cornelius Stribling, for their part, both led the remote and strategically unimportant East Gulf Squadron. Neither man ever had the resources or occasion to make their marks as squadron commanders. Instead of throwing their hands up in despair, however, they both used their few assets in creative ways to advance the Union war effort by maintaining the blockade and attacking rebel assets where they could. Finally, Henry Thatcher successfully cooperated with the Army in seizing Mobile, but the war ended before he had the chance to establish himself as an outstanding squadron commander. Although Welles should have used some of these officers more advantageously, he did nothing to undermine their positions once he assigned them to their positions, and in return they became good squadron commanders who provided solid service.

On the other hand, there were several squadron commanders whom Welles deliberately denied the opportunity to distinguish themselves. He did so because he alleged, rightly or wrongly, that these men, whatever their previous accomplishments, lacked the strength of character, rank, or ability to complete the missions that would have placed them at the forefront of Union naval heroes. This list includes Silas Stringham, Louis Goldsborough, Charles Davis, Phillips Lee, James Palmer, and John Rodgers. Stringham, Goldsborough, and Davis had all won battles—Stringham at Hatteras Inlet, Goldsborough at Roanoke Island, and Davis before Memphis—but subsequent events convinced Welles that he could not trust them to continue their duties and take on greater responsibilities. Goldsborough, Davis, and Stringham were all discredited by their supposed lack of aggressiveness, to which Stringham added the additional sin of constant complaining. Although Lee, Rodgers, and Palmer had no victories to their credit, they had all performed well as squadron or flotilla commanders. Even so, Welles opted to remove them anyway rather than entrust greater duties to them. In Rodgers' case, Welles did not believe that he had sufficient rank to lead the Western Flotilla. Palmer possessed the aggressiveness Welles sought in squadron commanders, so it is unclear why the navy secretary supplanted him, though it is quite possible that he always saw his tenure as temporary. As for Lee, Welles freely admitted that he had conducted the Wilmington blockade as well as could be expected. Nevertheless, he suspected that Lee lacked the killer instinct that seizing Fort Fisher required, so he replaced him with Porter. In most of these cases, Welles relieved these officers not because their records as squadron commanders warranted it, but rather because he questioned their potential to succeed in future operations. Whatever the veracity of his suspicions about these men, their actual records were generally creditable.

There were also two squadron commanders who failed in battle. Both Samuel Du Pont and John Dahlgren led Union naval forces in unsuccessful attacks on Charleston harbor's defenses. Du Pont established his wartime reputation as a winning officer by seizing Port Royal in November 1861. Although rebel defenses were weak and Du Pont's assault plan unraveled at the beginning of the engagement, his triumph came early in the conflict when Union victories were few and far between. As a result, he gained considerable clout in the Navy Department. Unfortunately, Welles and Fox overrated his talents, and expected too much of him when they assigned him the task of storming Charleston. Du Pont's serious and legitimate reservations about the operation's feasibility, his poor planning, and his inability to cooperate with the Army all contributed to his repulse on 7 April 1863. Worse yet, his subsequent efforts to shift the blame for his debacle onto the Navy Department caused strife and dissension in the service and destroyed his career. Despite his early achievement at Port Royal, Du Pont's overall record as a squadron commander was one of disappointment. This was equally true of Dahlgren, who also suffered defeat at Charleston for many of the same reasons. Unlike Du Pont, though, Dahlgren had no earlier victory on his record to counterbalance his setback. Worse yet, other officers resented his undeserved promotion to rear admiral. Dahlgren shared Du Pont's timidity in front of Charleston, and he too attempted to deflect the blame for his difficulties onto others. Dahlgren managed to hold onto his squadron command only because Welles considered it politically inconvenient to remove him. Both Du Pont and Dahlgren had their merits, but as battle commanders they never came up to par because while they were both willing to risk their lives to the Union cause, they were far more chary with their reputations.

Finally, there were several officers who contributed almost nothing to the Union war effort as squadron commanders. While it is true that Henry Bell, James Lardner, William McKean, William Mervine, and Charles Wilkes all lacked sufficient resources, they also made little effort to overcome these deficiencies through the kind of creative thinking that Bailey and Stribling demonstrated off the Florida coast. Instead, Mervine and Wilkes devoted much of their energies toward badgering the Navy Department for more warships, though it is doubtful that these additional assets would have made much of a difference. Bell, Lardner, and McKean were more reticent, but equally unsuccessful. Welles eventually removed Mervine and Wilkes from their positions because he demanded much more from them than they delivered. On the other hand, he was more forbearing toward Bell, Lardner, and McKean, seeing them as caretakers of whom he had few expectations. It is unfair to compare these men to awful Union generals such as John Pope, Benjamin Butler, and Nathaniel Banks who suffered major defeats that seriously damaged the Union cause. Instead, they were merely inadequate.

Despite the innumerable disadvantages under which he labored, Welles' choices for squadron command were generally fairly good. Welles had very little naval experience before becoming navy secretary. Although he had served as chief of the Bureau of Provisions during the Mexican War, he never had to make the kind of difficult, crucial, and complicated personnel decisions that the Civil War demanded of him. Moreover, he had to initially operate within the confines of a bureaucracy that prized conformity, routine, and especially seniority. Fortunately for the Union, Welles possessed many of the traits he looked for in his squadron commanders, including determination, single-mindedness, self-discipline, and complete devotion to the Union. This enabled him to hack his way through these bureaucratic weeds to find the good officers who led the Union Navy to victory. The key to Welles' success was his early recognition that the Navy's cherished seniority system could not provide the squadrons with enough quality leaders. Welles never completely overcame this obsession with seniority, but he managed to reduce its influence through legislation and force of will. Once he freed himself from the chains of seniority, he then depended upon his advisers and his own astute understanding of human nature to appoint the right men to the appropriate squadrons. In fact, Welles chose best when he relied primarily on his instincts and ignored other factors. He selected Farragut, Foote, and Rodgers almost entirely on the basis of perceived merit, and they all did good jobs. Of course, Welles' judgment was not foolproof. Most prominently, he overrated Du Pont. In terms of personnel selection, though, Welles got into the most trouble when he factored seniority and politics into his decisions, which explained Dahlgren's, McKean's, Mervine's, Stringham's, and Wilkes' unfortunate appointments to squadron command. These mistakes aside, Welles succeeded in choosing enough competent squadron commanders to justify Lincoln's praise.

Notes

Full citations for works cited by author and short title in the notes may be found in the bibliography. The abbreviation *OR* stands for *Official Records of the Union and Confederate Navies in the War of the Rebellion*. The names and citations of some archival sources used in the notes have been shortened as follows:

Bailey Papers	Theorodus Bailey Papers, Syracuse University Library, Syracuse, N.Y.
Bell diary	Papers of Henry Bell, National Archives and Records Administration, College Park, Md.
Bell papers	Papers of Henry Bell, National Archives and Records Administration, College Park, Md.
Blair papers	Blair Family Papers, Library of Congress, Washington, D.C.
Montgomery Blair Papers	Montgomery Blair Papers, Blair Family Papers, Library of Congress, Washington, D.C.
Du Pont Notes	Du Pont Notes, Samuel Francis Du Pont Papers, Hagley Museum and Library, Wilmington, Del.
Du Pont Papers	Samuel Francis Du Pont Papers, Hagley Museum and Library, Wilmington, Del.
Fox papers	Gustavus Fox Papers, Blair Family Papers, Library of Congress, Washington, D.C.
Virginia Fox diary	Virginia Woodbury Fox Diary, Levi Woodbury Family, Washington, D.C.
Goldsborough papers	Papers of Louis M. Goldsborough, Duke University Libraries, Durham, N.C.

Howard Papers	Mark Howard Papers, Connecticut Historical Society Museum, Hartford, Conn.
Lincoln Papers	Abraham Lincoln Papers, Library of Congress, Washington, D.C.
John Rodgers Papers	Rodgers Family Collection, John Rodgers Papers, Library of Congress, Washington, D.C.
Welles papers	Papers of Gideon Welles, National Archives and Records Administration, College Park, Md.
Wilkes papers	Papers of Charles Wilkes, Library of Congress, Washington, D.C.

Introduction

1. David Porter to Gideon Welles, 15 January 1865, *OR,* 11:435.
2. Welles, 16 September 1862, 21 September 1863, *Diary of Gideon Welles,* 1:134, 440.
3. Welles, "Farragut and New Orleans," 679.
4. Welles to S. Ledyard Phelps, 18 September 1862, Slagle, *Ironclad Captain,* 297–98.
5. The quote is from Welles, "Farragut and New Orleans," 679.
6. Welles, 25 May 1863, *Diary of Gideon Welles,* 1:312–13.

Chapter One. Scraping the Barnacles

1. The quotes and the story are from Welles, 3 March 1869, *Diary of Gideon Welles,* 3:539–40.
2. See, for example, James Dixon to Abraham Lincoln, 17 November 1860, Lincoln Papers; Samuel Austin to Lincoln, 17 December 1860, Lincoln Papers; William Buckingham to Lincoln, 25 December 1860, Lincoln Papers; Lafayette Foster et al. to Lincoln, 5 January 1861, Lincoln Papers; J. D. Baldwin to Lincoln, 7 January 1861, Lincoln Papers; Milo Holcomb to Lincoln, 14 January 1861, Lincoln Papers.
3. For a good discussion of Lincoln's decision to nominate Welles as his secretary of the navy, see Niven, *Gideon Welles,* 310–23. See also Hannibal Hamlin to Lincoln, 14, 29 December 1860, Lincoln Papers; George Fogg to Lincoln, 29 December 1860, 4 January 1861, Lincoln Papers; Henry Wilson to Lincoln, 5 January 1861, Lincoln Papers; Preston King to Lincoln, 11 January 1861, Lincoln Papers; Weed, *Life,* 1:606–7, 611; Hamlin, *Hannibal Hamlin,* 370, 373–75.
4. Quoted in Hamlin, *Hannibal Hamlin,* 487.
5. Welles, 15 September 1864, *Diary of Gideon Welles,* 2:145–46.
6. For opinions of Welles, see William Buckingham to Lincoln, 25 December 1860, Lincoln Papers; George Fogg to Lincoln, 29 December 1860, Lincoln Papers; Wilson to Lincoln, 5 January 1861, Lincoln Papers; Dana, *Recollections,* 170; Brooks,

Washington, D.C. in Lincoln's Time, 39–40; Porter, *Incidents and Anecdotes,* 114. See also Welles, "Farragut and New Orleans," 670; Gideon Welles to Mark Howard, 9 March 1861, Howard Papers, box 5, folder 8.

7. Welles, *Diary of Gideon Welles,* 1:19–20, 36; Welles, "Fort Sumter," 623–26, 629–30.

8. Welles, *Diary of Gideon Welles,* 1:15; Gustavus Fox to Winfield Scott, 8 February 1861, *OR,* 4:223; Welles, "Fort Sumter," 616–19; Lincoln to Fox, 1 May 1861, Fox, *Confidential Correspondence,* 1:43–44.

9. Andrew Foote to Welles, 4, 5 April 1861, *OR,* 4:236–37; Welles, "Fort Sumter," 628–30, 634; Porter, *Incidents and Anecdotes,* 13–15.

10. Welles, 16 September 1862, 20 December 1862, *Diary of Gideon Welles,* 1:133, 204; Welles, "Fort Sumter," 628–29.

11. Wilkes, *Autobiography,* 756.

12. Samuel Barron to Samuel Du Pont, 22 April 1861, Du Pont Papers, box 42, no. 120.

13. Franklin Buchanan to John Dahlgren, 22 April 1861, *OR,* 4:417; Du Pont to William Shubrick, 5 December 1860, Du Pont, *Samuel Francis Du Pont,* 1:3–4; George Magruder to Du Pont, 22 April 1861, Du Pont, *Samuel Francis Du Pont,* 1:59; Elizabeth Blair Lee to Samuel Phillips Lee, 12 February 1861, Lee, *Wartime Washington,* 35–36; Wilkes, *Autobiography,* 755–57; Buchanan to Du Pont, 20 May 1861, Du Pont Papers, box 42, no. 67.

14. Charles Steedman to [Unknown] Hudson, 21 April 1861, Steedman, *Memoir and Correspondence,* 273.

15. Percival Drayton to Lydig Hoyt, 19 May 1861, Drayton, *Naval Letters,* 3; Du Pont to Buchanan, Du Pont, *Samuel Francis Du Pont,* 1:72; Dewey, *Autobiography,* 44; John Missroon to Du Pont, 8 May 1861, Du Pont Papers, box 42, no. 40; Pollard, "Story of a Hero," 599–600.

16. Welles, *Diary of Gideon Welles,* 1:5, 10, 19; Welles, "Farragut and New Orleans," 679; Welles to Fox, 5 December 1870, Fox papers.

17. For examples of people who noticed the discontent and mistrust these Southern officers generated in and out of Washington, see Du Pont to Foote, 25 January 1861, Du Pont, *Samuel Francis Du Pont,* 1:27; Wilkes, *Autobiography,* 755–56; Missroon to Du Pont, 8 May 1861, Du Pont Papers, box 42, no. 40; Albert Smith to Welles, 15 April 1861, Welles papers, container 19; Robert Ritchie to Welles, 15 May 1861, Welles papers, container 19; Homer Adams to Du Pont, 30 April 1861, *OR,* 4:126; Hiram Paulding memo, 25 April 1869, Meade, *Life of Hiram Paulding,* 235–36. For Maury's and Magruder's resignations, see Welles, "Welles's Narrative," Welles papers, container 2.

18. Welles' quote is from Welles, "Welles's Narrative," Welles papers, container 2. See also Buchanan to Dahlgren, 22 April 1861, *OR,* 4:417; Dahlgren, 3 May 1861, *Memoir of John A. Dahlgren,* 330–31; Buchanan to Du Pont, 20 May 1861, Du Pont Papers, box 42, no. 67; *Washington National Intelligencer,* n.d., Du Pont

Papers, box 42, no. 20a; Welles to his son, 9 May 1861, Welles papers, container 19.

19. Welles, *Diary of Gideon Welles,* 1:5, 19–20; Welles, 27 June 1863, ibid., 1:346; Welles, "Farragut and New Orleans," 679; Welles to Fox, 5 December 1870, Fox papers; Welles Letter, 14 October 1861, Boynton, *History of the Navy,* 34–35.

20. Welles, "Fort Sumter," 616–17; Welles, "Welles' Narrative"; Meade, *Life of Hiram Paulding,* 219–24, 235–36.

21. Charles Davis to his wife, 31 May, 7 June 1861, Davis, *Life,* 122–23; Paulding to Du Pont, 4 May 1861, Du Pont Papers, box 42, no. 28; Paulding to his wife, 30 April, 2, 11, 15 May 1861, Meade, *Life of Hiram Paulding,* 241–43, 246–47.

22. Du Pont to Paulding, 25 April 1861, *OR,* 4:334.

23. Welles letter, 14 October 1861, Boynton, *History of the Navy,* 34–35.

24. The quote is from Welles to Howard, 6 October 1861, Howard Papers, box 5, folder 10. See also Du Pont to his wife, 13 January 1861, Du Pont, *Samuel Francis Du Pont,* 1:19–21; Wilkes, *Autobiography,* 757–59; Paulding memo, 25 April 1869, Meade, *Life of Hiram Paulding,* 235–36.

25. Cornelius Stribling to Welles, 10 July 1861, Welles papers, container 20.

26. Wilkes, *Autobiography,* 755–57; Charles Butler to Welles, 13 September 1861, Welles papers, container 20.

27. Dewey, *Autobiography,* 42–43.

28. For a good discussion of the Efficiency Board's history, see Weddle, "'Magic Touch of Reform,'" 471–504.

29. Welles to Louis Goldsborough, 25 November 1861, *OR,* 6:454; Welles to William Mervine, 2 October 1861, ibid., 16:695; Welles, 10 August, 20 September 1862, *Diary of Gideon Welles,* 1:76–77, 142; Welles, "Farragut and New Orleans," 679–81; Welles to S. Ledyard Phelps, 18 September 1861, Slagle, *Ironclad Captain,* 295–96.

30. Montgomery Blair to Fox, 26 April 1861, Fox, *Confidential Correspondence,* 1:37–38; Fox to Du Pont, 22 May 1861, Du Pont, *Samuel Francis Du Pont,* 1:71; Andrew Harwood to Du Pont, 2 May 1861, Du Pont Papers, box 42, no. 5; Welles, 23 August 1864, *Diary of Gideon Welles,* 2:117–18; Elizabeth Blair Lee to Samuel Phillips Lee, 6 July, 18 September 1861, Lee, *Wartime Washington,* 58, 80; Paulding to his wife, 30 April, 2, 11, 15 May 1861, Meade, *Life of Hiram Paulding,* 241–43, 246–47; Charles Davis to his wife, 22 May 1861, Davis, *Life,* 121–22.

31. Blair to Fox, 26 April 1861, *Confidential Correspondence,* 1:37–38; Lincoln to Fox, 1 May 1861, ibid., 1:43–44; Niven, *Gideon Welles,* 351–53; William Faxon to Howard, 13 February 1865, Howard Papers, box 6, folder 10; Faxon to Howard, 7 April 1861, Howard Papers, box 5, folder 8; Faxon to Howard, 10 April 1861, Howard Papers, box 5, folder 9.

32. Lincoln to Fox, 1 May 1861, Fox, *Confidential Correspondence,* 1:43–44; Fox to his wife, 25 July 1861, ibid., 1:363; Virginia Fox diary, 2 April 1861, reel 1.

33. Welles, 13 August 1863, *Diary of Gideon Welles,* 1:401; Welles, 30 January 1865, 17 January 1866, ibid., 2:233, 418; Charles Davis to his wife, 10 September 1861, Davis, *Life,* 132–33; Rodgers, "Du Pont's Attack at Charleston," 4:32–33; John Hay, 10 September 1863, Hay, *Inside Lincoln's White House,* 83; Du Pont to his wife, 10 September 1861, Du Pont, *Samuel Francis Du Pont,* 1:147; Du Pont to his wife, 30 June 1862, Du Pont, *Samuel Francis Du Pont,* 2:139; Elizabeth Blair Lee to Samuel Phillips Lee, 25 February 1864, Lee, *Wartime Washington,* 352–53; Sands, *From Reefer to Rear-Admiral,* 179; Brooks, *Washington, D.C. in Lincoln's Time,* 39; Villard, *Memoirs,* 1:172.

34. Welles, 17 January 1866, *Diary of Gideon Welles,* 2:418.

35. Brooks, *Washington, D.C. in Lincoln's Time,* 39.

36. Welles, 13 August 1863, *Diary of Gideon Welles,* 1:401; Welles, 26 November, 16 December 1864, 30 January, 21 February 1865, 17 January 1866, ibid., 2:183, 199–200, 232–33, 241, 418; Welles to his wife, 18 August 1861, Welles papers, container 20.

37. Welles, "Farragut and New Orleans," 679; Fox to Samuel Phillips Lee, 22 September 1862, Fox, *Confidential Correspondence,* 2:215.

38. Welles, 24, 25 September 1862, *Diary of Gideon Welles,* 1:147, 149–50; Du Pont to George Smith Blake, 27 September 1861, Du Pont, *Samuel Francis Du Pont,* 1:153–54; David Porter to Fox, 5 July 1861, Fox, *Confidential Correspondence,* 2:78; Welles, "Farragut and New Orleans," 680–81; Charles Davis to Du Pont, 15 June 1861, Du Pont Papers, box 42, no. 122; Charles Grimes to Welles, 14 May 1861, Welles papers, container 19; Welles, "Welles Narrative."

39. Welles, 2 December 1861, "Report of the Secretary of the Navy," 37th Cong., 2nd sess., *Congressional Globe,* 21; *An Act Providing for the Better Organization of the Military Establishment,* 3 August 1861, Acts of the Thirty-seventh Congress, Session 1, *U.S. Statutes at Large* 12 (1863), in American Memory's "A Century of Lawmaking for a New Nation," http://memory.loc.gov/ammem/hlawquery.html (accessed 8 May 2008), 290–92; *An Act to Further Promote the Efficiency of the Navy,* 21 December 1861, Acts of the Thirty-seventh Congress, Session 2, *U.S. Statutes at Large* 12 (1863), in American Memory's "A Century of Lawmaking for a New Nation," http://memory.loc.gov/ammem/hlawquery.html (accessed 8 May 2008), 329–30.

40. As things turned out, the legislation did not force Goldsborough into retirement. See Goldsborough to Fox, 15 December 1861, Fox, *Confidential Correspondence,* 1:218–20.

41. For examples of all these attitudes, see Du Pont to Fox, 24 September 1861, ibid., 1:54; Goldsborough to Fox, 15 December 1861, ibid., 1:218–20; Du Pont to Henry Winter Davis, 4, 29 September 1861, Du Pont, *Samuel Francis Du Pont,* 1:143, 161; Wilkes, *Autobiography,* 841–42; Adams to Wilkes, 20 December 1861, Wilkes papers, General Correspondence, container 13.

42. *An Act to Establish and Equalize the Grades of Line Officers in the United States Navy,* 16 July 1862, Acts of the Thirty-seventh Congress, Session 2, *U.S. Statutes at Large* 12 (1863), in American Memory's "A Century of Lawmaking for a New Nation," http://memory.loc.gov/ammem/hlawquery.html (accessed 8 May 2008), 583–87; Goldsborough to Welles, 27 June 1862, *OR,* 7:511; Foote to Fox, 2, 9, 13 November 1861, *OR,* 22:390, 393; 399; Du Pont to his wife, 14 April 1862, Du Pont, *Samuel Francis Du Pont,* 2:12; Fox to Du Pont, 14 September 1861, Du Pont Papers, box 42, no. 102.

43. The seven active-list rear admirals were David Farragut, Louis Goldsborough, Samuel Du Pont, Andrew Foote, Charles Davis, John Dahlgren, and David Porter. The original retired-list rear admirals were Charles Stewart, George Read, William Shubrick, Joseph Smith, George Storer, Francis Gregory, Elie Lavallette, Silas Stringham, Samuel Breese, and Hiram Paulding. Read died shortly after receiving his commission, giving Welles the opportunity to appoint Paulding.

44. Welles, 1 March 1864, *Diary of Gideon Welles,* 2:533–34.

45. Welles, 10 August 1862, ibid., 1:75–77; Du Pont to Fox, 14 August 1862, *OR,* 13:255; Samuel Phillips Lee to Fox, 27 February 1863, Fox, *Confidential Correspondence,* 2:248; James Grimes to Fox, 30 July 1862, Fox, *Confidential Correspondence,* 1:326; Samuel Phillips Lee to James Doolittle, 20 February 1865, "Letter from Samuel Phillips Lee to James Rood Doolittle," 121.

46. Charles Davis to his wife, 14 June 1861, Davis, *Life,* 124; Fox to Du Pont, 22 May 1861, Du Pont, *Samuel Francis Du Pont,* 1:71; Du Pont to Henry Winter Davis, 1 June 1861, Du Pont, *Samuel Francis Du Pont,* 1:75; Du Pont to his wife, 28 June 1861, Du Pont, *Samuel Francis Du Pont,* 1:85–86.

47. Welles, 23 August 1864, *Diary of Gideon Welles,* 2:117–18; Charles Davis to his wife, 22, 31 May, 14, 26 June 1861, Davis, *Life,* 121–22, 124–25; Du Pont to William Whetten, 3 May, 23 June 1861, Du Pont, *Samuel Francis Du Pont,* 1:67, 79; Fox to Du Pont, 22 May 1861, Du Pont, *Samuel Francis Du Pont,* 1:71; Du Pont to Alexander Bache, 30 May 1861, Du Pont, *Samuel Francis Du Pont,* 1:73–74; Du Pont to his wife, 28 June 1861, Du Pont, *Samuel Francis Du Pont,* 1:85–86.

48. The Strategy Board's six major reports are 5, 16, 26 July 1861, *OR,* 12:195–206; and 9 August, 3, 19 September 1861, ibid., 16:618–30, 651–54, 680–81. See also Du Pont to his wife, 28 June 1861, Du Pont, *Samuel Francis Du Pont,* 1:85–86; Charles Davis to his wife, 16 July, 29 August 1861, Davis, *Life,* 126, 129.

49. Welles to Howard, 23 August 1861, Howard Papers, box 5, folder 10.

Chapter Two. Atlantic Storms

1. For a good discussion of the capture of the Norfolk Navy Yard, see Musicant, *Divided Waters,* 28–40. See also Gideon Welles to Charles McCauley, 10, 16 April

1861, *OR,* 4:274, 277; McCauley to Welles, 25 April 1861, *OR,* 4:288; Hiram
Paulding to Welles, 23 April 1861, *OR,* 4:289–91.

2. Welles to Silas Stringham, 1 May 1861, *OR,* 5:621; Moses Grinnell to Abraham
Lincoln, 20 March 1862, Lincoln Papers; George Morgan to Welles, 2 May 1861,
Welles papers, container 19; Welles to Morgan, 31 October 1861, Welles papers,
container 20; Howard Barney to Welles, 21 February 1862, Welles papers, container 20; Welles to Gustavus Fox, 5 December 1870, Fox papers.

3. Fox to Samuel Du Pont, 13 December 1862, Fox, *Confidential Correspondence,*
1:169; J. P. Bankhead to Fox, 29 September 1861, ibid., 1:383–84; Du Pont to
George Smith Blake, 27 September 1861, Du Pont, *Samuel Francis Du Pont,* 1:155;
Thomas Looker to Lincoln, 9 May 1861, Lincoln Papers; Barney to Welles, 21 February 1862, Welles papers, container 20; Butler, *Butler's Book,* 286.

4. The quote is from Stringham to Fox, 16 August 1861, Fox, *Confidential Correspon-
dence,* 1:366. See also Welles to Stringham, 21 May 1861, *OR,* 5:661; Stringham to
Welles, 24, 30 May, 6, 20 June 1861, *OR,* 5:664–66, 681, 691, 732; Stringham to
Welles, 17, 30 July, 8, 15 August 1861, *OR,* 6:4, 42, 66, 83; Thomas Turner to Du
Pont, 9 August 1861, Du Pont Papers, box 42, no. 17.

5. Du Pont to Henry Winter Davis, 4 September 1861, Du Pont, *Samuel Francis Du
Pont,* 1:141–42; Du Pont to Blake, 27 September 1861, ibid., 1:154; Welles to his
wife, 1 September 1861, Welles papers, container 20; First Report of Conference
for the Consideration of Measures for Effectually Blockading the South Atlantic
Coast, 5 July 1861, *OR,* 12:100–200.

6. Butler, *Butler's Book,* 285–89.

7. Welles to Stringham, 2 September 1861, *OR,* 6:131.

8. Stringham to Welles, 2 September 1861, ibid., 6:158; Fox to Stringham, 14 September 1861, ibid., 6:210; Fox to Du Pont, 13 December 1862, Fox, *Confidential
Correspondence,* 1:169; Welles to Fox, 5 September 1861, Fox, *Confidential Correspondence,* 1:374; Du Pont to his wife, 15, 17 December 1861, Du Pont, *Samuel
Francis Du Pont,* 1:147–49; Du Pont to Blake, 27 September 1861, Du Pont,
Samuel Francis Du Pont, 1:153–54; Fox to Welles, 7 September 1861, Welles
papers, container 20; Welles to his wife, 15 September 1861, Welles papers, container 20; Welles to Morgan, 31 October 1861, Welles papers, container 20; Welles
to Fox, 5 December 1870, Fox papers; John Missroon to Du Pont, 21 November
1861, Du Pont Papers, box 42, no. 254.

9. Welles to Stringham, 18 September 1861, *OR,* 6:232.

10. Stringham to Welles, 16 September 1861, ibid., 6:216; Welles to Stringham, 18
September 1861, ibid., 6:232; Welles, 23 August 1864, *Diary of Gideon Welles,*
1:117–18; Du Pont to his wife, 15 September 1861, Du Pont, *Samuel Francis Du
Pont,* 1:147–48.

11. Ammen, *Campaigns,* 123.

12. Welles devoted considerable space in his diary trying to understand and explain Du Pont. See Welles, 2 October 1862, 20, 23, 25 May 1863, 4 June 1863, *Diary of Gideon Welles*, 1:160, 307, 309–12, 322–23; Welles, 23 August 1864, 2 September 1864, 23 June 1865, ibid., 2:117–18, 135, 320–21. See also Charles Davis to his wife, 8 November 1861, Davis, *Life*, 181; Rodgers, "Du Pont's Attack at Charleston," 4:33; Du Pont to his wife, 3 August 1861, Du Pont, *Samuel Francis Du Pont*, 1:123–24; Villard, *Memoirs*, 2:12; Osbon, *Sailor of Fortune*, 135, 149; Ammen, *Campaigns*, 123; Anne Rodgers to John Rodgers, 16 August 1862, John Rodgers Papers, box 21.

13. Du Pont to William Shubrick, 5 December 1860, Du Pont, *Samuel Francis Du Pont*, 1:4.

14. Du Pont to Henry Winter Davis, 29 September 1861, ibid., 1:161.

15. Welles to Du Pont, 27 April 1861, *OR*, 1:16; Welles to Stringham, 18 September 1861, ibid., 6:232; Welles to Du Pont, 3 August, 12 October 1861, 12:207, 214; Second Report of Conference for the Consideration of Measures for Effectually Blockading the South Atlantic Coast, 16 July 1861, ibid., 12:198; Du Pont to Fox, 24 September 1861, Fox, *Confidential Correspondence*, 1:54; Welles, 23 August 1864, *Diary of Gideon Welles*, 2:117–18; Du Pont to Blake, 27 September 1861, Du Pont, *Samuel Francis Du Pont*, 1:155.

16. Welles to Du Pont, 3 August, 12 October, 6 November 1861, *OR*, 12:207, 214, 259; Du Pont to his wife, 26 July, 1 August 1861, Du Pont,r *Samuel Francis Du Pont*, 1:114, 117; Du Pont to Henry Winter Davis, 8 October 1861, Du Pont, *Samuel Francis Du Pont*, 1:162–64.

17. Steedman, *Memoir and Correspondence*, 286.

18. Du Pont to Fox, 24, 29 September, 8 October, 11 November 1861, Fox, *Confidential Correspondence*, 1:53, 55–57; Welles, 23 August 1864, *Diary of Gideon Welles*, 2:117–18; Welles, 12 April 1863, *Diary of Gideon Welles*,1:268–69; Du Pont to Welles, 29 October, 11 November 1861, *OR*, 12:230, 265; Du Pont to his wife, 8, 15, 17 September 1861, Du Pont, *Samuel Francis Du Pont*, 1:143–49; Du Pont to Blake, 27 September 1861, Du Pont, *Samuel Francis Du Pont*, 1:155; Du Pont to Percival Drayton, 27 September 1861, Du Pont, *Samuel Francis Du Pont*, 1:156–57; Du Pont to Henry Winter Davis, 29 September 1861, Du Pont, *Samuel Francis Du Pont*, 1:160.

19. Du Pont to Welles, 6 November 1861, *OR*, 12:259; Du Pont to Henry Du Pont, 15 October 1861, Du Pont, *Samuel Francis Du Pont*, 1:165; Du Pont to his wife, 17, 30 October 1861, Du Pont, *Samuel Francis Du Pont*, 1:170–72, 202.

20. The quote is from Roswell Lamson to Flora Lamson, 4 November 1861, Lamson, *Lamson of the Gettysburg*, 41. See also Charles Davis to his wife, 8 November 1861, Davis, *Life*, 179–80; Du Pont to his wife, 5 May 1862, Du Pont, *Samuel Francis Du Pont*, 2:32–33; John Rodgers to Anne Rodgers, 9 November 1861, John Rodgers Papers, box 21.

21. Steedman, *Memoir and Correspondence*, 292.

22. The account of the Port Royal battle is based mostly on Robert M. Browning, *Success Is All That Was Expected,* 35–42, and Du Pont, "Detailed Report of Flag Officer Du Pont,"12:261–65. See also Charles Davis to his wife, 8 November 1861, Davis, *Life,* 179–80; Ammen, "Du Pont and the Port Royal Expedition," 1:686; Du Pont to his wife, 5 May 1862, Du Pont, *Samuel Francis Du Pont,* 2:32–33; Mahan, *From Sail to Steam,* 165; Osbon, *Sailor of Fortune,* 149.

23. Welles to Du Pont, 16 November 1861, *OR,* 12:294.

24. Du Pont to Welles, 11 November 1861, ibid., 12:265.

25. Welles to Du Pont, 13 November 1861, ibid., 12:290; Du Pont to Welles, 11 November 1861, ibid., 12:265; Du Pont to James Lardner, 9 November 1861, ibid., 12:286; Du Pont, "Detailed Report of Flag Officer Du Pont," ibid., 12:262; Du Pont to Fox, 11 November, 25 December 1861, Fox, *Confidential Correspondence,* 1:57, 83–84; Du Pont to his wife, 25 November 1861, Du Pont, *Samuel Francis Du Pont,* 1:259; Du Pont to James Stokes Biddle, 17 December 1861, Du Pont, *Samuel Francis Du Pont,* 1:282; Du Pont to his wife, 5 May 1862, Du Pont, *Samuel Francis Du Pont,* 2:32–33; Missroon to Du Pont, 21 November 1861, Du Pont Papers, box 42, no. 254.

26. Drayton to Lydig Hoyt, 22 April 1862, Drayton, *Naval Letters,* 16.

27. Welles to Du Pont, 28 July 1862, Du Pont, *Samuel Francis Du Pont,* 2:174.

28. Welles to Du Pont, 20, 25 January 1862, *OR,* 12:508, 521; Welles to Du Pont, 5 June 1862, ibid., 13:78; Raymond Rodgers to Du Pont, 18 July 1862, Du Pont, *Samuel Francis Du Pont,* 2:163; Welles to Mark Howard, 16 November 1861, Howard Papers, box 5, folder 10.

29. Du Pont to Fox, 11 January, 10 February 1862, *OR,* 12:477, 540; Du Pont to Welles, 3 April 1862, ibid., 12:706; Du Pont to Welles, 27 June, 7 August, 27 September 1862, ibid., 13:134, 239, 349; Du Pont to Fox, 29 July 1862, Fox, *Confidential Correspondence,* 1:139; Raymond Rodgers to Fox, 3 September 1862, Fox, *Confidential Correspondence,* 2:363–65; Du Pont to his wife, 12 December 1861, 10 April 1862, Du Pont, *Samuel Francis Du Pont,* 1:274, 405–8; Du Pont to George Hancock, 26 February 1862, Du Pont, *Samuel Francis Du Pont,* 1:342; Du Pont to his wife, 27 July 1862, Du Pont, *Samuel Francis Du Pont,* 2:168.

30. John Marchand, 24, 29 January, 3 February, 5 March, 5, 12 April, 8 May 1862, Marchand, *Charleston Blockade,* 85, 91, 95, 127–28, 141, 157, 264.

31. Du Pont to Welles, 6 March 1862, *OR,* 12:621.

32. Du Pont to his wife, 3, 28 January, 25 March, 10 April 1862, Du Pont, *Samuel Francis Du Pont,* 1:302–3, 320, 386, 407–8; Du Pont to his wife, 14 April, 1 May 1862, Du Pont, *Samuel Francis Du Pont,* 2:9, 23.

33. Du Pont to Lardner, 27 April 1862, *OR,* 12:781; Du Pont to Welles, 27 April 1862, ibid., 12:782.

34. The quote is from Du Pont to his wife, 25 December 1861, Du Pont, *Samuel Francis Du Pont,* 1:289. See also Du Pont to Biddle, 17 December 1861, ibid., 1:282; Du Pont to his wife, 10 April 1862, ibid., 1:408.

35. Du Pont to his wife, 14 April, 9, 10 August 1862, ibid., 2:12, 185–87; Raymond Rodgers to Du Pont, 19 July 1862, ibid., 2:166; Du Pont to Fox, 14 August 1862, *OR,* 13:255.

36. Welles, 20 September 1862, *Diary of Gideon Welles,* 1:142; Welles, 25 March 1869, ibid., 3:562; Du Pont to Fox, 24 September 1861, Fox, *Confidential Correspondence,* 1:55; Du Pont to his wife, 27 July 1862, Du Pont, *Samuel Francis Du Pont,* 2:169–70; William Frederick Keeler to his wife, 14 June 1862, Keeler, *Aboard the USS Monitor,* 155; Belknap, "Captain George Hamilton Perkins," 205; Franklin, *Memories of a Rear-Admiral,* 143–44; Welles to Fox, 24 March 1870, Fox papers; Saltonstall, *Reminiscences,* 20; John Rodgers to Anne Rodgers, [?] September 1862, John Rodgers Papers, box 21.

37. Du Pont to his wife, 15, 17 September 1861, Du Pont, *Samuel Francis Du Pont,* 1:147–48, 150.

38. Louis Goldsborough to Welles, 10 November 1861, *OR,* 6:421; Goldsborough to Fox, 9 November 1861, Fox, *Confidential Correspondence,* 1:204–5; Du Pont to his wife, 19 October 1861, Du Pont, *Samuel Francis Du Pont,* 1:175; Baird, "Narrative of Rear Admiral Goldsborough," 1023.

39. Goldsborough's report, 29 January 1862, *OR,* 6:536.

40. Welles to Goldsborough, 14 February 1862, ibid., 6:593; Fox to Goldsborough, 1 March 1862, ibid., 6:623; Goldsborough's report, 18 February 1862, ibid., 6:554; Welles to Goldsborough, 27 March 1862, ibid., 7:119; Goldsborough to Stephen Rowan, 22 March 1862, ibid.

41. Goldsborough to Welles, 23 March, 18, 19 April 1862, ibid., 7:165, 245–46; Goldsborough to Fox, 16 March 1862, Fox, *Confidential Correspondence,* 1:248–49; Baird, "Narrative of Rear Admiral Goldsborough," 1024–25, 1029; Goldsborough to his family, 13 April, 5 May 1862, Goldsborough papers; Saltonstall, *Reminiscences,* 28–29; John Rodgers to Anne Rodgers, 27 April 1862, John Rodgers Papers, box 21.

42. Welles, 20 September 1862, *Diary of Gideon Welles,* 1:142; Goldsborough to Fox, 9 November 1861, Fox, *Confidential Correspondence,* 1:218; Goldsborough to Welles, 17 October 1861, *OR,* 6:333; Welles to his son, 16 March 1862, Welles papers, container 21.

43. Saltonstall, *Reminiscences,* 28–29.

44. Du Pont to his wife, 25 March 1862, Du Pont, *Samuel Francis Du Pont,* 1:385; DiZerega, "Last Days of the Rebel Iron-Clad Merimac," 4.

45. Goldsborough to George McClellan, 27 June 1862, *OR,* 7:514.

46. Goldsborough to Welles, 27 June 1862, ibid., 7:511; Fox to Goldsborough, 15 April 1862, Goldsborough papers.

47. Welles, 20 September 1862, *Diary of Gideon Welles,* 1:142; Fox to Du Pont, 3 June 1862, Fox, *Confidential Correspondence,* 1:126–27; Goldsborough to Fox, 21 May 1862, Fox, *Confidential Correspondence,* 1:272–73; Fox to Goldsborough, 3

June 1862, Fox, *Confidential Correspondence,* 1:281–82; Du Pont to his wife, 27 July 1862, Du Pont, *Samuel Francis Du Pont,* 2:169–70; Keeler, 14, 25 June 1862, Keeler, *Aboard the USS Monitor,* 155, 162–63; Elizabeth Blair Lee to Samuel Phillips Lee, 3 July 1862, Lee, *Wartime Washington,* 160; Welles to Fox, 24 March 1870, Fox papers; John Rodgers to Anne Rodgers, 29 September 1862, John Rodgers Papers, box 26.

48. Osbon, *Sailor of Fortune,* 213–14; Dahlgren, 5 July 1862, Dahlgren, *Memoir of John A. Dahlgren,* 374; Orville Hickman Browning, 14 July 1862, *Diary of Orville Hickman Browning,* 1:558–59.

49. Welles, 10, 18 August 1862, 12 May 1863, *Diary of Gideon Welles,* 1:73, 86–87, 298; Wilkes, *Autobiography,* 223, 295, 867; Welles to Wilkes, 6 July 1862, *OR,* 7:548; Welles to Wilkes, 23 July 1862, Wilkes papers, General Correspondence, container 13.

50. Goldsborough to Fox, 23 May, 4 June 1862, Fox, *Confidential Correspondence,* 1:276, 282; Goldsborough to Welles, 11, 15 July 1862, *OR,* 7:565, 573; Welles to Goldsborough, 6 July 1862, *OR,* 7:547; Baird, "Narrative of Rear Admiral Goldsborough," 1031; Goldsborough to his family, 13 May, 10 June 1862, Goldsborough papers; Anne Rodgers to John Rodgers, 6 August, 11 September 1862, John Rodgers Papers, box 21.

51. Du Pont to his wife, 27 July 1862, Du Pont, *Samuel Francis Du Pont,* 2:169–70.

52. Welles to Goldsborough, 21 July 1862, *OR,* 7:574; Welles, 10 August 1862, *Diary of Gideon Welles,* 1:73; Raymond Rodgers to Du Pont, 18 July 1862, Du Pont, *Samuel Francis Du Pont,* 2:163; Goldsborough to Fox, 4 June 1862, Fox, *Confidential Correspondence,* 1:282; Goldsborough to his family, 13 April, 13 May 1862, Goldsborough papers.

Chapter Three. The Mighty Mississippi

1. Gideon Welles to Simon Cameron, 14 May, 12 June 1861, *OR,* 22:277, 285; Welles to John Rodgers, 16 May, 23 September 1861, ibid., 22:280, 349.

2. John Rodgers to Anne Rodgers, 7 September 1862, Johnson, *Rear Admiral John Rodgers,* 221.

3. The quote is from John Rodgers to Anne Rodgers, 31 October 1862, Johnson, *Rear Admiral Rodgers,* 200. See also Charles Davis to his wife, 26 June 1863, Davis, *Life,* 295–96; Rodgers, "Du Pont's Attack at Charleston," 4:36; Samuel Du Pont to James Stokes Biddle, 17 December 1861, Du Pont, *Samuel Francis Du Pont,* 1:282.

4. John Rodgers to [Unknown] Macomb, 11 July 1861, John Rodgers Papers, box 21.

5. Welles to John Rodgers, 11, 12, 17 June, 23 August 1861, *OR,* 22:284–86, 304; John Frémont to Montgomery Blair, 9 August 1861, ibid., 22:296; Andrew Foote to Welles, 13 September 1861, ibid., 22:322; John Rodgers to Welles, 7, 12 September 1861, ibid., 22:319, 334; John Rodgers to Anne Rodgers, 9 July 1861, John

Rodgers Papers, box 21; John Rodgers to William Hodge, 25 September 1861, John Rodgers Papers, box 26.

6. Welles to Foote, 30 August 1861, *OR,* 22:307; Welles to Foote, 17 June 1862, ibid., 23:213; Welles, 2 September 1864, *Diary of Gideon Welles,* 2:135; Welles, "Fort Sumter," 620, 634; Foote to Welles, 5, 15 June 1861, Welles papers, container 20.

7. Foote, "General Orders No. 6," 17 December 1861, *OR,* 22:466; Charles Davis to his wife, 23 June 1863, Davis, *Life,* 294; John Foote, "Notes on the Life of Admiral Foote," 1:347; Foote to John Dahlgren, 12 July 1861, Welles papers, container 29; Foote to Welles, 5 June 1861, Welles papers, container 20.

8. Welles, 18, 27 June 1863, *Diary of Gideon Welles,* 1:334, 346; Charles Davis to his wife, 23 June 1863, Davis, *Life,* 294; Eads, "Recollections of Foote and the Gun-Boats," 1:346; John Foote, "Notes on the Life of Admiral Foote," 1:347; Walke, "Gun-Boats at Belmont and Fort Henry," 1:359–60; Du Pont to his wife, 16 December 1862, Du Pont, *Samuel Francis Du Pont,* 2:302; Du Pont to his wife, 24 June 1863, Du Pont, *Samuel Francis Du Pont,* 3:180–81; Thomas Turner to Du Pont, 14 July 1863, Du Pont Papers, W9-15529; Junius Browne, *Four Years in Secessia,* 151–52.

9. John Rodgers to Welles, 7 September 1861, *OR,* 22:320.

10. Welles to John Rodgers, 30 August, 23 September 1861, ibid., 22:307, 349; Foote to Gustavus Fox, 8 September 1861, ibid., 22:320; Foote to Welles, 13 September 1861, ibid., 22:322; John Rodgers to Welles, 7, 12 September 1861, ibid., 22:318–20, 334; John Rodgers to Anne Rodgers, 30, 31 August, 3, 27 September 1861, John Rodgers Papers, box 21; John Rodgers to Hodge, 25 September 1861, John Rodgers Papers, box 26.

11. Foote to Welles, 2, 9, 13 November 1861, *OR,* 22:316, 390, 393, 399; Foote to Fox, 2 November, 28 December 1861, Fox, *Confidential Correspondence,* 2:9–10, 17; Seth Ledyard Phelps to Elisha Whittlesey, 10 October 1861, Slagle, *Ironclad Captain,* 146.

12. The quote is from Foote to Fox, 11 January 1862, Fox, *Confidential Correspondence,* 2:30. See also Foote to Welles, 28 November 1861, *OR,* 22:444; Foote to Fox, 4 November 1861, Fox, *Confidential Correspondence,* 2:11; Fox to Foote, 27 January 1862, Fox, *Confidential Correspondence,* 2:36–37; Foote to Phelps, 24 January 1862, Slagle, *Ironclad Captain,* 150; Phelps to Whittlesey, 26 February 1862, Slagle, *Ironclad Captain,* 192.

13. Walke, "Gun-Boats at Belmont and Fort Henry," 1:362; *Northwestern Advocate,* n.d., Welles papers, container 36.

14. Symmes Browne to Fannie Bassett, 10 February 1862, Henry Browne and Symmes Browne, *Fresh-Water Navy,* 25–26.

15. Browne to Bassett, 10 February 1862, ibid., 24; Eads, "Recollections of Foote and the Gun-Boats," 1:343; *Northwestern Advocate,* n.d., Welles papers, container 36.

16. The quote is from Foote to his wife, 9/12 March 1862, Slagle, *Ironclad Captain,* 196. See also Welles to Foote, 9, 13 February 1862, *OR,* 22:547; Foote to Welles, 3, 10 March 1862, *OR,* 22:642, 696; Foote to Fox, 5, 9 March 1862, Fox, *Confidential Correspondence,* 2:40–41, 46.

17. Eads, "Recollections of Foote and the Gun-Boats," 1:345.

18. The quote is from Foote to his wife, 17 March 1862, Slagle, *Ironclad Captain,* 200. See also Welles, 10 September 1862, *Diary of Gideon Welles,* 1:120; Eads, "Recollections of Foote and the Gun-Boats," 1:345.

19. Du Pont to Fox, 14 April 1862, Fox, *Confidential Correspondence,* 1:119.

20. Welles to Foote, 9 April 1862, *OR,* 22:723; Browne to Bassett, 15 April 1862, Henry Browne and Symmes Browne, *Fresh-Water Navy,* 61.

21. Welles to Foote, 23 April, 7 May, 17 June 1862, *OR,* 23:70, 84, 213; Foote to Welles, 27 March 1862, ibid., 22:702; Foote to Welles, 15, 27 April 1862, ibid., 23:62–63; Foote to Fox, 9 March 1862, Fox, *Confidential Correspondence,* 2:46; Fox to Charles Davis, 22 April 1862, Fox, *Confidential Correspondence,* 2:58; Phelps to Whittlesey, 6 May 1862, Slagle, *Ironclad Captain,* 217; Junius Browne, *Four Years in Secessia,* 164–65; Welles to Charles Davis, 22 April 1862, *OR,* 23:69.

22. Foote to Charles Davis, 9, 10 May 1862, *OR,* 23:85–86; Charles Davis to his wife, 9 May 1862, Davis, *Life,* 222–23; Junius Browne, *Four Years in Secessia,* 164–66.

23. Welles to Foote, 17 June 1862, *OR,* 23:213; Foote to Charles Davis, 10, 15, 28, 30 May 1862, ibid., 23:21, 86, 109, 110; Foote to Welles, 13 June 1862, ibid., 23:155; Raymond Rodgers to Du Pont, 18 July 1862, Du Pont, *Samuel Francis Du Pont,* 2:163–64; Foote to Welles, 22 May 1862, Welles papers, container 21; John Foote to Welles, 30 May 1862, Welles papers, container 21; Welles, 10 August 1862, 27 June 1863, *Diary of Gideon Welles,* 1:74–75, 346.

24. Elizabeth Blair Lee to Samuel Phillips Lee, 18 May 1862, Lee, *Wartime Washington,* 148.

25. Charles Davis to Welles, 10 May 1862, *OR,* 23:13; Fox to Charles Davis, 16 May 1862, ibid., 23:22; Du Pont to his wife, 3 January, 10 April 1862, Du Pont, *Samuel Francis Du Pont,* 1:302–3, 407; Du Pont to his wife, 14 April, 1 May 1862, Du Pont, *Samuel Francis Du Pont,* 2:9, 23; Welles, 1 October 1862, *Diary of Gideon Welles,* 1:157–58; Welles, 2 August 1865, *Diary of Gideon Welles,* 2:351; Wilkes, *Autobiography,* 395; Fox to Du Pont, 12 May 1862, Fox, *Confidential Correspondence,* 1:119; Charles Davis to his wife, 2 June 1862, Davis, *Life,* 236; Lodge, *Early Memories,* 195–97.

26. Du Pont to Welles, 11 November 1861, *OR,* 12:265; Foote to Welles, 27 March 1862, ibid., 22:702; Welles to William McKean, 20 January, 5 April 1862, ibid., 17:56, 210; McKean to Welles, 13 April 1862, ibid., 17:218; Charles Davis to Du Pont, 9 April 1862, Davis, *Life,* 212; Charles Davis to his wife, 9 May 1862, Davis, *Life,* 223; Fox to Charles Davis, 22 April 1862, Fox, *Confidential Correspondence,* 2:59; Charles Davis to Fox, 22, 25 April 1862, Fox, *Confidential Correspon-*

dence, 2:59, 61; Hiram Paulding to Fox, 23 October 1861, Fox, *Confidential Correspondence,* 1:387.

27. Charles Davis to Charles Ellet, 3 June 1862, *OR,* 23:42.

28. Charles Davis to his wife, 31 May 1862, Davis, *Life,* 235.

29. Charles Davis to his wife, 25 June 1862, ibid., 248.

30. Charles Davis to his wife, 8 June 1862, ibid., 244.

31. Welles to Charles Davis, 18 June 1862, *OR,* 23:137; Charles Davis to his wife, 30 May, 8, 19, 20, 25 June 1862, Davis, *Life,* 234–35, 244–48.

32. Welles to Charles Davis, 25 June 1862, *OR,* 18:585; Charles Davis to Welles, 28 June 1862, ibid., 18:589; Charles Davis to his wife, 30 June 1862, Davis, *Life,* 257–59; Bell diary, 1 July 1862, Bell papers, vol. 3; David Farragut diary, 2 July 1862, Farragut, *Life of David Farragut,* 282–83.

33. Strategy Board, 9 August 1861, *OR,* 16:618–30.

34. Fox to Du Pont, 13 December 1862, Fox, *Confidential Correspondence,* 1:169; Frederick Engle to Fox, 5 February 1862, ibid., 1:417; Welles, 10 August 1862, 25, 27 May 1863, *Diary of Gideon Welles,* 1:76, 312–15; Welles, 23 August 1864, *Diary of Gideon Welles,* 2:116; Porter, *Incidents and Anecdotes,* 37.

35. The quote is from Welles to William Mervine, 23 August 1861, *OR,* 16:644. See also Welles to Mervine, 14 May, 20 August 1861, ibid., 16:521, 642; Mervine to Welles, 17 July, 1 August 1861, ibid., 16:589, 598; Welles, 10 August 1862, *Diary of Gideon Welles,* 1:76; Welles, 23 August 1864, *Diary of Gideon Welles,* 2:116; David Porter to Fox, 5 July 1861, Fox, *Confidential Correspondence,* 2:74; Porter to Charles Davis, 22 July 1861, Welles papers, container 29; Paulding to his wife, 11 May 1861, Meade, *Life of Hiram Paulding,* 245–46.

36. Mervine to Welles, 17 July, 1 August, 5, 9, 11 September 1861, *OR,* 16:589, 598, 655, 662–64.

37. Welles, 23 August 1864, *Diary of Gideon Welles,* 2:116.

38. Welles to Mervine, 2 October 1861, *OR,* 16:695.

39. Mervine to Welles, 29 September 1861, ibid., 16:693; Welles to Mervine, 2 October 1861, ibid., 16:695; Maclay, *Reminiscences of the Old Navy,* 135; Welles, 10 August 1862, *Diary of Gideon Welles,* 1:76; Welles, 23 August 1864, *Diary of Gideon Welles,* 2:116.

40. Welles to McKean, 6 September 1861, *OR,* 16:651; Welles, 23 August 1864, *Diary of Gideon Welles,* 2:116; Du Pont to Henry Winter Davis, 4 September 1861, Du Pont, *Samuel Francis Du Pont,* 1:143; Fox to Du Pont, 22 August 1861, Du Pont Papers, box 42, no. 17.

41. Porter, *Incidents and Anecdotes,* 34–35; Schley, *Forty-Five Years,* 27–28; Steedman, *Memoir and Correspondence,* 120.

42. Welles to McKean, 6 September 1861, *OR,* 16:661; McKean to Welles, 22 September, 15 October 1861, ibid., 16:684, 703.

43. Theodorus Bailey to Edmund Bailey, 26 December 1861, Bailey Papers, box 1, correspondence.

44. Welles to McKean, 23 January 1862, *OR,* 6:528; McKean to Welles, 1 November 1861, ibid., 16:752; McKean to Fox, 30 January 1862, Fox, *Confidential Correspondence,* 1:413–14; Welles, 23 August 1864, *Diary of Gideon Welles,* 2:116.

45. Porter, "Opening of the Lower Mississippi," 2:23–24; Welles, "Farragut and New Orleans," 677; Porter, *Incidents and Anecdotes,* 63–65.

46. Porter, "Opening of the Lower Mississippi," 2:24.

47. Ibid., 2:23–26; Porter, *Incidents and Anecdotes,* 63–65; Welles, "Farragut and New Orleans," 676–77; Welles to Fox, 5 December 1870, Fox papers.

48. Paulding to Fox, 23 October 1861, Fox, *Confidential Correspondence,* 1:387.

49. Welles was always justifiably proud of his selection of Farragut, and commented extensively in his diary and postwar writings on the reasons behind his choice. See Welles, 12 September 1863, *Diary of Gideon Welles,* 1:431; Welles, 23 August, 2 September 1864, ibid., 2:116–18, 134–35; Welles, "Farragut and New Orleans," 679–82. See also Porter, "Opening of the Lower Mississippi," 2:26–28; Porter, *Incidents and Anecdotes,* 65.

50. Welles, 23 August, 2 September 1864, *Diary of Gideon Welles,* 2:116–17, 134; Porter, "Opening of the Lower Mississippi," 2:27–28; Welles, "Farragut and New Orleans," 679, 682–83; Virginia Fox diary, 21 December 1861, reel 1.

51. Mervine to Welles, 1 July 1861, *OR,* 16:565; Welles to McKean, 4 December 1861, ibid., 16:807; Welles to David Farragut, 9 January 1862, ibid., 18:5; Welles, "Farragut and New Orleans," 681.

52. Farragut, *Life of David Farragut,* 204.

53. Kinney, "Farragut at Mobile Bay," 4:383.

54. Porter, "Opening of the Lower Mississippi," 2:26–27; Bartlett, "'Brooklyn' at the Passage of the Forts," 2:56–58; Farragut, *Life of David Farragut,* 548; Dewey, *Autobiography,* 50, 52, 108–9; Schley, *Forty-Five Years,* 50–51; Franklin, *Memories of a Rear-Admiral,* 185.

55. Welles to David Farragut, 20, 25 January, 24 February 1862, *OR,* 18:8–9, 37; David Farragut to Welles, 30 January, [?] February, 18 March 1862, ibid., 18:11, 31, 71; David Farragut to Fox, 22 February 1862, ibid., 18:34; Welles to McKean and David Farragut, 12 March 1862, ibid., 17:189; Welles, 23 August, 2 September 1864, *Diary of Gideon Welles,* 2:116–17, 134; David Farragut to Fox, 30 January, 12, 17 February, 7 March 1862, Fox, *Confidential Correspondence,* 1:299–301, 303, 306.

56. Porter to Fox, 28 March 1862, Fox, *Confidential Correspondence,* 2:89–92.

57. Porter, "Opening of the Lower Mississippi," 2:26; Porter to Fox, 28 March, April 1862, Fox, *Confidential Correspondence,* 2:89–92, 97–98.

58. David Farragut, 10 March 1862, Farragut, *Life of David Farragut,* 217.

59. David Farragut, 25 March 1862, ibid., 212; David Farragut to Fox, 3 March 1862, *OR,* 18:43; David Farragut to Theodorus Bailey, 7 March 1862, Bailey Papers, box 1, 1862 correspondence.

60. Porter to Fox, 8 April 1862, Fox, *Confidential Correspondence,* 2:96.

61. David Farragut to Theodorus Bailey, 8 March 1862, Bailey Papers, box 1, 1862 correspondence.

62. Osbon, *Sailor of Fortune*, 185.

63. Bartlett, "'Brooklyn' at the Passage of the Forts," 2:58; George Perkins to his family, 4 April 1862, Perkins, *George Hamilton Perkins*, 109; Osbon, *Sailor of Fortune*, 179–80; Bell diary, 11 April 1862, Bell papers, vol. 2.

64. Osbon, *Sailor of Fortune*, 182–84.

65. Meredith, "Farragut's Capture of New Orleans," 2:70–71; Osbon, *Sailor of Fortune*, 180; Bell diary, 19, 20 April, Bell papers, vol. 2; David Farragut letter, 21 April 1862, Farragut, *Life of David Farragut*, 226–27.

66. Osbon, *Sailor of Fortune*, 195.

67. Ibid., 196–97.

68. Dewey, *Autobiography*, p. 54; Osbon, *Sailor of Fortune*, 193, 204; Farragut, *Life of David Farragut*, 228; Theodorus Bailey to Edmund Bailey, 30 January 1862, Bailey Papers, box 1, 1862 correspondence.

69. Bell diary, 29 April 1862, Bell papers, vol. 3; Perkins to his family, 27 April 1862, Perkins, *George Hamilton Perkins*, 120–22; Porter, *Incidents and Anecdotes*, 69–71.

70. Welles to Mark Howard, 26 April 1862, Howard Papers, box 5, folder 12.

71. Welles to David Farragut, 10 May 1862, *OR*, 18:245.

72. Porter to Fox, 10 May 1862, Fox, *Confidential Correspondence*, 2:100.

73. Fox to David Farragut, 12 May 1862, *OR*, 18:244; Percival Drayton to Lydig Hoyt, 22 June 1862, Drayton, *Naval Letters*, 18; Foote to Welles, 25 May 1862, Welles Papers, container 21; David Farragut to his family, 25 April 1862, Farragut, *Life of David Farragut*, 234.

74. David Farragut to McKean, 3 May 1862, *OR*, 18:462; Welles to Porter, 10 May 1862, ibid., 18:375; David Farragut to Theodorus Bailey, 27 April, 11 June 1862, ibid., 18:146, 550; David Farragut to Welles, 25, 29 April, 6 May 1862, ibid., 18:147, 152, 155, 157; David Farragut to Fox, 25 April 1862, ibid., 18:155; Elizabeth Blair Lee to Samuel Phillips Lee, 28 April 1862, Lee, *Wartime Washington*, 134; David Farragut to his family, 25 April 1862, Farragut, *Life of David Farragut*, 234; David Farragut letter, 2 June 1862, Farragut, *Life of David Farragut*, 270.

75. Porter to Fox, 2 June 1862, Fox, *Confidential Correspondence*, 2:114.

76. Elizabeth Blair Lee to Samuel Phillips Lee, 29 April, 9 May, 12 July 1862, Lee, *Wartime Washington*, 136, 140, 164.

77. Welles to David Farragut, 20 January 1862, *OR*, 18:8; David Farragut to Porter, 1 May 1862, ibid., 18:462; David Farragut to Welles, 6 May 1862, ibid., 18:470; David Farragut to Adam Bache, 1 July 1862, ibid., 18:663; Porter to Fox, 24 May 1862, Fox, *Confidential Correspondence*, 2:103; David Farragut diary, 29–30 April 1862, Farragut, *Life of David Farragut*, 261–62.

78. David Farragut to Welles, 3 June 1862, *OR*, 18:579; Bell diary, 8, 11, 14 May 1862, Bell papers, vol. 3.

79. James Autrey to Samuel Phillips Lee, 18 May 1862, *OR*, 18:491.

80. David Farragut to Samuel Phillips Lee, 24 May 1862, ibid., 18:509; David Farragut to Welles, 30 May 1862, ibid., 18:519; Bell diary, 19, 23 May 1862, Bell papers, vol. 3; Elizabeth Blair Lee to Samuel Phillips Lee, 26 July 1862, Lee, *Wartime Washington,* 170.

81. David Farragut letter, n.d., Farragut, *Life of David Farragut,* 270–71.

82. David Farragut to Welles, 30 May, 3 June 1862, *OR,* 18:519, 577; Bell diary, 24, 25, 28 May 1862, Bell papers, vol. 3.

83. The quote is from Fox to David Farragut, 17 May 1862, *OR,* 18:498. See also Fox to David Farragut, 16 May 1862, ibid., 18:498; Fox to David Farragut, 12 May, 10 July 1862, Fox, *Confidential Correspondence,* 1:313, 335.

84. Porter to Fox, 24 May 1862, Fox, *Confidential Correspondence,* 2:105–6.

85. Fox to Porter, 17 May 1862, ibid., 2:101–2; Porter to Fox, 24 May, 7, 30 June 1862, ibid., 2:103–11, 117–18, 122–25; Hearn, *Admiral David Dixon Porter,* 127.

86. David Farragut letter, 2 June 1862, Farragut, *Life of David Farragut,* 269.

87. David Farragut to Theodorus Bailey, 11 June 1862, *OR,* 18:550; David Farragut to Fox, 12 June 1862, Fox, *Confidential Correspondence,* 1:315–16; Bell diary, 6, 13 June 1862, Bell papers, vol. 3.

88. Farragut, *Life of David Farragut,* 364–65.

89. David Farragut to Welles, 28 June, 2, 6 July 1862, *OR,* 18:588, 610, 630; David Farragut to Charles Davis, 28 June 1862, ibid., 18:588; Franklin, *Memories of a Rear-Admiral,* 173; Phelps to Foote, 6 July 1862, Slagle, *Ironclad Captain,* 251–52; Bell diary, 27, 28 June, 1, 2 July 1862, Bell papers, vol. 3; Thomas Craven to David Farragut, 28 June 1862, *OR,* 18:597.

90. David Farragut to Welles, 7, 10, 11, 13 July 1862, *OR,* 18:594, 634, 638, 675; David Farragut to Bell, 10 July 1862, ibid., 18:632–33; David Farragut diary, 2 July 1862, Farragut, *Life of David Farragut,* 282–83; Phelps to Foote, 6 July 1862, Slagle, *Ironclad Captain,* 253–54; Bell diary, 1 July 1862, Bell papers, vol. 3.

91. David Farragut to Welles, 10 July 1862, *OR,* 18:675; David Farragut to Welles, 17 July 1862, ibid., 19:4; Phelps to Foote, 29 July 1862, Slagle, *Ironclad Captain,* 57; Charles Davis to his wife, 14 July 1862, Davis, *Life,* 263.

92. Charles Davis to his wife, 18 August 1862, Du Pont, *Samuel Francis Du Pont,* 3:29n.

93. David Farragut to Welles, 17 July 1862, *OR,* 19:4; David Farragut to Charles Davis, 15 July 1862, ibid., 19:7; David Farragut to Charles Davis, 16 July 1862, ibid., 23:236; Phelps to Foote, 29 July 1862, ibid., 19:57–58; Bell diary, 15, 17 July 1862, Bell papers, vol. 4; Charles Davis to his wife, 14 July 1862, Davis, *Life,* 264.

94. Charles Davis to David Farragut, 17 July 1862, *OR,* 19:11.

95. Charles Davis to his wife, 14, 18 July 1862, Davis, *Life,* 264–65; Charles Davis to his wife, 18 August 1862, Du Pont, *Samuel Francis Du Pont,* 3:29n; Bell diary, 16 July 1862, Bell papers, vol. 4; Charles Davis to David Farragut, 17, 20 July 1862, *OR,* 19:9–11, 15; David Farragut to Charles Davis, 18 July 1862, ibid., 19:13.

96. Charles Davis to his wife, 31 July 1862, Davis, *Life,* 267–72; Welles to David Farragut, 14 July 1862, *OR,* 18:595; Welles to David Farragut, 18, 21 July 1862, ibid., 19:18, 81; Welles to Charles Davis, 14 July 1862, ibid., 18:594.

97. Welles, 10 August 1862, *Diary of Gideon Welles,* 1:72; Welles to David Farragut, 25 July, 2 August 1862, *OR,* 19:5, 35.

98. Fox to Benjamin Butler, 17 November 1862, Fox, *Confidential Correspondence,* 2:446.

99. Welles, 28 January 1863, *Diary of Gideon Welles,* 1:230.

100. Welles to David Farragut, 25 July, 2 August, 12 December 1862, *OR,* 19:5, 35, 161; Raymond Rodgers to Du Pont, 18 July 1862, Du Pont, *Samuel Francis Du Pont,* 2:163; David Farragut to Bell, 24 November 1862, *OR,* 19:372; Butler to Fox, 2 November 1862, Fox, *Confidential Correspondence,* 2:422–23; David Farragut to his family, 29 July 1862, Farragut, *Life of David Farragut,* 290–91.

101. Welles, 1 October 1862, *Diary of Gideon Welles,* 1:157–58.

102. Welles to Charles Davis, 2 August 1862, *OR,* 19:7; Welles to Charles Davis, 16 August 1862, ibid., 23:294; Charles Davis to Welles, 19 August 1862, ibid., 23:305; Charles Davis to his wife, 25 June 1862, Davis, *Life,* 248; Fox to Charles Davis, 24 July, 9, 15, 22 September 1862, Fox, *Confidential Correspondence,* 2:61–66; Charles Davis to Fox, 28 September 1862, Fox, *Confidential Correspondence,* 2:68; Porter to Fox, 26 July 1862, Fox, *Confidential Correspondence,* 2:125; Elizabeth Blair Lee to Samuel Phillips Lee, 23 July 1862, Lee, *Wartime Washington,* 168.

103. Welles, 2 August 1865, *Diary of Gideon Welles,* 2:351; Welles, 25 March 1869, ibid., 3:562.

104. Fox to Charles Davis, 24 July, 9 September 1862, Fox, *Confidential Correspondence,* 2:61–63; Virginia Fox diary, 19 February 1863, reel 1.

105. Welles to David Farragut, 19 August 1862, *OR,* 19:161; Fox to David Farragut, 9 September 1862, 28 February 1863, ibid., 19:184, 639; David Farragut to Bell, 30 November 1862, ibid., 19:386; David Farragut to Fox, 11 October, 23 December 1862, Fox, *Confidential Correspondence,* 1:318–19, 322; Fox to Butler, 11 November 1862, Fox, *Confidential Correspondence,* 2:445; David Farragut to his family, 11 August 1862, Farragut, *Life of David Farragut,* 292–93.

106. David Farragut to his family, 21 September 1862, Farragut, *Life of David Farragut,* 294.

107. David Farragut to Bell, 4 December 1862, *OR,* 19:390; Bell to David Farragut, 27 February 1863, ibid., 19:638; David Farragut to Fox, 11 October 1862, Fox, *Confidential Correspondence,* 1:318; Edward Nichols to Fox, 3 October 1862, Fox, *Confidential Correspondence,* 2:398; David Farragut to his family, 29 July, 12, 21 August, 22 September 1862, 13, 17 February 1863, Farragut, *Life of David Farragut,* 290–95, 309–11.

108. Perkins to his family, n.d., Perkins, *George Hamilton Perkins,* 165, 171; Schley, *Forty-Five Years,* 29–31; Mahan, *From Sail to Steam,* 174, 188; Bell diary, 27 October 1862, Bell papers, vol. 5.

109. Farragut, *Life of David Farragut,* 304–6; Bell diary, 3, 9, 12 January 1863, Bell papers, vol. 5.

110. David Farragut to Welles, 12 May 1863, *OR,* 20:181.

111. Fox to Samuel Phillips Lee, 12 February 1863, ibid., 8:383; David Farragut to Welles, 18, 21 January 1863, ibid., 19:446, 552; Welles to Bell, 7 February 1863, ibid., 19:605; General Order No. 28, 7 January 1864, ibid., 19:463; David Farragut to Welles, 12 May 1863, ibid., 20:181; Welles to David Farragut, 6 April 1863, ibid., 20:121–22; Fox to Porter, 6 February 1863, ibid., 24:242; Welles, 12 January 1863, *Diary of Gideon Welles,* 1:220; Fox to Du Pont, 12 February 1863, Fox, *Confidential Correspondence,* 1:178–79; Fox to David Farragut, 6, 7 February 1863, Fox, *Confidential Correspondence,* 1:324–26; David Farragut to his wife, 13 January 1863, Farragut, *Life of David Farragut,* 307–8; Bell diary, 9 January 1863, Bell papers, vol. 5.

Chapter Four. Hammers on Anvils

1. Much of this information on Lee's background is from Cornish and Laas, *Lincoln's Lee.* See also Samuel Phillips Lee to James Doolittle, 20 February 1865, "Letters from Samuel Phillips Lee to James Doolittle," 112–13; Samuel Phillips Lee to Francis Blair Sr., 18 November 1860, Samuel Phillips Lee Letters, Blair papers, reel 8.

2. D. Stewart to Charles Wilkes, 18 November 1863, Wilkes papers, General Correspondence, container 15.

3. Roswell Lamson to Katie Buckingham, 10 March, 18 July 1863, Lamson, *Lamson of the Gettysburg,* 80, 122; Dewey, *Autobiography,* 118; Grattan, *Under the Blue Pennant,* 57; John Rodgers to Anne Rodgers, 11 September 1862, Johnson, *Rear Admiral John Rodgers,* 217; Welles, 16 May 1866, *Diary of Gideon Welles,* 3:504–5.

4. Welles, 7 October 1864, *Diary of Gideon Welles,* 2:172–73; Elizabeth Blair Lee to Samuel Phillips Lee, 25 June, 6, 13 July, 10 August 1861, 26 July, 9 August 1862, Lee, *Wartime Washington,* 51, 58, 59n–60n, 61, 170, 172.

5. Elizabeth Blair Lee to Samuel Phillips Lee, 21, 26, 31 July, 6, 7 August 1862, Lee, *Wartime Washington,* 166, 170–73.

6. The quote is from John Rodgers to Anne Rodgers, 11 September 1862, John Rodgers Papers, box 21, in response to Anne Rodgers to John Rodgers, 5, 6 September 1862, ibid., box 21. See also Gustavus Fox to Samuel Phillips Lee, 11 September 1862, Fox, *Confidential Correspondence,* 2:212; Samuel Phillips Lee to Doolittle, 20 February 1865, "Letters from Samuel Phillips Lee to James Doolittle," 118; Welles, 16 May 1866, *Diary of Gideon Welles,* 3:504–5; David Farragut to Gideon Welles, 30 May 1862, *OR,* 18:519–21; Welles to John Dahlgren, 14 October 1862, *OR,* 13:389–90.

7. These statistics are from Robert M. Browning, *From Cape Charles to Cape Fear,* 265–66.

8. Selfridge, *Memoirs,* 120–21.

9. The quote is from Lamson to Buckingham, 28 August 1863, Lamson, *Lamson of the Gettysburg,* 128. See also Welles to Samuel Phillips Lee, 30 June 1863, *OR,* 9:97; Samuel Phillips Lee to Welles, 6 June, 7 August 1863, *OR,* 9:64, 149.

10. Fox to Samuel Phillips Lee, 11 December 1862, Fox, *Confidential Correspondence,* 2:244–45.

11. Welles to Samuel Phillips Lee, 1 May 1863, *OR,* 8:834; Fox to Samuel Phillips Lee, 12 February, 26 March 1863, ibid., 8:383, 629; Samuel Phillips Lee to Welles, 5, 10, 20 May 1863, ibid., 9:3, 14, 33; Welles, 20 May 1863, *Diary of Gideon Welles,* 1:306–7; Samuel Du Pont to his wife, 26 January 1863, Du Pont, *Samuel Francis Du Pont,* 2:379; Percival Drayton to Du Pont, 8, 12 May 1863, Du Pont, *Samuel Francis Du Pont,* 3:94, 110–11; Samuel Phillips Lee to Fox, 11 December 1862, 29 March 1863, Fox, *Confidential Correspondence,* 2:239, 252; Fox to Samuel Phillips Lee, 11 December 1862, Fox, *Confidential Correspondence,* 2:244–45.

12. Samuel Phillips Lee to Fox, 27 February 1863, *OR,* 8:389; Elizabeth Blair Lee to Samuel Phillips Lee, 31 July 1862, 28 January, 17, 20, 26 February, 2, 3, 20 March 1863, Lee, *Wartime Washington,* 172, 235, 241–43, 246–48, 253–54.

13. Virginia Fox diary, 16 February 1863, reel 1.

14. Elizabeth Blair Lee to Samuel Phillips Lee, 3, 7, 20 March 1863, Lee, *Wartime Washington,* 247–48, 250, 253–54; Welles, 1 March 1864, *Diary of Gideon Welles,* 1:533–34.

15. Welles, 2 October 1862, *Diary of Gideon Welles,* 1:160; Fox to Du Pont, 1 October 1862, *OR,* 13:353; Rodgers, "Du Pont's Attack at Charleston," 4:33; Raymond Rodgers to Fox, 3 September 1862, Fox, *Confidential Correspondence,* 2:363–65.

16. Fox to Du Pont, 20 February 1863, Fox, *Confidential Correspondence,* 1:181.

17. Fox to Du Pont, 3 June 1862, ibid., 1:126.

18. Welles to Du Pont, 6, 31 January 1863, *OR,* 13:503, 571; Welles, 26 September 1862, 5 January, 16 February 1863, *Diary of Gideon Welles,* 1:153, 216, 236–37; Fox to Du Pont, 12 May, 3 June 1862, Fox, *Confidential Correspondence,* 1:119, 126; Fox to Alexander Rhind, 6 September 1862, Fox, *Confidential Correspondence,* 2:368; Fox to George Morgan, 18 December 1862, Fox, *Confidential Correspondence,* 2:471; Virginia Fox diary, 19 February 1863, reel 1; Salmon Chase diary, 13 September 1862, Chase, *Papers,* 1:383–84; Chase to Hiram Barney, 26 October 1862, Chase, *Papers,* 3:306; Du Pont to his wife, 18 October 1862, Du Pont, *Samuel Francis Du Pont,* 2:248–49.

19. Du Pont to his wife, 22 June 1862, Du Pont, *Samuel Francis Du Pont,* 2:129.

20. For a long and dreary record of Du Pont's reservations, hesitations, and obscurations, see Du Pont to Theodorus Bailey, 30 October 1862, *OR,* 13:423; Du Pont to Welles, 24, 28 January, 8, 9 February, 6, 7 March 1863, ibid., 13:535, 543, 651, 655, 716, 728; Du Pont to Fox, 2 March 1863, ibid., 13:712; Du Pont to Fox, 31 May 1862, 25 February, 2, 7 March 1863, Fox, *Confidential Correspondence,* 1:121–22, 182, 186, 190–91; Du Pont to his wife, 16, 18, 20, 22 October,

27 December 1862, 2 March 1863, Du Pont, *Samuel Francis Du Pont,* 2:247–49, 259, 324, 461; Du Pont to Charles Davis, 4 January 1863, Du Pont, *Samuel Francis Du Pont,* 2:340; Du Pont to William Whetten, 17 March 1863, Du Pont, *Samuel Francis Du Pont,* 2:489; Du Pont to James Biddle, 25 March 1863, Du Pont, *Samuel Francis Du Pont,* 2:511.

21. Fox to Du Pont, 6 March 1863, Fox, *Confidential Correspondence,* 1:189–90.

22. For Welles', Fox's, and Lincoln's unhappiness with Du Pont, see Welles to Du Pont, 31 January, 27 March 1863, *OR,* 13:751, 786; Fox to Du Pont, 2 April 1863, ibid., 13:803; Welles, 12, 16 February, 12 March, 2, 7, 9 April 1863, *Diary of Gideon Welles,* 1:236–37, 247, 259, 263–65; Fox to Du Pont, 12 February, 3, 26 March 1863, Fox, *Confidential Correspondence,* 1:178–79, 188, 196; Elizabeth Blair Lee to Samuel Phillips Lee, 27 March 1863, Lee, *Wartime Washington,* 256; Dahlgren, 29 March 1863, Dahlgren, *Memoir of John A. Dahlgren,* 389.

23. Edward Pierce to Chase, 2 April 1863, Chase, *Papers,* 3:4–5.

24. The quote is from Du Pont to Biddle, 25 March 1863, Du Pont, *Samuel Francis Du Pont,* 2:512. See also Rodgers, "Du Pont's Attack at Charleston," 4:34; Du Pont to his wife, 6 March 1863, Du Pont, *Samuel Francis Du Pont,* 2:470; Du Pont to Biddle, 25, 27, 30 March 1863, Du Pont, *Samuel Francis Du Pont,* 2:511–12; 526n–27n; Du Pont to Charles Davis, 31 March 1863, Du Pont, *Samuel Francis Du Pont,* 2:530–31; Du Pont to his wife, 6 April 1863, Du Pont, *Samuel Francis Du Pont,* 2:553.

25. Du Pont to Welles, 6 March 1863, *OR,* 13:716; Du Pont to Fox, 19 November 1862, Fox, *Confidential Correspondence,* 1:168; Drayton to Alexander Hamilton, 11 February, 15 April 1863, Drayton, *Naval Letters,* 26, 34; Rodgers, "Du Pont's Attack at Charleston," 4:33–34; Du Pont to Charles Davis, 4 January 1863, Du Pont, *Samuel Francis Du Pont,* 2:339; Du Pont to his wife, 7 March, 3, 6 April 1863, Du Pont, *Samuel Francis Du Pont,* 2:472, 540–41, 550–51; Villard, *Memoirs,* 2:12–13; Raymond Rodgers to Charles Steedman, 16 February 1863, Steedman, *Memoir and Correspondence,* 360; John Rodgers to Anne Rodgers, 30 March 1863, John Rodgers Papers, box 22.

26. Rodgers, "Du Pont's Attack at Charleston," 4:37.

27. Du Pont to Welles, 8, 15, 22 April, 27 May 1863, *OR,* 14:3, 6, 52, 66; Rodgers, "Du Pont's Attack at Charleston," 4:37, 39–40; Du Pont to his wife, 5, 6 April 1863, Du Pont, *Samuel Francis Du Pont,* 2:545, 550–51; Du Pont to his wife, 8, 10 April 1863, Du Pont, *Samuel Francis Du Pont,* 3:3–4, 9–10, 17; Du Pont to William McKean, 29 April 1863, Du Pont, *Samuel Francis Du Pont,* 3:66; Osbon, *Sailor of Fortune,* 230, 236; Ammen, *Campaigns,* 372; John Rodgers to William Hodge, [?] April 1863, John Rodgers Papers, box 26.

28. Rodgers, "Du Pont's Attack at Charleston," 4:40.

29. Du Pont to Welles, 8, 15, 22 April, 22 May 1863, *OR,* 14:3, 7, 52, 55, 66; Rodgers, "Du Pont's Attack at Charleston," 4:37–41; Du Pont to his wife, 8, 10 April 1863, Du Pont, *Samuel Francis Du Pont,* 3:3–4, 14; Villard, *Memoirs,* 2:45;

John Rodgers to Anne Rodgers, 15 April, 2 May 1863, John Rodgers Papers, box 22.

30. For Welles' reaction to the repulse, see Welles to Du Pont, 11 April 1863, *OR*, 14:123; Fox to Montgomery Blair, 10 April 1863, ibid., 14:38; Du Pont to Welles, 8 April 1863, ibid., 14:112; Welles, 10, 12, 15, 19, 20, 21 April, *Diary of Gideon Welles*, 1:266–69, 273, 276–77; Dahlgren, 21 April 1863, Dahlgren, *Memoir of John A. Dahlgren*, 390; Virginia Fox diary, 10 April 1863, reel 1.

31. Du Pont to Drayton, 16 May 1863, Du Pont, *Samuel Francis Du Pont*, 3:131–32.

32. For examples of Du Pont's changing views of and growing unhappiness with the Navy Department, see Du Pont to Welles, 22 April, 12, 27 May, 3 June 1863, *OR*, 14:51, 59, 65–66, 70–71; Rodgers, "Du Pont's Attack at Charleston," 4:45; Du Pont to his wife, 10, 25 April, 2, 10, 17 May, 5 June 1863, Du Pont, *Samuel Francis Du Pont*, 3:17–18, 57–58, 74, 99–101, 118–20, 161–62; Du Pont to Charles Davis, 26 April 1863, Du Pont, *Samuel Francis Du Pont*, 3:62; Du Pont to Henry Winter Davis, 3, 11 May 1863, Du Pont, *Samuel Francis Du Pont*, 3:75, 102–3; Du Pont to Drayton, 4 May 1863, Du Pont, *Samuel Francis Du Pont*, 3:87–88. See also Welles, 14 May 1863, *Diary of Gideon Welles*, 1:302; Fox to Welles, 13 May 1863, Du Pont, *Samuel Francis Du Pont*, 3:116n.

33. John Rodgers to Du Pont, 14 April 1863, Du Pont, *Samuel Francis Du Pont*, 3:24; Du Pont to his wife, 2, 8 May 1863, ibid., 3:73–74, 91; Du Pont to Biddle, 4 May 1863, ibid., 3:86–87; Cornelius Stribling to Du Pont, n.d., Du Pont Papers, box 42, no. 174; William Shubrick to Du Pont, 16 April 1863, Du Pont Papers, box 42, no. 175; Drayton to Du Pont, 26 April 1863, Du Pont Papers, box 42, no. 192; Thomas Turner to Du Pont, 4 May 1863, Du Pont Papers, box 42, no. 198; Andrew Foote to Du Pont, 14 April 1863, Du Pont Papers, box 42, no. 187; Steedman to his wife, 3 May 1863, Steedman, *Memoir and Correspondence*, 372.

34. Drayton to Du Pont, 2 June 1863, Du Pont, *Samuel Francis Du Pont*, 3:151; Charles Davis to Du Pont, 16 April 1863, Du Pont Papers, box 42, no. 172.

35. Welles, 8, 25 May 1863, *Diary of Gideon Welles*, 1:295–96, 311–12; Drayton to Hamilton, 15 April 1863, Drayton, *Naval Letters*, 34; Charles Davis to his wife, 30 May 1863, Davis, *Life*, 293; Du Pont to Drayton, 4 May 1863, Du Pont, *Samuel Francis Du Pont*, 3:87–88; Drayton to Du Pont, 8, 12 May, 2 June 1863, Du Pont, *Samuel Francis Du Pont*, 3:92, 110–11, 151; Dahlgren, 6 May 1863, Dahlgren, *Memoir of John A. Dahlgren*, 391; Shubrick to Du Pont, 8 May 1863, Du Pont Papers, box 42, no. 203.

36. Henry Winter Davis to Du Pont, 3 May 1863, Du Pont, *Samuel Francis Du Pont*, 3:80–83.

37. Du Pont to his wife, 10, 17, 27 May, 5, 6 June 1863, ibid., 3:101, 121–22, 141, 161–62, 164–65; Du Pont to Drayton, 4, 19 May 1863, ibid., 3:87–88, 131–32.

38. Du Pont to Welles, 16 April 1863, *OR*, 14:140.

39. For an evolution of Welles' thinking about Du Pont, see Welles, 19, 30 April, 8, 23, 25 May, 3 June 1863, *Diary of Gideon Welles*, 1:276, 288, 295–96, 309–13,

320–21. See also Elizabeth Lee Blair to Samuel Phillips Lee, 10, 29 August 1863, Lee, *Wartime Washington,* 297, 303–4; Du Pont to Welles, 16 April 1863, *OR,* 14:140; Charles Fulton to Montgomery Blair, 7 August 1863, Montgomery Blair Papers.

40. Du Pont to Welles, 5 July 1863, *OR,* 14:310.

41. Du Pont to his wife, 21, 25 January 1863, Du Pont, *Samuel Francis Du Pont,* 2:372, 378; Du Pont to his wife, 8, 9 June 1863, ibid., 3:167–68; Welles to Du Pont, 26 June, 15 July 1863, *OR,* 14:282, 342; Dahlgren, 4 July 1863, Dahlgren, *Memoir of John A. Dahlgren,* 396.

42. Welles, 26 June 1863, *Diary of Gideon Welles,* 1:344; Welles, 9, 18 April, 13 May 1864, 23 June 1865, ibid., 2:7–8, 14, 30, 320–21; Du Pont to Shubrick, 17 July 1863, Du Pont, *Samuel Francis Du Pont,* 3:202; Du Pont to Welles, 22 October 1863, Du Pont, *Samuel Francis Du Pont,* 3:253; Welles to Du Pont, 4 November 1863, Du Pont, *Samuel Francis Du Pont,* 3:258–76; Raymond Rodgers to Du Pont, 16 November, 26 December 1863, Du Pont, *Samuel Francis Du Pont,* 3:282–84, 300; Du Pont to Henry Winter Davis, 15 January, 20 May 1864, 11 June 1864, Du Pont, *Samuel Francis Du Pont,* 3:304, 346–47, 354–55; Hodge to Du Pont, 2 March 1864, Du Pont, *Samuel Francis Du Pont,* 3:314; Charles Davis to Du Pont, 5 June 1864, Du Pont, *Samuel Francis Du Pont,* 3:353; Du Pont to Whetten, 15 August 1864, Du Pont, *Samuel Francis Du Pont,* 3:367–68; Du Pont to William King Hall, 13 October 1864, Du Pont, *Samuel Francis Du Pont,* 3:402; Charles Davis to Du Pont, 31 July, 12 November 1863, 11 June 1864, Du Pont Papers, W9-15734, W9-15928, W9-16334; Shubrick to Du Pont, 14 July 1863, Du Pont Papers, W9-15697.

43. Welles, 2 October 1862, 23 May, 26 June 1863, *Diary of Gideon Welles,* 1:160, 310–12, 344; Du Pont to Henry Winter Davis, 15 January 1864, Du Pont, *Samuel Francis Du Pont,* 3:304.

44. John Rodgers to Anne Rodgers, 26 April 1863, Johnson, *Rear Admiral John Rodgers,* 248; John Rodgers to Du Pont, 8 August 1863, Du Pont Papers, W9-15752; Welles, 26, 29 June, 16 July 1863, *Diary of Gideon Welles,* 1:344, 351, 373.

45. For a good biography of Dahlgren, see Schneller, *Quest for Glory,* on which part of this paragraph is based.

46. John Hay, 11 May 1862, Hay, *Lincoln's Journalist,* 262.

47. Du Pont Notes, W9-18475, box 47, 73–75.

48. Welles, *Diary of Gideon Welles,* 1:62; Welles, 1, 9 October 1862, 22 February 1863, ibid., 1:158, 163–64, 239–40; Rodgers, "Du Pont's Attack at Charleston," 4:34; Hay, 25 April 1861, Hay, *Inside Lincoln's White House,* 11; Hay, 16 December 1861, Hay, *Lincoln's Journalist,* 169; Hay, *At Lincoln's Side,* 133; Du Pont Notes, box 47, 73–75; Dahlgren, 5 July 1862, Dahlgren, *Memoir of John A. Dahlgren,* 374; Fox to Du Pont, 7 October 1862, Fox, *Confidential Correspondence,* 1:158.

49. Welles to Dahlgren, 14 October 1862, *OR*, 13:389–90; Welles, *Diary of Gideon Welles*, 1:62; Welles, 1, 9 October 1862, 19, 22 February 1863, *Diary of Gideon Welles*, 1:158, 163–64, 238–40; Dahlgren, 4 August, 15, 30 November 1861, Dahlgren, *Memoir of John A. Dahlgren*, 341, 349–50; Dahlgren to Abraham Lincoln, 10 July 1862, Lincoln Papers; William Seward to Lincoln, July 1862, Lincoln Papers; Welles to Fox, 5 December 1870, Fox papers.

50. David Farragut to Du Pont, 20 April 1863, Du Pont, *Samuel Francis Du Pont*, 3:49.

51. John Rodgers to Anne Rodgers, 29 June 1863, Johnson, *Rear Admiral John Rodgers*, 259.

52. Welles, 23 June 1863, *Diary of Gideon Welles*, 1:341; Du Pont to his wife, 3 August 1861, Du Pont, *Samuel Francis Du Pont*, 1:119; Raymond Rodgers to Du Pont, 18 July 1862, Du Pont, *Samuel Francis Du Pont*, 2:163–64; Du Pont to his wife, 7 March 1863, Du Pont, *Samuel Francis Du Pont*, 2:473–74; Stribling to Du Pont, 16 October 1863, Du Pont Papers, W9-15928; Theodorus Bailey to Edmund Bailey, 6 December 1863, Bailey Papers, box 1, 1863 correspondence; Wilkes diary, 2 May 1861, Wilkes papers, reel 1; John Rodgers to Anne Rodgers, 18 July 1862, John Rodgers Papers, box 21; Hodge to John Rodgers, 3 March 1863, John Rodgers Papers, box 26.

53. Du Pont to Fox, 8 October 1862, Fox, *Confidential Correspondence*, 1:160–61.

54. Dahlgren to Welles, 1, 11, 25 October 1862, *OR*, 13:353, 377, 416; Welles to Dahlgren, 8, 14 October, 1 November 1862, ibid., 13:376, 389–90, 426; Welles, 1, 9 October 1862, *Diary of Gideon Welles*, 1:158, 163–65; Fox to Du Pont, 7 October 1862, Fox, *Confidential Correspondence*, 1:158; Du Pont to Fox, 8 October 1862, Fox, *Confidential Correspondence*, 1:160–61; Du Pont to James Wilson Grimes, 8 August 1863, Du Pont, *Samuel Francis Du Pont*, 3:221–23.

55. Welles, 19 April, 25, 27 May 1863, *Diary of Gideon Welles*, 1:276, 312–14; Davis, *Life*, 311; Dahlgren, 28 May 1862, Dahlgren, *Memoir of John A. Dahlgren*, 391–92; Welles, "Fort Sumter," 616–17.

56. Welles, 23, 29, 31 May, 8, 9, 18, 27 June 1863, *Diary of Gideon Welles*, 1:312–13, 317–18, 325–26, 334, 346–47; Charles Davis to his wife, 30 May 1863, Davis, *Life*, 293; Drayton to Du Pont, 8 May, 2 June 1863, Du Pont, *Samuel Francis Du Pont*, 3:92–95, 151; Dahlgren, 9 June 1863, Dahlgren, *Memoir of John A. Dahlgren*, 394; Foote to Du Pont, 20 March 1863, Du Pont Papers, box 42; Foote to Du Pont, 12 May 1863, Du Pont Papers, box 42, no. 210; Anne Rodgers to John Rodgers, 3 April 1863, John Rodgers Papers, box 22.

57. Welles, 27 May 1863, *Diary of Gideon Welles*, 1:314–15.

58. Welles, 27, 28 May 1863, ibid., 1:314–16; Dahlgren, 21 April, 13 May 1863, Dahlgren, *Memoir of John A. Dahlgren*, 390–91; Turner to Du Pont, 14 July 1863, Du Pont Papers, W9-15529.

59. Welles, 29 May 1863, *Diary of Gideon Welles*, 1:317–18; Dahlgren, 28, 29 May, 2, 3 June 1863, Dahlgren, *Memoir of John A. Dahlgren*, 391–93; Welles to Foote, 3

June 1863, Letters Sent by the Secretary of the Navy to Chiefs of Navy Bureaus, 1842–1886, National Archives and Records Administration, roll 2, vol. 4.

60. The quote is from Bell diary, 18 July 1863, Bell papers, vol. 6. See also Hiram Paulding to Welles, 16 June 1863, *OR,* 2:286; Welles, 19, 20, 21, 27 June 1863, *Diary of Gideon Welles,* 1:335–36, 346–47; Charles Davis to his wife, 27 June 1863, Davis, *Life,* 298; Dahlgren, 25 June 1863, Dahlgren, *Memoir of John A. Dahlgren,* 395; Drayton to Du Pont, 18 June 1863, Du Pont Papers, box 42, no. 326.

61. Welles, 21, 23 June 1863, *Diary of Gideon Welles,* 1:337–38, 341; Du Pont to his wife, 25 June 1863, Du Pont, *Samuel Francis Du Pont,* 3:182–83; Dahlgren, 21, 22, 24 June 1863, Dahlgren, *Memoir of John A. Dahlgren,* 395.

62. James Palmer to David Porter, 6, 7 July 1863, *OR,* 20:253–55; Rodgers, "Du Pont's Attack at Charleston," 4:45; Du Pont to his wife, 11, 25, 28 June 1863, Du Pont, *Samuel Francis Du Pont,* 3:172, 182–83, 186–87; Grimes to Du Pont, 1 July 1863, Du Pont, *Samuel Francis Du Pont,* 3:191; Dahlgren, 4 July 1863, Dahlgren, *Memoir of John A Dahlgren,* 396; Drayton to Du Pont, 18 June 1863, Du Pont Papers, box 42, no. 326; John Missroon to Du Pont, 25 July 1863, Du Pont Papers, W9-15715; Turner to Du Pont, 14 July 1863, Du Pont Papers, W9-15529; John Rodgers to Anne Rodgers, 29 June 1863, Johnson, *Rear Admiral John Rodgers,* 259; Charles Fulton to Montgomery Blair, 7 August 1863, Montgomery Blair Papers.

63. The quote is from Fulton to Montgomery Blair, 7 August 1863, Montgomery Blair Papers. See also Dahlgren, 5 June 1863, Dahlgren, *Memoir of John A Dahlgren,* 393; Dahlgren to Welles, 6 July 1863, *OR,* 14:311.

64. The quote is from Dahlgren, 28 August 1863, Dahlgren, *Memoir of John A. Dahlgren,* 411. See also Welles, 3 October 1863, *Diary of Gideon Welles,* 1:466–67; Dahlgren to Fox, 20 July, 24 September 1863, *OR,* 14:376, 671; Rodgers, "Du Pont's Attack at Charleston," 4:43; Hay, 20, 28 October 1863, 24 January 1864, Hay, *Inside Lincoln's White House,* 96, 101–2, 146; Dahlgren, 22 July, 4, 15, 26 August, 2 September 1863, Dahlgren, *Memoir of John A. Dahlgren,* 404, 406–7, 410, 412.

65. Dahlgren, 23 July, 22, 26 August, 1, 2 September 1863, Dahlgren, *Memoir of John A. Dahlgren,* 404–5, 410–12; Du Pont to his wife, 7 March 1863, Du Pont, *Samuel Francis Du Pont,* 2:473–74; Hay, 20 October 1863, Hay, *Inside Lincoln's White House,* 96; Welles, 21 June 1863, *Diary of Gideon Welles,* 1:337.

66. For examples of Dahlgren's thinking, see Dahlgren to Fox, 24 September 1863, *OR,* 14:671; Dahlgren to Welles, 1, 10 October, 22 January 1863, ibid., 15:4, 31, 250–51; Dahlgren to Lincoln, 23 January 1863, ibid., 15:252; Dahlgren, 22 October, 29 November 1863, 1 January 1864, Dahlgren, *Memoir of John A. Dahlgren,* 419, 428, 434.

67. Welles, 20 April, 15 July, 16, 17 September, 1, 3 October 1863, *Diary of Gideon Welles,* 1:276–77, 372, 434–35, 449, 466–67, 474–75; Welles to Dahlgren, 9

October 1863, *OR,* 15:26; Hay, 20, 28 October 1863, Hay, *Inside Lincoln's White House,* 96, 101–2; Fox to Porter, 16 July 1863, Fox, *Confidential Correspondence,* 2:185; Elizabeth Blair Lee to Samuel Phillips Lee, 10, 29 August 1863, Lee, *Wartime Washington,* 297, 303–4; Dahlgren, 21 January 1864, Dahlgren, *Memoir of John A. Dahlgren,* 436.

68. Dahlgren, 4 March 1864, Dahlgren, *Memoir of John A. Dahlgren,* 443.

69. Welles, 3 February, 2, 26 March 1864, *Diary of Gideon Welles,* 1:520, 534, 547; Dahlgren, 5, 6, 25, 29 March, 17, 28 April 1864, Dahlgren, *Memoir of John A. Dahlgren,* 443–44, 448, 450–51.

70. Welles, 7 March 1864, *Diary of Gideon Welles,* 1:536; Welles, 8 April 1864, ibid., 2:7.

71. Welles, 30 August 1864, ibid., 2:128–29; Dahlgren to Welles, 14, 21 May, 4 June 1864, *OR,* 15:430, 438, 468; Hay, 24 April 1864, Hay, *Inside Lincoln's White House,* 99; Dahlgren, 25, 29 March, 10, 12, 17 May, 2 July 1864, Dahlgren, *Memoir of John A. Dahlgren,* 448, 453–54, 461.

72. Welles, 30 August 1864, *Diary of Gideon Welles,* 2:128–29; Fox to Porter, 25 May 1864, *OR,* 26:325; Welles to Dahlgren, 15 July 1864, *OR,* 15:569; Dahlgren, 22 July, 28 September 1864, Dahlgren, *Memoir of John A. Dahlgren,* 470, 473–74.

73. Dahlgren, 15 January 1865, Dahlgren, *Memoir of John A. Dahlgren,* 492.

74. Dahlgren, 1 March 1865, ibid., 503–4.

75. Much of this information on Dahlgren's subsequent career is from Schneller, *Quest for Glory,* 321–62.

Chapter Five. The Peripheries

1. Vail, *Three Years on the Blockade,* 20; William Frederick Keeler, 14 April 1865, Keeler, *Aboard the USS Florida,* 208–9; C. Marion Dodson, 1, 2 May 1864, Dodson, *Yellow Flag,* 31; Percival Drayton to Alexander Hamilton, 13 January 1864, Drayton, *Naval Letters,* 39; Theodorus Bailey to Edmund Bailey, 27 August 1863, Bailey Papers, box 1, 1863 correspondence.

2. Welles, 23 August 1864, *Diary of Gideon Welles,* 2:116; Gustavus Fox to Samuel Du Pont, 3 June 1862, Fox, *Confidential Correspondence,* 1:127; David Porter to Fox, 28 March, 24 May 1862, *Confidential Correspondence,* 2:93, 110–11; William McKean to L. G. Arnold, 23 April 1862, *OR,* 17:220; Gideon Welles to McKean, 20 January 1862, *OR,* 17:56; Welles to McKean and David Farragut, 12 March 1862, *OR,* 17:189; McKean to Welles, 11 March, 3, 25 April, 14, 26 May 1862, *OR,* 17:187, 200, 218, 238, 244.

3. McKean to Welles, 13, 19 April 1862, *OR,* 17:217–18; Welles, "Farragut and New Orleans," 682; Welles to McKean, 5 April 1862, *OR,* 17:219; Anne Rodgers to John Rodgers, 22 July 1862, Rodgers Papers, box 21.

4. Du Pont, 11 November 1861, "Detailed Report of Flag Officer Du Pont," 12:262; Louis Goldsborough to Welles, 12 May 1862, ibid., 7:343; Davis, 9 April 1862, Davis, *Life*, 212.

5. Welles, 22 May, 1 June 1863, *Diary of Gideon Welles*, 1:309, 319; Evans, *Sailor's Log*, 61–62; John B. Marchand, 12 April 1862, Marchand, *Charleston Blockade*, 141; DiZerega, "Last Days of the Rebel Iron-Clad Merimac," 4; Mervine, 24 November 1863, 28 February 1864, Mervine, "Jotings on the Way," 263.

6. Du Pont to his wife, 16 December 1862, Du Pont, *Samuel Francis Du Pont*, 2:300; Welles to James Lardner, 10 June, 8, 15 September, 4 November 1862, *OR*, 17:263, 309, 312, 325; Fox to Lardner, 9 September 1862, *OR*, 17:310; Lardner to Welles, 29 June, 1 August, 12 October 1862, *OR*, 17:264, 294, 315; Lardner to Fox, 1 November 1862, Fox, *Confidential Correspondence*, 2:421–22.

7. Stephen Rowan to Fox, 5 November 1862, Fox, *Confidential Correspondence*, 2:427.

8. Welles to Theodorus Bailey, 4 November 1862, *OR*, 17:325; Fox to David Farragut, 1 November 1862, ibid., 19:338; Theodorus Bailey to Fox, 15 October, 17 November 1862, Fox, *Confidential Correspondence*, 2:406–7, 444; William Mervine to Theodorus Bailey, 21 September 1861, Bailey Papers, box 1, 1861 correspondence; Theodorus Bailey to Edmund Bailey, 30 June 1862, Bailey Papers, box 1, 1862 correspondence.

9. Theodorus Bailey to Edmund Bailey, 26 December 1861, Bailey Papers, box 1, 1861 correspondence; Mervine to Theodorus Bailey, 21 September 1861, ibid.; Theodorus Bailey to Edmund Bailey, 28 May 1864, ibid., box 1, 1864 correspondence.

10. Theodorus Bailey, 9 December 1862, *OR*, 17:329.

11. Theodorus Bailey to Welles, 6 January 1864, ibid., 17:621; Drayton to Hamilton, 13 January 1864, Drayton, *Naval Letters*, 39; Theodorus Bailey to Fox, 17 November 1862, Fox, *Confidential Correspondence*, 2:444; Theodorus Bailey to Edmund Bailey, 20, 29 April, 1 July, 17 August, 6 December 1863, Bailey Papers, box 1, 1863 correspondence; Theodorus Bailey to Edmund Bailey, 17 March, 11 July 1864, Bailey Papers, box 1, 1864 correspondence.

12. Theodorus Bailey to Edmund Bailey, 29 March, 20, 29 April, 24 July, 16 October, 24 November, 6, 29 December 1863, Bailey Papers, box 1, 1863 correspondence; Theodorus Bailey to Edmund Bailey, 11 February, 28 May, 1 June, 14 July 1864, ibid., box 1, 1864 correspondence; William Temple to Theodorus Bailey, 28 June, 7 August 1864, ibid.; Temple to Theodorus Bailey, 12 February 1865, ibid., box 1, 1865 correspondence.

13. Welles to Theodorus Bailey, 19 July, 12 September 1864, *OR*, 17:732, 753; Theodorus Bailey to Welles, 6 September 1863, 24 March, 27 July 1864, ibid., 17:548, 672, 737; Welles, 16 September 1864, *Diary of Gideon Welles*, 2:148;

Theodorus Bailey to David Farragut, 30 July 1864, Farragut, *Life of David Far-ragut,* 412; Theodorus Bailey to Edmund Bailey, 2 April, 28 May, 24 June, 11, 14 July 1864, Bailey Papers, box 1, 1864 correspondence.

14. Porter to Theodorus Bailey, 22 October 1864, Bailey Papers, box 1, 1864 corre-spondence.

15. Belknap, "Captain George Hamilton Perkins," 205.

16. Cornelius Stribling to Du Pont, n.d., 6 June, 16 October 1863, Du Pont Papers, box 42, nos. 174, 317, W9-15928; Stribling to Welles, 10 July 1861, Welles papers, container 20; Charles Butler to Welles, 13 September 1861, Welles papers, container 20; Temple to Theodorus Bailey, 1 June 1864, Bailey Papers, box 1, 1864 correspondence; Langley, *Social Reform,* 244.

17. William McClure to Abraham Lincoln, 16 September 1863, Welles papers, con-tainer 29.

18. Welles to Stribling, 23 September, 19 November 1864, 14 July 1865, *OR,* 17:757, 778, 861; Stribling to Welles, 14 November 1864, 27 March, 30 May 1865, ibid., 17:775, 829, 854; Robert Handy to Welles, 3 September 1864, ibid., 17:752.

19. For the controversy surrounding Preble's actions, see Welles to George Preble, 20 September 1862, ibid., 1:434; Welles to Lincoln, 10 February 1863, ibid., 457–59; David Farragut to Welles, 8 September, 18 October 1862, ibid., 1:431, 455; Welles, 10 August, 19, 20 September, 7, 9 October, 4 December 1862, *Diary of Gideon Welles,* 1:74, 140–41, 162–63, 188–91; Fox to Joseph Fay, 25 November 1862, Fox, *Confidential Correspondence,* 2:455; James Grimes to Fox, 10 November 1862, Fox, *Confidential Correspondence,* 2:469; George Hamilton Perkins, 13 Octo-ber 1862, Perkins, *George Hamilton Perkins,* 146; Charles Adams to William Seward, 24 October 1862, *Foreign Relations of the United States, 1862,* 225; Eliza-beth Blair Lee to Samuel Phillips Lee, 12 December 1862, Lee, *Wartime Letters,* 214n.

20. Welles to Charles Wilkes, 8 September 1862, *OR,* 1:470; Fox to David Farragut, ibid., 19:184; Fox to Du Pont, 3 April 1862, Fox, *Confidential Correspondence,* 1:115; Welles, 26 December 1863, *Diary of Gideon Welles,*1:497; Fowler, *Under Two Flags,* 298; Anderson, *By Sea and by River,* 213.

21. Welles, 10, 18 August, 5 September 1862, 16 May, 4 June, 18 December 1863, 19 February 1864, *Diary of Gideon Welles,* 1:73, 86–87, 110, 304–5, 322–23, 489, 528; Welles, 2 August 1865, ibid., 2:351; Drayton to Lydig Hoyt, 18 January 1862, Drayton, *Naval Letters,* 13; D. Fairfax, "Captain Wilkes's Seizure of Mason and Slidell," 2:142; Du Pont to his wife, 25 December 1861, Du Pont, *Samuel Francis Du Pont,* 1:290; Du Pont to William Whetten, 28 December 1861, Du Pont, *Samuel Francis Du Pont,* 1:295; Wilkes, *Autobiography,* 223, 295; Franklin, *Memories of a Rear-Admiral,* 186; Virginia Fox diary, 12 January 1863, reel 1; Wilkes diary, 2 May 1861, Wilkes papers, reel 1; John Rodgers to Anne Rodgers, 8, 13, 15, 18 July, 5, 7 August, 2 September 1862, John Rodgers Papers, box 21;

Anne Rodgers to John Rodgers, 5 December 1861, 11 September 1862, John Rodgers Papers, box 21.

22. Welles to Wilkes, 7 August 1862, *OR,* 7:634; Welles, 10, 18 August, 4 September 1862, *Diary of Gideon Welles,* 1:73; 86–87, 109.

23. Welles, 16 September 1862, 12 May 1863, *Diary of Gideon Welles,* 1:134, 298–99.

24. Welles to Wilkes, 8, 20 September 1862, *OR,* 1:470, 476; Wilkes to Welles, 29 September 1862, ibid., 1:483; Welles to Goldsborough, 4 September 1862, ibid., 7:697; Welles, 6, 11, 12 September 1862, *Diary of Gideon Welles,* 1:111, 122–24; Wilkes, *Autobiography,* 776–78, 833; Wilkes to Welles, 12 September 1862, Wilkes papers, General Correspondence, container 14.

25. The quote is from Wilkes to Welles, 11 December 1863, *OR,* 2:567. See also Welles to Wilkes, 15 December 1862, ibid., 1:587; Welles to Wilkes, 15 December 1863, ibid., 2:569–70; Wilkes to Welles, 20 October, 11, 25 November, 14, 31 December 1862, ibid., 1:513, 542, 558, 586, 606; Wilkes to Welles, 2, 23 January, 3, 16 April 1863, ibid., 2:4, 7, 54, 141, 160; Welles, 6 July 1863, *Diary of Gideon Welles,* 1:363; Fox to Du Pont, 11 March 1863, Fox, *Confidential Correspondence,* 1:192; Wilkes, *Autobiography,* 784–85, 787, 805–7.

26. Wilkes to Welles, 16 June 1863, *OR,* 2:352; David Farragut to Welles, 21 February, 23 June 1863, ibid., 19:535, 623; David Farragut to Wilkes, 6 February 1863, ibid., 19:602.

27. Welles to Charles Baldwin, 27 January 1863, ibid., 2:60; Baldwin to Welles, 30 March 1863, ibid., 2:183; Welles to Wilkes, 15 December 1863, ibid., 2:569–70; Wilkes to Welles, 11 December 1863, ibid., 2:567–68; Welles, 29 May 1863, *Diary of Gideon Welles,* 1:316; Wilkes, *Autobiography,* 785–87.

28. The quote is from Welles, 12 May 1863, *Diary of Gideon Welles,* 1:298. See also Welles to Wilkes, 8 September 1862, *OR,* 1:470; Wilkes to Welles, 29 September, 12 October 1862, *OR,* 1:484, 502–3; Wilkes to Welles, 2, 24 January, 2 April 1863, *OR,* 2:7–8, 14, 139; Welles, 6 January, 4 June 1863, *Diary of Gideon Welles,* 1:217, 322–23; Du Pont to Whetten, 14 November 1862, Du Pont, *Samuel Francis Du Pont,* 2:283–84; Lord Lyons to Seward, 24 November, 5, 29, 31 December 1862, 20 January, 7 March, 9 April 1863, *Foreign Relations of the United States, 1863,* 1:408–9, 415–17, 422–23, 427, 440–41, 465, 502; Welles to Seward, 6 January 1863, *Foreign Relations of the United States, 1863,* 1:442–43; Wilkes, *Autobiography,* 783–84, 787–88, 791, 834.

29. Charles Fleming to Welles, 25 March 1863, *OR,* 2:135; Fleming to Wilkes, 28 March 1863, ibid., 2:135; Lyons to Seward, 29 April, 7 May 1863, *Foreign Relations of the United States, 1863,* 1:520, 526; Seward to Lyons, 7 May 1863, *Foreign Relations of the United States, 1863,* 1:529.

30. Welles to Lardner, 1 June 1863, *OR,* 2:250–51; Welles, 12, 16, 22, 29 May, 4, 8, 10 June 1863, *Diary of Gideon Welles,* 1:298–99, 304–5, 309, 316, 322–23, 325, 327.

31. D. Stewart to Wilkes, 18 November 1863, Wilkes papers, General Correspondence, container 15; Wilkes to Welles, 30 August 1863, ibid.; Wilkes to Welles, 11 December 1863, *OR,* 2:567; Welles, 4 June 1863, *Diary of Gideon Welles,* 1:323; Wilkes, *Autobiography,* 834.

32. Wilkes to Welles, 11 December 1863, *OR,* 2:567–68; Welles to Wilkes, 15 December 1863, ibid., 2:569–70; Welles, 4 June, 18 December 1863, *Diary of Gideon Welles,* 1:322–23, 489; Wilkes, *Autobiography,* 833–34.

33. Welles, 18, 19, 21 December 1863, 7, 26 January, 19, 24 February, 19 March 1864, *Diary of Gideon Welles,* 1:489–91, 505–6, 515, 528, 530, 543–44; Welles, 7, 30 April 1864, ibid., 2:6, 21–22; William Hodge to Du Pont, 2 March 1864, Du Pont, *Samuel Francis Du Pont,* 3:315; Wilkes, *Autobiography,* 835–36, 838–39; Welles to Wilkes, 19, 21 December 1863, Wilkes papers, General Correspondence, container 15; Wilkes to Welles, 19, 22 December 1863, Wilkes papers, General Correspondence, container 15; Report of the Lighthouse Board, 13 January 1864, Wilkes papers, General Correspondence, container 15; Wilkes, 9 March 1864, Wilkes papers, General Correspondence, container 15.

34. Welles, 24 February 1864, *Diary of Gideon Welles,* 1:530; Welles, 20 December 1864, ibid., 2:203; Smith Payne to Lincoln, 5 May 1864, Lincoln Papers; Britton Hill to Lincoln, 5 May 1864, Lincoln Papers; Du Pont to Wilkes, 21 May 1864, Du Pont, *Samuel Francis Du Pont,* 3:349; Wilkes, *Autobiography,* 840.

35. Welles to Charles Bell, 8 November 1864, *OR,* 3:351; Welles, 12, 22 May, 1 June 1863, *Diary of Gideon Welles,* 1:299, 309, 319; Welles, 23 August 1864, *Diary of Gideon Welles,* 2:116; Hiram Paulding to Fox, 23 October 1861, Fox, *Confidential Correspondence,* 1:387.

36. Welles to Lardner, 1 June, 22 September, 13 October, 4, 16 November 1863, *OR,* 2:250–51, 460, 471, 490, 497; Lardner to Welles, 16 November 1863, ibid., 2:497; Welles to Lardner, 1 September 1864, ibid., 3:196; Lardner to Welles, 17 June, 22 August, 3 October 1864, ibid., 3:55, 192, 249; Evans, *Sailor's Log,* 61, 63–66; Mervine, 15 December 1863, 15, 28 February 1864, Mervine, "Jotings on the Way," 251, 261, 263.

Chapter Six. Turning the Tide

1. For the story of Porter's selection, see Louis Goldsborough to Gideon Welles, 1 August 1862, *OR,* 7:607; Welles, 2 April 1863, *Diary of Gideon Welles,* 1:259; David Porter to Gustavus Fox, 26 July, 5 August, 1, 10 September 1862, Fox, *Confidential Correspondence,* 2:125, 127, 133, 136; Fox to Porter, 6 September 1862, Fox, *Confidential Correspondence,* 2:135; Porter, *Incidents and Anecdotes,* 114–22.

2. Virginia Fox diary, 5 January 1863, reel 1.

3. Welles, 20 August, 23 September, 1, 10 October 1862, 17 March, 14 April, 13 July 1863, *Diary of Gideon Welles,* 1:87–88, 145, 157–58, 167, 249, 273, 369;

Porter to Fox, 10 November 1862, Fox, *Confidential Correspondence,* 2:148; James Grimes to Fox, 24 October 1862, Fox, *Confidential Correspondence,* 2:410; Selfridge, *Memoirs,* 116–17; Walke, *Naval Notes,* 410; Dana, *Recollections,* 85–86; Virginia Fox diary, 5 January 1863, reel 1.

4. Welles to John Dahlgren, 8 October 1862, *OR,* 13:376; Welles, 20 August, 21 September, 1, 10 October 1862, 17 March, 2, 14 April, 13 July 1863, *Diary of Gideon Welles,* 1:87–88, 145, 157–58, 167, 249, 259, 273, 369; Welles, 13 March 1865, *Diary of Gideon Welles,* 2:255.

5. Charles Davis to Seth Ledyard Phelps, 27 September 1862, Slagle, *Ironclad Captain,* 299.

6. Dahlgren to Welles, 1 October 1862, *OR,* 13:353; Welles to Dahlgren, 8 October 1862, ibid., 13:376; Virginia Fox diary, 5 January 1863, reel 1.

7. Welles to Phelps, 18 September 1862, Slagle, *Ironclad Captain,* 296.

8. Charles Davis to Welles, 29 August 1862, *OR,* 23:324; Welles, 1, 10 October 1862, *Diary of Gideon Welles,* 1:158, 167; Fox to Charles Davis, 9 September 1862, Fox, *Confidential Correspondence,* 2:63; Grimes to Fox, 2 September 1862, Fox, *Confidential Correspondence,* 2:362–63; Fox to Grimes, 11 September 1862, Fox, *Confidential Correspondence,* 2:375; Phelps to Andrew Foote, 7 September 1862, Slagle, *Ironclad Captain,* 289; Phelps to Elisha Whittlesey, 9 September 1862, Slagle, *Ironclad Captain,* 290–91; Andrew Foote to Phelps, 12 September 1862, Slagle, *Ironclad Captain,* 292; Welles to Phelps, 18 September 1862, Slagle, *Ironclad Captain,* 295–98; Charles Davis to Phelps, 27 September 1862, Slagle, *Ironclad Captain,* 299.

9. Welles, 4 November 1862, *Diary of Gideon Welles,* 1:180; Welles to Porter, 21 October 1862, *OR,* 23:429; Fox to Porter, 24 October, 8 November 1862, *OR,* 23:443, 469; Porter to Welles, 21 October, 5 November 1862, *OR,* 13:428, 465; Porter to Fox, 8 November 1862, *OR,* 23:469; Porter to Fox, 17, 29 October, 10 November 1862, *Confidential Correspondence,* 2:140–42, 144, 148, Slagle, *Ironclad Captain,* 304–10.

10. Selfridge, *Memoirs,* 116–17.

11. Quoted in Shelby Foote, *Civil War,* 2:78.

12. Welles, 10 October 1862, *Diary of Gideon Welles,* 1:167.

13. Welles, 1 October 1862, ibid., 1:157–58.

14. Porter to D. P. Heap, 20 January 1863, *OR,* 24:185; Porter to Welles, 7, 22, 27 February, 26 March, 11 April 1863, ibid., 24:322, 382, 388, 479, 540; Porter to David Farragut, 26 March 1863, ibid., 20:28; Porter to Fox, 16 January, 3 March 1863, Fox, *Confidential Correspondence,* 2:153–54, 158–59, 161, Franklin, *Memories of a Rear-Admiral,* 194.

15. Welles to Samuel Du Pont, 27 March 1863, *OR,* 13:786; Welles to Porter, 2 April 1863, ibid., 20:44; Welles to Porter, 19, 28 January, 2 March 1863, ibid., 24:127, 181, 388; Welles, 17, 31 March, 2, 14, 17 April 1863, *Diary of Gideon Welles,*

1:249, 251, 259, 273–74; Fox to Porter, 6 February, 6 April 1863, *OR*, 24:242, 532; Dahlgren, 29 March, 21 April 1863, Dahlgren, *Memoir of John A. Dahlgren*, 389–90; Anne Rodgers to John Rodgers, 2 February 1863, John Rodgers Papers, box 22.

16. Fox to Samuel Phillips Lee, 28 February 1863, *OR*, 8:577; Fox to David Farragut, 28 February 1863, ibid., 19:639; Henry Bell to David Farragut, 27 February 1863, ibid., 19:637–38; David Farragut to Bell, 4 December 1862, ibid., 19:390; David Farragut to Welles, 29 December 1862, 3, 21 January 1863, ibid., 19:434, 481, 551; David Farragut to Fox, 23 December 1862, Fox, *Confidential Correspondence*, 1:322; Fox to Benjamin Butler, 17 November 1862, Fox, *Confidential Correspondence*, 2:445; David Farragut to his wife, 13 February 1863, Farragut, *Life of David Farragut*, 309.

17. The quote is from Farragut, *Life of David Farragut*, 307. See also David Farragut to his wife, 17 February 1863, ibid., 310–11; George Perkins to his family, 3 January 1863, Perkins, *George Hamilton Perkins*, 149.

18. David Farragut to Bell, 5 March 1863, *OR*, 19:651; David Farragut to Fox, 28 February 1863, Fox, *Confidential Correspondence*, 1:327–28; Farragut, *Life of David Farragut*, 317.

19. Farragut, *Life of David Farragut*, 319.

20. Welles to David Farragut, 2 April, 15 June 1863, *OR*, 20:43, 83; Fox to David Farragut, 2 April 1863, ibid., 20:44; David Farragut to Porter, 25, 27 March 1863, ibid., 20:24, 33; David Farragut to Theodorus Bailey, 22 April 1863, ibid., 20:157; David Farragut to Welles, 20 June 1863, ibid., 20:367; Fox to Du Pont, 26 March 1863, Fox, *Confidential Correspondence*, 1:196; David Farragut to Fox, 27 March 1863, Fox, *Confidential Correspondence*, 1:329–30; Fox to David Farragut, 10 July 1863, Fox, *Confidential Correspondence*, 1:335; Fox to Porter, 6 April 1863, Fox, *Confidential Correspondence*, 2:164–65; David Farragut to Du Pont, 20 April 1863, Du Pont Papers, box 42, no. 185.

21. Porter to David Farragut, 26 March 1863, *OR*, 20:28; Welles to Porter, 2 April 1863, ibid., 20:44; Hearn, *Admiral David Dixon Porter*, 212.

22. Welles, 7 July 1863, *Diary of Gideon Welles*, 1:364–65; Brooks, *Washington, D.C. in Lincoln's Time*, 81–82.

23. Porter to Fox, 16 August 1863, Fox, *Confidential Correspondence*, 2:187–88.

24. Welles to Porter, 15 April 1863, *OR*, 20:55; Welles to Porter, 5 May 1863, ibid., 24:565; Welles to David Farragut, 13 July 1863, ibid., 25:109–11; Welles, 25 April, 13 July 1863, *Diary of Gideon Welles*, 1:284, 369; Fox to Porter, 16 July 1863, *OR*, 25:306; Porter to Welles, 13 July 1863, *OR*, 25:277; Porter to Phelps, 27 July 1863, Slagle, *Ironclad Captain*, 336.

25. Welles to David Farragut, 6 April, 3 June, 17 July 1863, *OR*, 20:121, 290, 396; David Farragut to Porter, 27 March 1863, ibid., 20:33; David Farragut to Welles, 30 June, 16 July, 10 August 1863, ibid., 20:367, 395, 442; David Farragut to

Theodorus Bailey, 22 April 1863, ibid., 20:157; David Farragut to Bell, 3, 28 July 1863, ibid., 20:386, 423; David Farragut to James Palmer, 27 July 1863, ibid., 20:421; David Farragut to Fox, 27 March 1863, Fox, *Confidential Correspondence,* 1:329–30; David Farragut to his family, 11 May 1863, Farragut, *Life of David Farragut,* 365; David Farragut to Du Pont, 20 April 1863, Du Pont Papers, box 42, no. 185.

26. David Farragut to Bell, 28, 29 July, 28 August 1863, *OR,* 20:423–24, 489; Welles to David Farragut, 6 April 1863, ibid., 20:121–22; Welles to Bell, 10 August 1863, ibid., 20:440; David Farragut to Welles, 12 May 1863, ibid., 20:181; David Farragut to his family, 25 April, 29 July, 21 August 1862, Farragut, *Life of David Farragut,* 234, 290–91, 293; Farragut, *Life of David Farragut,* 306; Bell diary, 18 July 1863, Bell papers, vol. 6.

27. Bell diary, 31 July 1863, Bell papers, vol. 6.

28. The quote is from Pollard, "Story of a Hero," 599–600. See also Bell diary, 1 July 1862, Bell papers, vol. 3; Bell diary, 22 April, 31 July 1863, Bell papers, vol. 6.

29. David Farragut to Welles, 12 May 1863, *OR,* 20:181; Porter, *Incidents and Anecdotes,* 69–71.

30. Franklin, *Memories of a Rear-Admiral,* 193–94; Bell to Theodorus Bailey, 18 December 1863, Bailey Papers, box 1, 1863 correspondence; Bell to Welles, 18, 19 August, 4 September, 6, 16, 30 October 1863, *OR,* 20:464, 468, 509, 611, 631, 650; Bell to Henry Thatcher, 3 October 1863, *OR,* 20:609; William Shock to Bell, 11 December 1863, *OR,* 20:717; Bell to Robert Townsend, 28 August 1863, *OR,* 25:390.

31. Bell to Welles, 4 September 1863, *OR,* 20:515.

32. Welles to Bell, 2, 9 October 1863, ibid., 20:538; Bell to Welles, 13 September 1863, ibid., 20:521; Bell to John Madigan, 11 September 1863, ibid., 20:520; Welles, 22 September 1863, *Diary of Gideon Welles,* 1:441–42.

33. Farragut, *Life of David Farragut,* 384–88; Welles to David Farragut, 11 August 1863, *OR,* 20:443; David Farragut to Welles, 17 August 1863, *OR,* 20:464; David Farragut to Bell, 15 October 1863, *OR,* 20:629; David Farragut to Fox, 19 October 1863, Fox, *Confidential Correspondence,* 1:337; Salmon Chase, 11 January 1864, Chase, *Papers,* 1:516; Homer Adams to Charles Wilkes, 24 November 1863, Wilkes papers, General Correspondence, container 15; Welles, 11 August, 11, 12 September 1863, *Diary of Gideon Welles,* 1:396, 431.

34. Welles to David Farragut, 3, 15 June 1863, *OR,* 20:83, 290; David Farragut to Welles, 28 May, 17 August 1863, ibid., 20:277, 464; David Farragut to Bell, 30 October 1863, ibid., 20:652; David Farragut to Bell, 17 January 1864, ibid., 21:39; David Farragut to Porter, 17 January 1864, ibid., 21:39; David Farragut to Welles, 12 August 1864, ibid., 21:420; Percival Drayton to Alexander Hamilton, 5 December 1863, Drayton, *Naval Letters,* 38–40; David Farragut to Fox, 19 October 1863, Fox, *Confidential Correspondence,* 1:337.

35. David Farragut to his family, 4 April 1864, Farragut, *Life of David Farragut,* 391.

36. For information on Farragut's preparations, see David Farragut to Welles, 7 February, 15 July 1864, *OR,* 21:90, 374; Drayton to Hamilton, 26 January, 19 February, 20 March, 14 April, 19 August 1864, Drayton, *Naval Letters,* 41, 44, 47, 50, 67; David Farragut to Fox, 5 March 1864, Fox, *Confidential Correspondence,* 1:348; David Farragut to Loyall Farragut, n.d., Farragut, *Life of David Farragut,* 390.

37. Quoted in Shelby Foote, *Civil War,* 3:493–94.

38. David Farragut to Porter, 17 January 1864, *OR,* 21:39; David Farragut to Welles, 20, 22 January, 7 February, 3, 9 May, 15 July 1864, ibid., 21:45, 52, 90, 242, 268, 374; David Farragut to Theodorus Bailey, 19 February, 26 May 1864, ibid., 21:94, 298; Drayton to Lydig Hoyt, 14 July 1864, Drayton, *Naval Letters,* 63; Porter to Welles, 13 June 1864, *OR,* 26:387; David Farragut to Fox, 18 January, 28 February 1864, Fox, *Confidential Correspondence,* 1:341–42, 345; Theodorus Bailey to David Farragut, 30 July 1864, Farragut, *Life of David Farragut,* 411–12.

39. The quote is from David Farragut to his wife, 4 August 1864, Farragut, *Life of David Farragut,* 405. See also Kinney, "Farragut at Mobile Bay," 4:383; David Farragut, General Orders No. 10, 12 July 1864, *OR,* 21:397.

40. Kinney, "Farragut at Mobile Bay," 4:389.

41. For good descriptions of the battle, see Musicant, *Divided Waters,* 314–19, and Shelby Foote, *Civil War,* 3:496–503. See also Kinney, "Farragut at Mobile Bay," 4:391; J. Crittenden Watson, 6 September 1880, Watson, "Lashing of Admiral Farragut in the Rigging," 4:406–7; Farragut, *Life of David Farragut,* 413.

42. Welles, 9 August 1864, *Diary of Gideon Welles,* 2:100; Charles Steedman to his wife, 30 September 1864, Steedman, *Memoir and Correspondence,* 385; Edwin Morgan to Abraham Lincoln, 21 December 1864, Lincoln Papers.

43. Welles to David Farragut, 15 August 1864, *OR,* 21:542; David Farragut to Welles, 27 August, 13, 18 October, 22 November 1864, ibid., 21:612, 682, 690, 735; Drayton to Hamilton, 19 February, 19 August, 8 September 1864, Drayton, *Naval Letters,* 44, 67, 69; David Farragut to Fox, 5 September 1864, Fox, *Confidential Correspondence,* 1:249–50; Drayton to Du Pont, 8 September 1864, Du Pont, *Samuel Francis Du Pont,* 3:380; Farragut, *Life of David Farragut,* 421; George Perkins to his family, 24 August 1864, Perkins, *George Hamilton Perkins,* 201.

44. Farragut, *Life of David Farragut,* 475–76.

45. Welles, 14 January 1865, *Diary of Gideon Welles,* 2:223; Drayton to Hamilton, 13, 15 February 1865, Drayton, *Naval Letters,* 77–78.

46. David Farragut to Welles, 13 October, 22 November 1864, *OR,* 21:682, 735; David Farragut to his family, 29 July, 21 August 1862, Farragut, *Life of David Farragut,* 290–91, 293; Farragut, *Life of David Farragut,* 364–65; Franklin, *Memories of a Rear-Admiral,* 173.

47. Palmer to Welles, 17 November 1861, *OR,* 1:208–9; Palmer to Porter, 7 July 1863, ibid., 20:255; Palmer to Porter, 8 February 1864, ibid., 25:744; David Far-

ragut to Welles, 6 July 1862, ibid., 18:630; Fox to Silas Stringham, 6 September 1861, ibid., 6:180; J. P. McKinstry to Fox, 19 December 1861, ibid., 1:251–52; Drayton to Hamilton, 26 January 1864, Drayton, *Naval Letters,* 41; Farragut, *Life of David Farragut,* 364–65; —— Dixon to Welles, 16 February 1862, Welles papers, container 20; Du Pont Notes, box 47, W9-18475.

48. Thatcher to Welles, 3 May 1865, *OR,* 22:174.

49. Palmer to Welles, 7, 16, 28 December 1864, ibid., 21:750, 761, 774; Palmer to Welles, 8 February 1865, ibid., 22:16; Edward Canby to Palmer, 3 May 1865, ibid., 22:194.

50. Porter to Welles, 28 January 1865, ibid., 11:452.

51. David Farragut to Welles, 5 February 1864, ibid., 21:11; Welles to Thatcher, 24 January 1865, ibid., 22:20; Elizabeth Blair Lee to Samuel Phillips Lee, 11 February 1865, Lee, *Wartime Washington,* 475; Porter to Thatcher, 17 January 1865, *OR,* 11:611; Drayton to Hamilton, 26 January 1865, Drayton, *Naval Letters,* 76; Preble, *Henry Knox Thatcher,* 13–17.

52. Welles to Thatcher, 26 January, 29 April 1865, *OR,* 22:21, 96; Thatcher to Welles, 21 February, 3 April 1865, ibid., 22:47, 70.

53. Welles to Edwin Stanton, 2 January 1864, ibid., 9:384; Samuel Phillips Lee to Welles, 22 December 1863, ibid., 9:370–72; Ammen, *Old Navy and the New,* 384.

54. The quote is from Samuel Phillips Lee to Welles, 22 December 1863, *OR,* 9:372. The statistics are from Cornish and Laas, *Lincoln's Lee,* 122–23. See also Samuel Phillips Lee to Welles, 17, 22 December 1863, 20 February 1864, *OR,* 9:358, 370–72, 497; Samuel Phillips Lee to Welles, 24 June, 6 September 1864, *OR,* 10:207, 432; Samuel Phillips Lee to James Doolittle, 20 February 1865, Mowry, *American Civil War,* 116; Samuel Phillips Lee to Fox, 20 February 1864, Fox, *Confidential Correspondence,* 2:274–76.

55. Welles, 1 March 1864, *Diary of Gideon Welles,* 1:533–34.

56. Welles to Samuel Phillips Lee, 19 March 1864, *OR,* 9:556; Fox to Samuel Phillips Lee, 7 December 1863, ibid., 9:340; Fox to Samuel Phillips Lee, 17 November 1863, Fox, *Confidential Correspondence,* 2:266–67; Elizabeth Blair Lee to Samuel Phillips Lee, 28, 30 August, 30 November 1863, 13 January, 8 June 1864, Lee, *Wartime Washington,* 302–3, 304–5, 320, 339, 390.

57. The quote is from Welles' letter to Samuel Phillips Lee, 19 July 1864, *OR,* 10:284. See also D. L. Braine to Samuel Phillips Lee, 13 July 1864, ibid., 10:265; Welles to Samuel Phillips Lee, 14 July 1864, ibid., 10:271–72; Samuel Phillips Lee to Welles, 14 July 1864, ibid., 10:272; Grattan, *Under the Blue Pennant,* 113–14.

58. Welles to Samuel Phillips Lee, 21 July, 26 August 1864, *OR,* 10:295, 402; Elizabeth Blair Lee to Samuel Phillips Lee, 26 July, 2, 15, 16 August 1864, Lee, *Wartime Washington,* 410, 416, 422–23; Samuel Phillips Lee to Doolittle, 20 February 1865, Mowry, *American Civil War,* 110–13.

59. Welles, 30 August 1864, *Diary of Gideon Welles,* 2:127.

60. The statistics come from Musicant, *Divided Waters,* 416. See also Samuel Phillips Lee to Welles, 6, 8 September 1864, *OR,* 10:432, 441–44; Welles to David Farragut, 5 September 1864, *OR,* 10:430; Welles to Lincoln, 28 October 1864, *OR,* 11:3; Welles, 30 August, 15 September 1864, *Diary of Gideon Welles,* 2:127–28, 146–47; Ulysses Grant to Fox, 10 September 1864, Grant, *Papers,* 12:141; Henry Halleck to Grant, 1 September 1864, Grant, *Papers,* 12:141n; Fox to Grant, 3 September 1864, Grant, *Papers,* 12:142n; William Temple to Theodorus Bailey, 28 June 1864, Bailey Papers, box 1, 1864 correspondence.

61. Welles, 30 August, 15, 27 September, 7 October 1864, *Diary of Gideon Welles,* 2:127, 145–47, 161–62, 172–73; Welles, 16 May 1866, ibid., 3:504–5; Elizabeth Blair Lee to Samuel Phillips Lee, 9 November 1864, Lee, *Wartime Washington,* 439–40; Grant to Butler, 4 August 1864, Grant, *Papers,* 11:313–14; Ammen, *Old Navy and the New,* 382, 384.

62. Welles to David Farragut, 5, 22 September, 1 October 1864, *OR,* 10:430, 473, 512; Welles to Dahlgren, 9 September 1864, ibid., 10:449; David Farragut to Welles, 22 September, 18 October 1864, ibid., 21:655–56, 690; Welles, 30 August, 2, 15 September 1864, *Diary of Gideon Welles,* 2:127–28, 133–34, 146–47; David Farragut to Fox, 5 September, 19 October 1864, Fox, *Confidential Correspondence,* 1:350, 352.

63. Welles, 30, 31 August, 15 September, 6 October 1864, *Diary of Gideon Welles,* 2:128–29, 146-47, 172; Welles, 25 March 1869, ibid., 3:562.

64. Welles to Samuel Phillips Lee, 17 September 1864, *OR,* 10:467; Welles to Samuel Phillips Lee, 17 October 1864, ibid., 26:693; Welles, 15, 27 September 1864, *Diary of Gideon Welles,* 2:147, 161–62.

65. Elizabeth Blair Lee to Samuel Phillips Lee, 17 September 1864, Lee, *Wartime Washington,* 430–31.

66. Welles to Samuel Phillips Lee, 17 September 1864, *OR,* 10:467; Samuel Phillips Lee to Welles, 6 September, 1 October 1864, ibid., 10:432, 513; Elizabeth Blair Lee to Samuel Phillips Lee, 9, 17, 30 September, 9 November 1864, Lee, *Wartime Washington,* 427, 430–31, 436, 439–40; William Frederick Keeler, 7 August 1864, Keeler, *Aboard the USS Florida,* 188; Welles, 27 September 1864, *Diary of Gideon Welles,* 2:161; Katie Buckingham, 13 October 1864, Lamson, *Lamson of the Gettysburg,* 204–5.

67. Samuel Phillips Lee to Doolittle, 20 February 1865, Mowry, *American Civil War,* 111–12, 120; Roswell Lamson to Katie Buckingham, 13 October 1864, Lamson, *Lamson of the Gettysburg,* 204–5.

68. Welles, 21 February 1865, *Diary of Gideon Welles,* 2:241; Samuel Phillips Lee to Welles, 22 January 1865, *OR,* 27:28; Elizabeth Blair Lee to Samuel Phillips Lee, 20 February 1865, Lee, *Wartime Letters,* 471n; Samuel Phillips Lee to Doolittle, 20 February 1865, Mowry, *American Civil War,* 111–21; Cornish and Laas, *Lincoln's Lee,* 250.

69. For a summary of Lee's postwar life, see Cornish and Laas, *Lincoln's Lee*, 156–87.

70. Quoted in Shelby Foote, *Civil War*, 3:78.

71. Welles to Porter, 31 May 1864, *OR*, 10:160; Porter to Welles, 2 March 1864, ibid., 26:7; Selfridge, *Memoirs*, 109.

72. Elizabeth Blair Lee to Samuel Phillips Lee, 17 September 1864, Lee, *Wartime Letters*, 430; Lamson to Buckingham, 6, 25 November, 5 December 1864, Lamson, *Lamson of the Gettysburg*, 208, 211–12; Porter to Fox, 15 October 1864, Merrill, "Fort Fisher," 464; Porter, 12 October 1864, *OR*, 10:558; Welles, 26 April, 9 May, 31 August, 6 October 1864, *Diary of Gideon Welles*, 2:18, 26–27, 129, 172.

73. Porter, *Incidents and Anecdotes*, 269–70.

74. See also Porter to Welles, 13 October 1864, *OR*, 10:563; Welles to Lincoln, 28 October 1864, ibid., 11:3; John Hay, 25 June, September 1864, Hay, *Inside Lincoln's White House*, 232; Porter, *Incidents and Anecdotes*, 212, 267; Porter to Fox, 15, 19 October 1864, Merrill, "Fort Fisher," 464, 466; Welles, 25 December 1864, *Diary of Gideon Welles*, 2:209.

75. Porter, *Incidents and Anecdotes*, 272.

76. Elizabeth Blair Lee to Samuel Phillips Lee, 31 December 1864, Lee, *Wartime Letters*, 457.

77. Porter to Welles, 27 December 1864, *OR*, 11:261.

78. Porter to Welles, 24, 27 December 1864, ibid., 11:253, 261; Welles to Grant, 29 December 1864, ibid., 11:391; Grant to Porter, 30 December 1864, ibid., 11:394; Porter, 30 December 1864, "General Orders No. 75," ibid., 11:252; Welles to Porter, 31 December 1864, ibid., 11:396; Welles, 29, 30, 31 December 1864, *Diary of Gideon Welles*, 2:213–16; Selfridge, "Navy at Fort Fisher," 4:657.

79. Grattan, *Blue Pennant*, 172.

80. W. H. Merriam to Butler, 1 February 1865, Butler, *Private and Official Correspondence*, 5:532.

81. Welles to Porter, 17 January 1865, *OR*, 11:457; Porter to Welles, 28 January 1865, ibid., 11:452–54; N. G. Upham to Butler, 19 January 1865, Butler, *Private and Official Correspondence*, 5:532; Anne Rodgers to John Rodgers, 21 January 1865, John Rodgers Papers, box 23.

Conclusion

1. Porter, *Incidents and Anecdotes*, 283.

2. Welles, 21 September 1863, *Diary of Gideon Welles*, 1:440.

3. Anne Rodgers to John Rodgers, 17 June 1862, John Rodgers Papers, box 21.

4. Welles, 21 September 1863, *Diary of Gideon Welles*, 1:440.

Bibliography

Archival Sources

Connecticut Historical Society Museum, Hartford, Conn.
    Mark Howard Papers

Department of Special Collections, University of Notre Dame, Notre Dame, Ind.
    Manuscripts of the American Civil War
        Pugh Family Papers (accessed at http://rarebooks.nd.edu/digital/civil_
        war/letters/pugh

Duke University Libraries, Durham, N.C.
    Papers of Louis M. Goldsborough

Hagley Museum and Library, Wilmington, Del.
    Samuel Francis Du Pont Papers
    Du Pont Notes

Library of Congress, Washington, D.C.
    Blair Family Papers
        Montgomery Blair Papers
        Gustavus Fox Papers
    Papers of Andrew H. Foote
    Abraham Lincoln Papers
    Rodgers Family Collection, John Rodgers Papers
    Papers of Charles Wilkes
    Levi Woodbury Family Papers
        Virginia Woodbury Fox Diary

National Archives and Records Administration, College Park, Md.
    Papers of Henry Bell
        Journals of Henry Bell

Letters Received by the Secretary of the Navy from Chiefs of Navy Bureaus, 1842–1885 (Microcopy No. 518)

Letters Sent by the Secretary of the Navy to Chiefs of Navy Bureaus, 1842–1886 (Microcopy No. 480)

Papers of Stephen Rowan

Papers of Gideon Welles

New York Public Library, New York, N.Y.
Louis M. Goldsborough Papers

Syracuse University Library, Syracuse, N.Y.
Theodorus Bailey Papers

Primary Sources

Ammen, Daniel. "Du Pont and the Port Royal Expedition." In Johnson and Buel, *Battles and Leaders,* 1:671–91.

———. *The Old Navy and the New.* Philadelphia: J. B. Lippincott, 1891.

Baird, Henry Carey. "Narrative of Rear Admiral Goldsborough, U.S. Navy." *Proceedings of the United States Naval Institute* 59 (1933): 1023–31.

Bartlett, John Russell. "The 'Brooklyn' at the Passage of the Forts." In Johnson and Buel, *Battles and Leaders,* 2:22–55.

Batten, John M. *Reminiscences of Two Years in the United States Navy.* Lancaster, Pa.: Inquirer Printing and Publishing, 1881.

Boyer, Samuel Pellman. *Naval Surgeon: Blockading the South, 1862–1866.* Edited by Elinor Barnes and James A. Barnes. Bloomington: Indiana University Press, 1963.

Brooks, Noah. *Washington, D.C. in Lincoln's Time.* Edited by Herbert Mitgang. Athens: University of Georgia Press, 1989.

Browne, Henry, and Symmes Browne. *From the Fresh-Water Navy: 1861–64.* Edited by John D. Milligan. Annapolis, Md.: United States Naval Institute, 1970.

Browne, Junius. *Four Years in Secessia.* New York: Arno Press, 1970.

Browning, Orville Hickman. *The Diary of Orville Hickman Browning,* vols. 1–2. Edited by Theodore Calvin Peade and James Randall. Springfield: Illinois State Historical Library, 1927, 1933.

Butler, Benjamin. *Butler's Book: Autobiography and Personal Reminiscences of Major-General Benjamin F. Butler.* Boston: A. M. Thayer, 1892.

———. *Private and Official Correspondence of Gen. Benjamin F. Butler.* 5 vols. Norwood, Mass.: Plimpton Press, 1917.

Chase, Salmon. *The Salmon P. Chase Papers.* Edited by John Niven. 4 vols. Kent, Ohio: Kent State University Press, 1993–97.

Clark, Charles E. *My Fifty Years in the Navy.* Boston: Little, Brown, 1917.

Dahlgren, Madeleine Vinton. *Memoir of John A. Dahlgren: Rear Admiral United States Navy.* Boston: James R. Osgood, 1882.

Dana, Charles. *Recollections of the Civil War: With the Leaders at Washington and in the Field in the Sixties.* New York: D. Appleton, 1902.

Davis, Charles H., Jr. *Life of Charles Henry Davis, Rear Admiral, 1807–1877.* Boston: Houghton Mifflin, 1899.

Dewey, George. *Autobiography of George Dewey: Admiral of the Navy.* New York: Charles Scribner's Sons, 1913.

DiZerega, Alfred L. B. "The Last Days of the Rebel Iron-Clad Merimac and Occupation of Norfolk, as Seen from the USS *Susquehanna.*" War Papers series. Military Order of the Loyal Legion of the United States (MOLLUS). http://suvcw.org/mollus/warpapers/DCv1p447.htm (accessed 8 May 2008).

Dodson, C. Marion. *Yellow Flag: The Civil War Journal of Surgeon's Steward C. Marion Dodson.* Edited by Charles Albert Earp. Baltimore: Maryland Historical Society, 2002.

Drayton, Percival. *Naval Letters from Captain Percival Drayton, 1861–1865.* New York, 1906.

Du Pont, Samuel Francis. "Detailed Report of Flag Officer Du Pont." In *Official Records,* 12:262–65.

———. *From Private Journal-Letters of Captain S. F. Du Pont, While in Command of the Cyane, during the War with Mexico, 1846–1848.* Wilmington, Del.: Ferris Bros., 1885.

———. *Samuel Francis Du Pont: A Selection from His Civil War Letters.* Edited by John D. Hayes. 3 vols. Ithaca, N.Y.: Cornell University Press, 1969.

Eads, James. "Recollections of Foote and the Gun-Boats." In Johnson and Buel, *Battles and Leaders,* 1:338–46.

Evans, Robley D. *A Sailor's Log: Recollections of Forty Years of Naval Life.* New York: D. Appleton, 1902.

Fairfax, MacNeill. "Captain Wilkes's Seizure of Slidell and Mason." In Johnson and Buel, *Battles and Leaders,* 2:135–42.

Farragut, Loyall. *The Life of David Farragut, First Admiral of the United States Navy, Embodying His Journal and Letters.* New York: D. Appleton, 1879.

Foote, John. "Notes on the Life of Admiral Foote." In Johnson and Buel, *Battles and Leaders,* 1:347–48.

Foreign Relations of the United States, 1863. Pt. 1. Washington, D.C.: Government Printing Office, 1864.

Foreign Relations of the United States, 1862. Washington, D.C.: Government Printing Office, 1862.

Fox, Gustavus. *Confidential Correspondence of Gustavus Vasa Fox: Assistant Secretary of the Navy, 1861–1865.* Edited by Robert Means Thompson and Richard Wainwright. 2 vols. New York: Naval Historical Society, 1918.

Franklin, S. R. *Memories of a Rear-Admiral Who Has Served for More than Half a Century in the Navy of the United States.* New York: Harper and Brothers, 1898.

Grant, Ulysses. *The Papers of Ulysses S. Grant,* vols. 4–13. Edited by John Simon. Carbondale: Southern Illinois University Press, 1972–85.

Grattan, John. *Under the Blue Pennant, or, Notes of a Naval Officer.* Edited by Robert Schneller. New York: John Wiley and Sons, 1999.

Hay, John. *At Lincoln's Side: John Hay's Civil War Correspondence and Selected Writings.* Edited by Michael Burlingame. Carbondale: Southern Illinois University Press, 2000.

———. *Inside Lincoln's White House: The Complete Civil War Diary of John Hay.* Edited by Michael Burlingame and John R. Turner Ettlinger. Carbondale: Southern Illinois University Press, 1997.

———. *Lincoln's Journalist: John Hay's Anonymous Writings for the Press, 1860–1864.* Edited by Michael Burlingame. Carbondale: Southern Illinois University Press, 1998.

Higgins, Josiah Parker. *Yeoman in Farragut's Fleet: The Civil War Diary of Josiah Parker Higgins.* Edited by E. C. Herrmann. Carmel, Calif.: Guy Victor Publications, 1999.

History Set Right: Attack on New Orleans and Its Defenses by the Fleet under Admiral Farragut. New York: Office of the Army and Navy Journal, 1869.

Hunter, Alvah Folsom. *A Year on a Monitor and the Destruction of Fort Sumter.* Edited by Craig L. Symonds. Columbia: University of South Carolina Press, 1987.

Johnson, Robert Underwood, and Clarence Clough Buel, eds. *Battles and Leaders of the Civil War.* 4 vols. New York: Thomas Yoseloff, 1956.

Keeler, William Frederick. *Aboard the USS Florida: 1863–65.* Edited by Robert W. Daly. Annapolis, Md.: United States Naval Institute, 1968.

———. *Aboard the USS Monitor: 1862.* Edited by Robert W. Daly. Annapolis, Md.: United States Naval Institute, 1964.

Kinney, John Coddington. "Farragut at Mobile Bay." On Johnson and Buel, *Battles and Leaders,* 4:379–400.

Lamson, Roswell H. *Lamson of the Gettysburg: The Civil War Letters of Lieutenant Roswell H. Lamson, U.S. Navy.* Edited by James M. McPherson and Patricia R. McPherson. Oxford, U.K.: Oxford University Press, 1997.

Lee, Elizabeth Blair. *Wartime Washington: The Civil War Letters of Elizabeth Blair Lee.* Edited by Virginia Jeans Laas. Urbana: University of Illinois Press, 1991.

"Letters from Samuel Phillips Lee to James Doolittle, February 20, 1865." In Mowry, *American Civil War,* 111–22.

Library of Congress. "A Century of Lawmaking for a New Nation." American Memory. http://memory.loc.gov/ammem/hlawquery.html (accessed 8 May 2008).

Lodge, Henry Cabot. *Early Memories.* New York: Charles Scribner's Sons, 1913.

Maclay, Edgar Stanton, ed. *Reminiscences of the Old Navy: From the Journals and Private Papers of Captain Edward Trenchard, and Rear-Admiral Stephen Decatur Trenchard.* New York: G. P. Putnam's Sons, 1898.

Mahan, Alfred Thayer. *From Sail to Steam: Recollections of Naval Life.* New York: Da Capo Press, 1968.

Marchand, John B. *Charleston Blockade: The Journals of John B. Marchand, US Navy, 1861–1862.* Edited by Craig L. Symonds. Newport, R.I.: Naval War College, 1976.

Meredith, William. "Farragut's Capture of New Orleans." In Johnson and Buel, *Battles and Leaders,* 2:70–73.

Merrill, James, ed. "The Fort Fisher and Wilmington Campaign: Letters from Rear Admiral David D. Porter." *North Carolina Historical Review* 35, no. 4 (October 1958): 461–75.

Mervine, William. "Jotings on the Way, part 2." *Pennsylvania Magazine of History and Biography* 71 (July 1947): 242–82.

Mowry, Duane, ed. *American Civil War: Letters and Diaries.* Alexandria, Va.: Alexandria Street Press, 2003.

Official Records of the Union and Confederate Navies in the War of the Rebellion. 30 vols. Washington, D.C.: Government Printing Office, 1894–1921.

Osbon, B. S. *A Sailor of Fortune: Personal Memoirs of Captain B. S. Osbon.* New York: McClure, Phillips, 1906.

Perkins, George Hamilton. *George Hamilton Perkins, Commodore, U.S.N.: His Life and Letters.* Edited by Carroll Storrs Alden. Boston: Houghton Mifflin, 1914.

Porter, David D. *Incidents and Anecdotes of the Civil War.* New York: D. Appleton, 1885.

———. "The Opening of the Lower Mississippi." In Johnson and Buel, *Battles and Leaders,* 2:22–55.

Rodgers, Raymond. "Du Pont's Attack at Charleston." In Johnson and Buel, *Battles and Leaders,* 4:32–47.

Saltonstall, William Gurdon. *Reminiscences of the Civil War and Autobiography of William Gurdon Saltonstall.* N.p.: privately printed, 1913.

Sands, Benjamin. *From Reefer to Rear-Admiral.* New York: Frederick A. Stokes, 1899.

Schley, Winfield Scott. *Forty-five Years under the Flag.* New York: D. Appleton, 1904.

Selfridge, Thomas O. *Memoirs of Thomas O. Selfridge.* New York: G. P. Putnam's Sons, 1924.

———. "The Navy at Fort Fisher." In Johnson and Buel, *Battles and Leaders,* 4:655–61.

Steedman, Charles. *Memoir and Correspondence of Charles Steedman.* Edited by Amos Lawrence Mason. Cambridge, Mass.: Riverside Press, 1912.

Vail, I. E. *Three Years on the Blockade: A Naval Experience.* New York: Abbey Press, 1902.

Villard, Henry. *Memoirs of Henry Villard, Journalist and Financier, 1835–1900.* 2 vols. Boston: Houghton Mifflin, 1904.

Walke, Henry. "The Gun-Boats at Belmont and Fort Henry." In Johnson and Buel, *Battles and Leaders,* 1:358–67.

———. *Naval Notes and Reminiscences of the Civil War in the United States, on the Southern and Western Waters, during the Years 1861, 1862, and 1863.* New York: F. R. Reed, 1877.

Watson, J. Crittenden. "The Lashing of Admiral Farragut in the Rigging." In Johnson and Buel, *Battles and Leaders,* 4:406–7.

Weed, Thurlow. *Life of Thurlow Weed.* 2 vols. Boston: Houghton Mifflin, 1883–84.

Welles, Gideon. "Admiral Farragut and New Orleans." *Galaxy* 12, no. 5 (November 1871): 669–83.

———. *Diary of Gideon Welles.* 3 vols. Boston: Houghton Mifflin, 1911.

———. "Fort Sumter." *Galaxy* 10, no. 5 (November 1870): 613–38.

Wilkes, Charles. *Autobiography of Rear Admiral Charles Wilkes, U.S. Navy, 1798–1877.* Washington, D.C.: Department of the Navy, Naval History Division, 1978.

Secondary Sources

Ammen, Daniel. *Campaigns of the Civil War: The Atlantic Coast.* New York: Thomas Yoseloff, 1963.

Anderson, Bern. *By Sea and by River: The Naval History of the Civil War.* New York: Alfred A. Knopf, 1962.

Beale, Howard. "Is the Printed Diary of Gideon Welles Reliable?" *American Historical Review* 30, no. 3 (April 1925): 547–52.

Belknap, George. "Captain George Hamilton Perkins, USN." *Bay State Monthly* 1, no. 4 (April 1884): 201–30.

Boynton, Charles. *The History of the Navy during the Rebellion.* New York: D. Appleton, 1867.

Bradford, James, ed. *Command under Sail: Makers of the American Naval Tradition, 1775–1850.* Annapolis, Md.: United States Naval Institute, 1985.

Browning, Robert M., Jr. *From Cape Charles to Cape Fear: The North Atlantic Blockading Squadron during the Civil War.* Tuscaloosa: University of Alabama Press, 1993.

———. *Success Is All That Was Expected: The South Atlantic Blockading Squadron during the Civil War.* Washington, D.C.: Brassey's, 2002.

Buker, George. *Blockaders, Refugees, & Contrabands: Civil War on Florida's Gulf Coast, 1861–1865.* Tuscaloosa: University of Alabama Press, 1993.

Carrison, Daniel J. *The Navy from Wood to Steel, 1860–1890.* New York: Franklin Watts, 1965.

Chaitin, Peter M. *The Coastal War: Chesapeake Bay to Rio Grande.* Alexandria, Va.: Time-Life Books, 1984.

Chisholm, Donald. *Waiting for Dead Men's Shoes: Origins and Development of the US Navy's Officer Personnel System, 1793–1941.* Stanford, Calif.: Stanford University Press, 2001.

Coombe, Jack D. *Gunfire around the Gulf: The Last Major Naval Campaigns of the Civil War.* New York: Bantam, 1999.

———. *Thunder Along the Mississippi: The River Battles That Split the Confederacy.* New York: Sarpedon, 1996.

Cornish, Dudley Taylor, and Virginia Jeans Laas. *Lincoln's Lee: The Life of Samuel Phillips Lee, United States Navy, 1812–1897.* Lawrence: University Press of Kansas, 1986.

Dudley, William. *Going South: U.S. Navy Officer Resignations and Dismissals on the Eve of the Civil War.* Washington, D.C.: Naval Historical Foundation, 1981.

Ellicott, John Morris. *The Life of John Ancrum Winslow, Rear-Admiral, United States Navy.* New York: G. P. Putnam's Sons, 1905.

Fonvielle, Chris E., Jr. *The Wilmington Campaign: Last Rays of Departing Hope.* Campbell, Calif.: Savas Publishing, 1997.

Foote, Shelby. *The Civil War: A Narrative.* 3 vols. New York: Vintage, 1986.

Fowler, William F., Jr. *Under Two Flags: The American Navy in the Civil War.* New York: W. W. Norton, 1990.

Hagan, Kenneth J., ed. *In Peace and War: Interpretations of American Naval History, 1775–1984.* Westport, Conn.: Greenwood Press, 1984.

Hamersly, Lewis. *The Records of Living Officers of the U.S. Navy and Marine Corps, with a History of Naval Operations during the Rebellion of 1861–5, and a List of the Ships and Officers Participating in the Great Battles.* Philadelphia: J. B. Lippincott, 1870.

Hamlin, Charles Eugene. *The Life and Times of Hannibal Hamlin.* N.p.: privately printed, 1899.

Hearn, Chester. *Admiral David Dixon Porter: The Civil War Years.* Annapolis, Md.: Naval Institute Press, 1996.

Hooper, Edwin Bickford. *United States Naval Power in a Changing World.* New York: Praeger, 1988.

Johnson, Robert E. *Rear Admiral John Rodgers, 1812–1882.* Annapolis, Md.: United States Naval Institute, 1967.

Karsten, Peter. *The Naval Aristocracy: The Golden Age of Annapolis and the Emergence of Modern American Navalism.* New York: Free Press, 1972.

Langley, Harold D. *Social Reform in the United States Navy, 1798–1862.* Chicago: University of Illinois Press, 1967.

Lewis, Charles Lee. *David Glasgow Farragut: Our First Admiral.* Annapolis, Md.: United States Naval Institute, 1943.

Macartney, Clarence. *Mr. Lincoln's Admirals.* New York: Funk and Wagnalls, 1956.

Martin, Christopher. *Damn the Torpedos! The Story of America's First Admiral: David Glasgow Farragut.* London: Abelard-Schuman, 1970.

McPherson, James. *Battle Cry of Freedom: The Civil War Era.* Oxford, U.K.: Oxford University Press, 1988.

Meade, Rebecca Paulding. *Life of Hiram Paulding, Rear-Admiral, U.S.N.* New York: Baker and Taylor, 1910.

Merrill, James M. *Du Pont: The Making of an Admiral.* New York: Dodd, Mead, 1986.

———. *The Rebel Shore: The Story of Union Sea Power in the Civil War.* Boston: Little, Brown, 1957.

Milligan, John D. *Gunboats down the Mississippi.* Annapolis, Md.: United States Naval Institute, 1965.

Musicant, Ivan. *Divided Waters: The Naval History of the Civil War.* New York: HarperCollins, 1995.

Niven, John. *Gideon Welles: Lincoln's Secretary of the Navy.* New York: Oxford University Press, 1973.

Paullin, Charles Oscar. *Paullin's History of Naval Administration, 1775–1911.* Annapolis, Md.: United States Naval Institute, 1968.

———. "President Lincoln and the Navy." *American Historical Review* 14, no. 2 (January 1909): 284–301.

Pollard, Edward. "The Story of a Hero." *Galaxy* 6, no. 5 (November 1868): 598–606.

Pratt, Fletcher. *Civil War on Western Waters.* New York: Henry Holt, 1956.

Preble, George. *Henry Knox Thatcher, Rear Admiral, U.S. Navy.* Boston: privately printed, 1882.

Reed, Rowena. *Combined Operations in the Civil War.* Annapolis, Md.: Naval Institute Press, 1978.

Ringle, Denis J. *Life in Mr. Lincoln's Navy.* Annapolis, Md.: Naval Institute Press, 1998.

 Roberts, William H. *Civil War Ironclads: The U.S. Navy and Industrial Mobilization.* Baltimore: Johns Hopkins University Press, 2002.

Schneller, Robert J., Jr. *A Quest for Glory: A Biography of Rear Admiral John A. Dahlgren.* Annapolis, Md.: Naval Institute Press, 1996.

Slagle, Jay. *Ironclad Captain: Seth Ledyard Phelps and the U.S. Navy, 1841–1864.* Kent, Ohio: Kent State University Press, 1996.

Sloan, Edward William, III. *Benjamin Franklin Isherwood, Naval Engineer: The Years as Engineer in Chief, 1861–1869.* Annapolis, Md.: United States Naval Institute, 1965.

Soley, James Russell. *Campaigns of the Civil War: The Blockade and the Cruisers.* New York: Thomas Yoseloff, 1963.

Sprout, Harold, and Margaret Sprout. *The Rise of American Naval Power, 1776–1918.* Princeton, N.J.: Princeton University Press, 1939.

Sullivan, William. "Gustavus Vasa Fox and Naval Administration, 1861–1866." PhD diss., Catholic University of America, 1977.

Sweetman, Jack, ed. *Great American Naval Battles*. Annapolis, Md.: Naval Institute Press, 1998.

Thompson, Kenneth E., Jr. *Civil War Commodores and Admirals: A Biographical Directory of All Eighty-eight Union and Confederate Officers Who Attained Commissioned Flag Rank during the War*. Portland, Maine: Thompson Group, 2001.

Thomson Gale. Biography Resources Center. http://gale.cengage.com/ BiographyRC.

Time-Life Books, eds. *The Blockade: Runners and Raiders*. Alexandria, Va.: Time-Life Books, 1983.

Tucker, Spencer C. *Andrew Foote: Civil War Admiral on Western Waters*. Annapolis, Md.: Naval Institute Press, 2000.

Underwood, Rodman L. *Waters of Discord: The Union Blockade of Texas during the Civil War*. Jefferson, N.C.: McFarland, 2003.

Valle, James E. *Rocks and Shoals: Order and Discipline in the Old Navy, 1800–1861*. Annapolis, Md.: Naval Institute Press, 1980.

van Duyn Southworth, John. *The Age of Steam: The Story of Engine-Powered Naval Warfare, 1783–1936*. Pt. 1. New York: Twayne Publishers, 1970.

Waugh, John C. *Reelecting Lincoln: The Battle for the 1864 Presidency*. New York: Crown Publishers, 1997.

Weddle, Kevin. "'The Magic Touch of Reform': Samuel Francis Du Pont and the Efficiency Board of 1855." *Journal of Military History* 68, no. 2 (2004): 471–504.

West, Richard, Jr. *Gideon Welles: Lincoln's Navy Department*. New York: Bobbs-Merrill, 1943.

———. "(Private and Confidential) My Dear Fox – ." *Proceedings of the United States Naval Institute* 63 (May 1937): 694–98.

Wise, Stephen. *Lifeline of the Confederacy: Blockade Running during the Civil War*. Columbia: University of South Carolina Press, 1988.

Index

Alabama, 183, 186, 187, 188, 190, 191–92, 195

Arkansas, 7, 54

Arkansas, 109–11, 112, 113, 197

Arkansas Post, Arkansas, 204, 206, 208

Army, Union: competence of officers, 255; defeats of, 134–35, 182, 223–24; general in chief position, 15; Grant as general in chief, 165, 223–24; importance of in Civil War, xiii–xiv, 255; interservice cooperation, 23–24, 28, 127, 130, 155, 159–60, 161–62, 163–64, 167–68, 252–53; personnel selection, xiii–xiv, 262; Roanoke Island operations, 52, 53–54; success of, 54, 163; Vicksburg operations, 112, 119; Wilmington operations, 127

Army of the Potomac: evacuation from James River peninsula, 61; Richmond campaign, 54, 57–58, 125, 134, 196, 223, 251; Seven Days' Battles, 58, 196. *See also* McClellan, George

Atlanta, Georgia, 168, 223–24, 239–40, 245

Atlantic Blockading Squadron, 32–35, 38. *See also* North Atlantic Squadron; South Atlantic Squadron

Bailey, Theodorus: appearance of, 176; blockades, purpose of, 176; character of, 175, 176; competence of, 181, 263–64; death of, 179; Florida operations, 176–77, 181; health of, 102, 175, 177, 178, 262; life and career of, 175–76; lobbyists for, 260; message delivery to Washington, 106; New Orleans operations, 96, 97, 98–99, 100, 175; praise for following New Orleans operations, 102, 103; prize money, 176, 177, 178; rear admiral appointment, 175, 177; relief of, 178–79; shortages experienced by, 176; squadron commander appointment, 175, 257, 261; Wilmington operations, 178

Banks, Nathaniel, 209, 213, 215, 218, 222, 245–46, 255, 266

Barron, Samuel, 5, 6, 10, 16–17, 254

Baton Rouge, Louisiana, 104, 111

Beaufort, North Carolina, 53, 127

Bell, Charles, 87, 177, 193–94

Bell, Henry: appearance of, 217; character of, 217; competence of, 265–66; death of, 219; East India Squadron command, 219; Galveston operations, 117, 118, 217; Gulf Coast blockade, 218; health of, 104; life and career of, 216–17; loyalty to the Union, 118, 216, 217, 258; Mississippi River operations, 104; Mobile operations, 115, 217; New Orleans operations, 96, 98, 99, 100, 217; opinions of others about, 106–7, 216; praise for following New

About the Author

S TEPHEN R. TAAFFE is the author of *MacArthur's Jungle War: The 1944 New Guinea Campaign, The Philadelphia Campaign,* and *Commanding the Army of the Potomac.* He was born and raised in Canfield, Ohio, and attended Grove City College in Pennsylvania for his undergraduate degrees, and Ohio University in Athens, Ohio, for his graduate work. He currently teaches American military and diplomatic history at Stephen F. Austin State University in Nacogdoches, Texas, where he resides with his wife, Cynthia, and their three children.

The **Naval Institute Press** is the book-publishing arm of the U.S. Naval Institute, a private, nonprofit, membership society for sea service professionals and others who share an interest in naval and maritime affairs. Established in 1873 at the U.S. Naval Academy in Annapolis, Maryland, where its offices remain today, the Naval Institute has members worldwide.

Members of the Naval Institute support the education programs of the society and receive the influential monthly magazine *Proceedings* or the colorful bimonthly magazine *Naval History* and discounts on fine nautical prints and on ship and aircraft photos. They also have access to the transcripts of the Institute's Oral History Program and get discounted admission to any of the Institute-sponsored seminars offered around the country.

The Naval Institute's book-publishing program, begun in 1898 with basic guides to naval practices, has broadened its scope to include books of more general interest. Now the Naval Institute Press publishes about seventy titles each year, ranging from how-to books on boating and navigation to battle histories, biographies, ship and aircraft guides, and novels. Institute members receive significant discounts on the Press's more than eight hundred books in print.

Full-time students are eligible for special half-price membership rates. Life memberships are also available.

For a free catalog describing Naval Institute Press books currently available, and for further information about joining the U.S. Naval Institute, please write to:

Member Services
U.S. Naval Institute
291 Wood Road
Annapolis, MD 21402-5034
Telephone: (800) 233-8764
Fax: (410) 571-1703
Web address: www.usni.org